International Library of Ethics, Law, and the New Medicine

Volume 58

The book series International Library of Ethics, Law and the New Medicine comprises volumes with an international and interdisciplinary focus. The aim of the Series is to publish books on foundational issues in (bio) ethics, law, international health care and medicine. The 28 volumes that have already appeared in this series address aspects of aging, mental health, AIDS, preventive medicine, bioethics and many other current topics. This Series was conceived against the background of increasing globalization and interdependency of the world's cultures and governments, with mutual influencing occurring throughout the world in all fields, most surely in health care and its delivery. By means of this Series we aim to contribute and cooperate to meet the challenge of our time: how to aim human technology to good human ends, how to deal with changed values in the areas of religion, society, culture and the self-definition of human persons, and how to formulate a new way of thinking, a new ethic. We welcome book proposals representing the broad interest of the interdisciplinary and international focus of the series. We especially welcome proposals that address aspects of 'new medicine', meaning advances in research and clinical health care, with an emphasis on those interventions and alterations that force us to re-examine foundational issues.

More information about this series at http://www.springer.com/series/6224

Jitendra Kumar Trivedi • Adarsh Tripathi
Editors

Mental Health in South Asia: Ethics, Resources, Programs and Legislation

Second Edition

Springer

Editors
Jitendra Kumar Trivedi
King George's Medical University
Lucknow
India

Adarsh Tripathi
King George's Medical University
Lucknow
India

ISSN 1567-8008
ISBN 978-94-017-9016-1 ISBN 978-94-017-9017-8 (eBook)
DOI 10.1007/978-94-017-9017-8
Springer Dordrecht Heidelberg New York London

Library of Congress Control Number: 2014944066

Printed on acid-free paper

Springer is part of Springer Science+Business Media (www.springer.com)

Introduction

This book on "Mental Health in South Asia: Ethics, Resources, Programs and Legislation" is a readable, timely resource for the relevant mental health related issues from this socio-politically and culturally varied region of the world. The field of mental health has seen dramatic and rapid advances in the recent years, making it difficult for many regions of the world to keep abreast with relatively more developed parts of the world. Ethical and legal issues, mental health program and community psychiatry practices in this part of world are in a stage of flux with further rapid changes are expected. While sharing a number of similarities like scarcity of resources, financial limitations and increasing adversities, there is a growing consensus among the regional countries in South Asia that the delivery of mental health care can be improved with better cooperation and collaboration. This is an important time to put the mental health picture of the South Asian region in right perspective so as to enable us to grow further rapidly. This book is likely to be of great use to academic psychiatrists, educators, students of mental health professionals as well as of law and policy makers of South Asian region and others interested in the developments in this part of the world like advocacy groups etc.

Despite the great diversity in this region, a common thread of a large burden of mental and behavioural disorders, huge treatment gap due to scarcity of available mental health resources, inequities in their distribution, and inefficiencies in their use persist. It has been our guiding principle to trace the common links between these countries mental health issues pertaining to ethical, legal and community psychiatry related issues and to suggest wherever possible ways for further growth and developments. We have a galaxy of scholarly authors prominent in the South Asian region or consistently working actively in collaboration with these countries so that they have developed an authority on the topics on which they have wrote chapters and countries they worked. They are leaders in their fields and represent a broad spectrum of institutions and countries. We are indebted to these outstanding academicians and researchers who have contributed to this volume. Each chapter is written with a specific focus. There are minor repetitions in few places as may be expected from a multi-author book but each chapter can stand alone. Use of headings and subheadings make it easy to use book.

The first section starts with ethics. This is indeed most important part of the book. In the current scenario, a greater awareness and interest in the ethics is expected and this region should be discussed with even more cautiously. Psychiatry as a medical science has been under constant scrutiny. Considering the various varied dimensions of psychiatry, ethics plays a crucial role in safeguarding psychiatry as a profession. Ethics helps psychiatrists to be transparent and accountable in their practice and also helps us to protect the rights of the persons with mental illness. These ethical issues are all the more pertinent in South Asia, which consists of countries with great cultural and linguistic diversity. This section has five chapters. These chapters include important discussions on human rights related issues, ethical principles in practice, research and its implications in psychological interventions. Considerations related to spirituality and religiosity is becoming increasingly important in psychiatry. These are even more important in countries and societies where religion and spirituality play a big role and it is inevitable that these values may well carry more weight than legal frameworks. Finally, a chapter makes relevant comparisons of differences in ethical practises and standards in east and west as a whole. We have been deliberate in choices of the topics and sequence of the chapters so as to develop a smooth flow of information in a particular way.

The second section deals with the resources and opportunities pertaining to mental health services in this geographical area. Mental health care is by no means is standardised and is extremely varied in perception and practices. This section has been again divided into 2 subsections. The first subsection deals with mental health program and policies. First chapter looked into mental health programs in South Asian countries and have discussed the areas that need urgent attention both by the governments as well as by other mental health services providers. The second chapter continues the discussion with special opportunities and major obstacles faced in planning, implementation and supervision. In order to address the huge and largely unmet burden of mental health disorders in the region, it is essential to scale-up evidence-based interventions by progressively strengthening existing mental health systems. This also mentions unique endeavours made for improvement in mental health scenarios by NGOs and private sectors. The last two chapters are written with a more specific comparative approach. These discuss USA's mental health policies, programs, ethical, legal and human right related issues with the aim to examine what South Asian countries, India as an example, can learn.

The second subsection deals with status of community psychiatry in South Asia comprehensively. The chapters discuss community mental health programs and mental health resources including professionals. Public health sector has failed to deal with the huge treatment gap in South Asian countries. Private sector psychiatry has emerged as a potential service provider. Same way Non Government Organizations (NGO) has also played an active role in many countries of the region. Chapters on private sector and NGOs discuss pertinent issues related to them. Another chapter in the sequence attempt to answer the question of what is needed to correct situation of mental health care mainly community psychiatry in South Asia. Transcultural comparisons taking the differences and commonalities into consideration in the fields of mental health and ill mental health have always been a focus of scientific interest. The ‘East’ and ‘West’ comparison is one of the most widely deliberated. Beyond the inequalities and discrepancies in resources, the `East and West` dualism has an impact on mental health theory and praxis in a many different ways. The range, content, patterns and expression of psychopathology and psychiatric symptomatology; the explanatory models of mental disorders; the diagnostic and classification systems; the psychopharmacological and psychotherapeutical interventions; the role of family and psychosocial support systems; basic conceptualizations of mental health and disease are some of the main areas of diversity. Next chapter discuss theses aspects in greater details.

The third and last section deals with the mental health legislation. Due to the common heritage and historical roots, the current scenario of mental health legislation has many commonalities in countries of the South Asian region. Many of the mental health acts in the region still have an institutional and custodial philosophy and lack human rights and community based mental health care approach. Besides ambiguities and lacunae in the acts themselves, there are major flaws in their implementation too. Therefore, mentally ill people continue to be vulnerable to various types of abuse and violation of their rights in the region. The first chapter describes a type of law governing involuntary treatment that is based on decision-making capacity and not on risk of harm to self or others. It is consistent with the legal and ethical principles followed in general medicine, and non-discriminatory against people with a mental illness. It is argued that this type of law could be adapted to the needs of many countries in South Asia without sacrificing its underlying principles and this approach could better serve need of these countries. Other chapters aims to discuss major shortcomings, difficulties faced during implementations and possible solutions for these problems. Finally comparisons between legislation of south Asian countries and western countries have been made with the intention to identify lacunae and plan comprehensive strategies for further improvements. Major areas of concern includes include poor constitutional guarantees for proper health services, inadequate attention to socioeconomic rights, civil and political rights and different cultural norms in comparison to the industrialized countries. Reform of mental health legislation will have to go hand in hand with increased trained manpower, improved resource allocation and improvement of services.

We acknowledge the help of Mr. Christopher Wilby and Prof. Weisstub for their continuous support and encouragement during this arduous journey. We would also like to acknowledge with thanks the feedback of our colleagues, trainees and students, who have been essential in the modification of the content of this volume. Finally, we also like to thank our patients and their families whose unheard voices motivated us to undertook this project and complete it so that it might help in the betterment of the mental health services in South Asia.

Jitendra Kumar Trivedi
Adarsh Tripathi

Acknowledgments

Prof J. K. Trivedi

I am very much aggrieved to inform the readers about the sad and untimely demise of Prof JK Trivedi on 16th September, 2013. Prof Trivedi was born on 15th March 1952.

Charismatic, disciplinarian, eminently fair, honest to the point of bluntness, punctual, hard working and dedicated to the cause of psychiatry, is the simplest description that comes to my mind when I think about of Prof J.K Trivedi. He was passionate about treatment and care of his patients and enjoyed teaching clinical skills in psychiatry to his students. Hundreds of his former students who are working in India and around the globe adore and respect him for his teaching and training skills.

He held many official positions in various psychiatric organizations. Important ones of them are mentioned below. He was Vice-President and President of Central Zone Indian Psychiatric Society respectively in 1996 and 1997. He became Editor, Indian Journal of Psychiatry (1997–2003), Vice Precident and President of Indian Psychiatric Society in 2003–2004 and 2004–05 respectively, and WPA Zonal Representative for Zone 16- Southern Asia (2005–08). At the time of his demise, he was President-Elect of Indian Association for Social Psychiatry, member of Executive Council of Preventive Psychiatry Section of World Psychiatric Association and Chairman- Preventive Psychiatry Section of World Association of Social Psychiatry. He was also Fellow of the Royal College of Psychiatrists UK and Fellow of the American Psychiatric Association. He was one of the foremost researchers in Psychiatry in India today and received funding from his researches and conducted researches with National Institute of Mental Health, USA; John Hopkins Institute, USA; Indian Council of Medical Research, World Health Organization, World Psychiatric Association and World Association of Social Psychiatry etc. He has more than 300 publications in national and international Journals. He received many prestigious awards in Indian psychiatry including Tilak Venkoba Rao Oration, Marfatia Award, Dr. D. L. N. Murthy Rao Oration, Dr.V. N. Bagadia Oration and NN DE oration among many others. In addition, Prof J K Trivedi was a teacher, academician and educator of par excellence. He was also very well known internationally.

His sudden and untimely demise is a big loss to fraternity of psychiatrists in our country and to the civil society in general. He will live in our memory forever.

Adarsh Tripathi

Contents

Part I Ethical Issues

1 **Human Rights and Psychiatry in South Asia** ... 3
Ajit Avasthi and Shubhmohan Singh

2 **Ethical Issues in Psychiatry in Southeast Asia: Research and Practice** ... 19
Suresh Bada Math, Biju Viswanath and Ami Sebastian Maroky

3 **Ethical Handling of Religious and Spiritual Issues: South Asian Perspective** ... 35
Gurvinder S. Kalra, Nilesh Shah, Nicholas Deakin and Dinesh Bhugra

4 **Ethics and Psychiatry: East Versus West** ... 49
T. Maniam and Fifa Rahman

5 **Ethics in Counselling and Psycho-Therapy: What We Need to Learn—South Asian Perspective** ... 63
Arun Kumar Gupta and Anuradha Menon

Part II Resources and Opportunities/Mental Health Programs and Policies (their current status and future prospects)

6 **Programs of Mental Health and Policies in South Asia: Origin and Current Status** ... 81
M. A. Javed

7 **Mental Health Programs and Policies in South Asia: Initiatives and Obstacles** ... 95
Rahul Shidhaye, Siham Sikander, Mark Jordans, Prasantha De Silva and Sudipto Chatterjee

8 **Mental Health Services in USA: Policies and Programs —What can India Learn from Western Models?** ... 113
Anand K. Pandurangi, Antony Fernandez and Jagannathan Srinivasaraghavan

9 **Mental Health Services in USA: Ethical and Legal Aspects and Human Rights—What India can Learn from Western Models** 133
Jagannathan Srinivasaraghavan, Antony Fernandez and Anand K. Pandurangi

Part III Resources and Opportunities/Community Psychiatry

10 **Mental Health Programs at Community Level in South Asian Countries: Progress, Problems and Prospects** 155
R. Srinivasa Murthy

11 **Mental Health Resources in South Asia** 191
Anju Kuruvilla and K. S. Jacob

12 **Allied Mental Health Professionals: Clinical Psychologists, Psychiatric Nurses and Psychiatric Social Workers: Availability and Competency** 221
R. K. Chadda and R. Prashanth

13 **Private Sector Psychiatry: What it has to Offer to Mentally Ill Persons in South Asia?** 233
Charles Pinto and Malay Dave

14 **NGOs and Mental Health: Initiatives and Progress** 247
R. Thara

15 **Roadmap for the Future: What is Needed in the Region?** 263
Pratap Sharan and T. Sivakumar

16 **Mental Health in the Context of East and West: Beyond Resources and Geographical Realities** 279
Levent Küey and ElifKırmızı Alsan

Part IV Mental Health Legislation 289

17 **A Mental Health Law for Low and Middle Income Countries** 291
George Szmukler, Adarsh Tripathi, Rowena Daw and John Dawson

18 **Mental Health Legislation in South Asian Countries: Shortcomings and Possible Solutions** 319
Mohan Isaac

19 **Mental Health Legislation: Evolution and Implementation in South Asian Countries** 333
Vijoy K Varma, Harischandra Ghambeera and Nitin Gupta

20 **Mental Health Legislation: Comparison of South Asian and Western Countries** 343
Jitendra Kumar Trivedi, Adarsh Triapthi and Sujit Kumar Kar

Erratum E1

Contributors

ElifKırmızı Alsan Department of Psychiatry, Kocaeli University, Kocaeli, Turkey

Ajit Avasthi Department of Psychiatry, Postgraduate Institute of Medical Education and Research, Chandigarh, India

Dinesh Bhugra Institute of Psychiatry, London, England

R. K. Chadda Department of Psychiatry, All India Institute of Medical Sciences, New Delhi, India

Sudipto Chatterjee School of Population Health, University of Melbourne and Sangath and Parivartan, Goa, India

Malay Dave Gayatri Polyclinic, Mumbai, India

Rowena Daw Royal College of Psychiatrists, London, UK

John Dawson University of Otago, Dunedin, New Zealand

Prasantha De Silva Family Wellness Unit, HEMAS Hospital, Wattala, Sri Lanka

Nicholas Deakin Institute of Psychiatry, London, England

Antony Fernandez Department of Psychiatry, West Hospital, Richmond, VA, USA

Mc Guire V.A. Medical Center, Richmond, VA, USA

Harischandra Ghambeera National Institute of Mental health, Angoda, Sri Lanka

Arun Kumar Gupta Northumberland, Tyne & Wear NHS Foundation Trust, Cherry Knowle Hospital, Ryhope, Sunderland, UK

Nitin Gupta Government Medical College & Hospital, Chandigarh, India

Mohan Isaac Fremantle Hospital, Fremantle, WA, Australia

K. S. Jacob Christian Medical College, Vellore, India

M. A. Javed Fountain House, Lahore, Pakistan

Mark Jordans Research and Development Department, HealthNet TPO and Institute of Psychiatry, Amsterdam, the Netherlands

Gurvinder S. Kalra Institute of Psychiatry, London, England

Sujit Kumar Kar Department of Psychiatry, King George's Medical University, Lucknow, UP, India

Levent Küey Istanbul Bilgi University, Istanbul, Turkey

Anju Kuruvilla Christian Medical College, Vellore, India

T. Maniam National University of Malaysia (UKM) Medical Centre, Kuala Lumpur, Malaysia

Ami Sebastian Maroky Department of Psychiatry, National Institute of Mental Health and Neuro Sciences, Bangalore, India

Suresh Bada Math Department of Psychiatry, National Institute of Mental Health and Neuro Sciences, Bangalore, India

Anuradha Menon Psychoanalytic Psychotherapy & Adult Psychiatry, Leeds and York NHS Partnerships Foundation Trust, Leeds, Southfield House, 40 Clarendon Road, UK

R. Srinivasa Murthy The Association for the Mentally Challenged, Bangalore, India

Anand K. Pandurangi Virginia Commonwealth University, Richmond, VA, USA

Department of Psychiatry, Virginia Commonwealth University, Richmond, VA, USA

Charles Pinto Topiwala National Medical College & BYL Nair Charitable Hospital, Mumbai, India

R. Prashanth Department of Psychiatry, All India Institute of Medical Sciences, New Delhi, India

Fifa Rahman Malaysian AIDS Council, Kuala Lumpur, Malaysia

Nilesh Shah Institute of Psychiatry, London, England

Pratap Sharan Department of Psychiatry, AIIMS, New Delhi, India

Rahul Shidhaye Centre for Chronic Conditions and Injuries, Public Health Foundation of India, New Delhi, India

Siham Sikander Research Division, Human Development Research Foundation, Islamabad, Pakistan

Shubhmohan Singh Department of Psychiatry, Postgraduate Institute of Medical Education and Research, Chandigarh, India

T. Sivakumar Department of Psychiatry, NIMHANS, Bengaluru, India

Jagannathan Srinivasaraghavan V. A. Medical Center, Marion, IL, USA

Southern Illinois University, School of Medicine & V.A. Medical Center, Marion, IL, USA

George Szmukler Institute of Psychiatry, King's College, London, UK

R. Thara Schizophrenia Research Foundation, Chennai, India

Adarsh Triapthi Department of Psychiatry, King George's Medical University, Lucknow, UP, India

Jitendra Kumar Trivedi Department of Psychiatry, King George's Medical University, Lucknow, UP, India

Vijoy K Varma Indiana University School of Medicine, Indianapolis, IN, USA

Postgraduate Insititute of Medical Education & Research, Chandigarh, India

Biju Viswanath Department of Psychiatry, National Institute of Mental Health and Neuro Sciences, Bangalore, India

Part I
Ethical Issues

Chapter 1
Human Rights and Psychiatry in South Asia

Ajit Avasthi and Shubhmohan Singh

1.1 Introduction

There is no single universally accepted definition of what constitutes South Asia. For the discussion that follows in this chapter, we have taken South Asia to include the countries of Bangladesh, Bhutan, India, Nepal, Pakistan and Sri Lanka. These countries have had a shared socio-cultural milieu from remote times to a more recent similar colonial experience. Of these countries, Bhutan and Nepal remained nominally independent of British sovereignty but continued to be influenced by British policies and expansionism for most part. Bangladesh, India and Pakistan were for all practical purposes a single colony made up of directly British administered and nominally independent kingdoms until 1947 after which there was a partition of the Indian subcontinent into India and the two wings of Pakistan. The current state of nations was arrived at in 1971 when East Pakistan became independent and Bangladesh was born. Sri Lanka became independent in 1948 and a republic in 1972.

The post-colonial history of South Asia has been marked by socio-economic upheaval, progress and strife. All countries of this region have experienced and continue to experience civil strife of varying degrees. India and Pakistan have fought at least four major wars and are today nuclear powers. On the other hand, all countries have also made rapid progress after independence. There is an increased assertiveness on the part of the traditionally down-trodden sections of societies. This has led to a churning of populations, increased urbanization and according to some, a greater socio-economic divide than has ever existed before.

A. Avasthi (✉) · S. Singh
Department of Psychiatry, Postgraduate Institute of Medical Education and Research, Chandigarh, India
e-mail: drajitavasthi@yahoo.co.in

S. Singh
e-mail: shubhmohan@gmail.com

J. K. Trivedi, A. Tripathi (eds.), *Mental Health in South Asia: Ethics, Resources, Programs and Legislation,* International Library of Ethics, Law, and the New Medicine 58,
DOI 10.1007/978-94-017-9017-8_1

India, Pakistan and Bangladesh are also among the most populous countries of the world. In fact, India is likely to overtake China as the most populous country of the world by the end of the next decade. South Asia is already the repository of about a fourth of the population of the world. These peoples are also amongst the poorest in the world. Generally, health has remained an area of low priority for policy makers in most countries of South Asia.

It is with this background that we discuss the practice and infrastructure of the practice of psychiatry and it's relation to the human rights of people with mental illness or disability who come in contact with these services and also of those who cannot, for various reasons access these resources.

Data in this area is vast but very unevenly available across different countries and different aspects of the discussion. Therefore, we will try to present an overview rather than an exhaustive review of the data that is available to derive inferences, trends and make recommendations.

1.2 The Concept of Human Rights

The idea that men and women have certain rights and privileges has probably been around ever since humans have existed as rational and thinking organisms. However this idea has undergone several mutations. From the times of antiquity, rules and regulations and by corollary, rights and obligations have been enshrined in cultures and civilizations as is exemplified in the Code of Hammurabi or the edicts of King Asoka to name a few examples. However, these rights were for a privileged few, and the exploitation of humans by fellow humans in form of practices such as slavery was accepted and encouraged. It was only in the early modern period that concept of Natural Laws was developed which proposed that all people are naturally free and equal (Locke 1963; Tuckness 2011). Further developments of thought and history led to a concept of human rights as fundamental, inalienable rights to which a person is entitled to simply because he or she is a human being (Beitz 2009). The atrocities committed in the Second World War underlined the need for human rights and that human rights needed to have a legal sanctity that is not bound by sovereignties of nation states. Subsequent developments have conceptualized human rights as universal and egalitarian, and have stressed the need to enforce these laws by giving a legal framework (Beitz 2009). In light of the experiences of the second world war and learning from its experiences, the most important step of the adoption of the Universal Declaration of Human Rights in 1948 by the General Assembly of the United Nations and its subsequent ratification by a sufficient number of countries was undertaken (United Nations General Assembly 1948). This declaration thus took on the form of an international law by which it became obligatory on the signatories to ensure that this was followed in their own countries and also to make available for scrutiny the situation in their country to international scrutiny. Another important step is the United Nations International Covenant on Economic, Social and Cultural rights (ICESCR) in 1966. This gives a holistic and all encompassing framework of the concept, extent and interdependence of various human rights in the modern world (United Nations 1966). Therefore, human rights are now understood as inalienable rights of an individual that have an international legal sanctity. Not only is it immoral, but also illegal to impinge upon these rights.

1.3 Health as a Human Right

The World Health Organization (WHO) defines health as a state of complete physical, mental and social well being and not simply an absence of disease (WHO 1986). Article 12 of the ICESCR refers to and builds upon this concept of health (United Nations 1966). It declares that the highest attainable state of health is a human right. Any obstruction to this attainment of an optimum or highest attainable state of health, due to any discrimination, or any other reason, is a human rights violation (Committee on Economic and Social and Cultural Rights 2000). Health care professionals thus need to be aware of the human rights aspects of their work, prevent or minimize human rights violations, and promote and advocate principles and policies that are strengthen the human rights of the population (Gruskin and Tarantola 2007). The recognition of health as a human right changes the emphasis of governments and policy from one of welfare provisions to a moral and legal obligation. Health and human rights interact in various ways. The delivery of human rights or the lack thereof can have positive or negative impacts on health, health can have a direct bearing on the human rights of people, and public health policies can influence delivery of human rights (Mann et al. 1994). Legal recourse is increasingly being taken in various countries to enforce the right to health as a human right (Singh et al. 2007).

1.4 Psychiatry and Human Rights

Psychiatry is a medical specialty concerned with the study, diagnosis, treatment and prevention of behavior disorders (Campbell 2004). Psychiatry thus involves itself with all facets of mental illness as manifested by behavioral disorders. Psychiatry also cannot function in vacuum and is interlinked with factors of policy, health setup, legislative controls and financial inputs among other factors.

Though the practice and scientific basis of psychiatry have grown immensely over the past century, its relationship with human rights has not been entirely cordial. When psychiatry and its relationship with human rights is examined, it is immediately obvious that it is not a psychiatrist in isolation, but the system and milieu in which he or she practices that defines the state of human rights and their violations. This was initially due to poor understanding of the nature of mental illness and the stigma associated with it. This was exemplified in the pitiable state of the inmates of various asylums in Europe in the middle ages (Foucault 2012). In relatively recent times, there are various instances of psychiatry being used as an instrument of repression and human rights violations, most notably in Nazi Germany and the former Soviet Union (Cohen 1954; Adler and Gluzman 1993).

The examples cited above are the more extreme cases. The more usual violations that have been described can occur in a setting of official systems or the organized sector and outside the official system or in the non-organized sector. These people suffer from denial or lack of access to health services, exclusion, marginalization, and discrimination in the community, denial or restriction of employment rights and opportunities, physical abuse/violence, Sexual abuse/violence, arbitrary detention, denial of opportunities for marriage/right to found a family, lack of means to enable people to live independently in the community, denial of access to general health/ medical services and finally financial exploitation. These abuses can take place in any setting, particularly the community, health care settings and in relation to law enforcement agencies (Natalie Drew et al. 2007).

Broadly the major steps that have been taken across the world for promotion and protection of human rights of mentally ill has been the provision of mental health services that has taken the form of national mental health programmes and the neccessary legislations to regulate these services in form of mental health acts. Along with the other declarations for human rights as discussed above, the United Nations also provided principles for protection of people with mental illness and improvement of mental health care in 1991. In 1996, the World Health Organization also developed 'Guidelines for promotion of human rights of persons with mental disorders' as a guide to countries for development of mental health laws (Mishra 2008). These are particularly illuminating and can serve as a guideline. These are as follows:

1. Promotion of mental health and prevention of mental disorders
2. Access to basic mental health care
3. Mental health assessments in accordance with internationally accepted principles
4. Provision of least restrictive type of mental health care
5. Self-determination
6. Right to be assisted in the exercise of self-determination
7. Availability of review procedure
8. Automatic periodic review mechanism
9. Qualified decision-maker (acting in official capacity or surrogate)
10. Respect of the rule of law

Psychiatry in particular and mental health in general can interact with human rights issues in various ways. Broadly, these are as follows.

1.4.1 Psychiatry as a Branch of Medicine that Promotes Human Rights

Mental diseases are common and exact a heavy toll from the sufferers, their friends and family members and the society in which they live. WHO also gives an equal emphasis on the mental aspect of well being (WHO 1986). Psychiatry as a branch of medicine that works to ameliorate the suffering of these people can thus be taken as force that promotes the right to health and human rights. In the same vein, it can be argued that the human rights of people who are suffering due to mental illness and who are not able to derive benefit from psychiatry due to treatment gap, ignorance, lack of resources or stigma are being trampled upon. Psychiatry also has a role in the promotion of rights of people disabled due to mental illness and in the rehabilitation of such people. This can be achieved by making psychiatry available to all those who need it, effective mental health programs, legislations, advocacy and removal of treatment barriers.

Psychiatry also has a role to play in the description and management of psychiatric morbidity arising as a result of human rights violations as is commonly seen after torture, or in refugees.

1.4.2 Psychiatry as a Branch of Medicine that is Used to Violate Human Rights

On the other hand, psychiatry has been used to deliberately trample upon the human rights of political prisoners, dissidents etc. Psychiatry has also come in for criticism as a system that tramples upon the right of a person to refuse treatment, the right to refuse treatments perceived as unpleasant or barbaric, as a system that impinges upon the liberty of people by institutionalizing them, and by medicalization of normal variations in behavior. Psychiatry survivors groups all over the world have also advocated for protection of human rights of the mentally ill (Zungu-Dirwayi et al. 2004). These survivor groups have a heterogeneous outlook ranging from mental illness deniers to those advocating for protection of rights and autonomy of those with mental illness.

We will examine the state of psychiatry and human rights in South Asia from these facets.

1.5 Human Rights and Psychiatry in South Asia

For psychiatry to be a potent force in the service of human rights, it is necessary that it be available to those who need it and that there are checks and balances within the system to prevent human rights violations in form of coercion or forced incarceration or prolonged incarceration in mental hospitals beyond that deemed necessary.

1.5.1 The Influence of Psychiatry on the Human Rights of the Mentally Ill

In South Asia, the care of the mentally ill is carried out predominantly by families. A mentally ill person can approach and be treated in the unorganized sector as described below or by regular psychiatric services by qualified psychiatrists and mental health professionals. Sometimes, the mentally ill people get no treatment at all and if the illness is severe enough are either kept locked and chained in their homes or are often found abandoned. This choice of treatment setting is often dictated by convenience, finances, traditional beliefs or availability. We will examine both of these scenarios separately.

1.5.2 The Influence of Psychiatry on the Human Rights of the Mentally Ill in the Unorganized Sector

A person with mental illness can seek treatment from the unorganized sector as exemplified by faith healers, magico-religious practitioners, or inadequately qualified people who practice in rural and remote areas. Often, these organizations also undertake to keep chronically mentally ill patients for long durations or indefinitely in exchange of some monetary consideration. Here, the situation of those with mental illness is often pitiable. Mentally ill people also make up a significant proportion of the homeless population. These people have ended up on the streets due to their families abandoning them or due to poverty and neglect. Here they are subjected to further abuse or fall prey to drug addiction and other infectious or non-infectious illnesses. For these groups of people, the following developments have taken place.

i. A host of data has been generated on the issues of the epidemiological data of mental illness, pathways and barriers to care and the treatment gap that exists between need and provision of mental health services (Thara et al. 1998; Thara and Srinivasan 2000; Thara et al. 2004; Patel 2007).

It has been shown that mental illness have a tremendous burden on the society and the country, there is a vast proportion of mentally ill people who are unable to derive benefit from mental health services and are often left to their own devices (Patel 2007). The untreated people in the community have chronic illness and provision of mental health services is inadequate and needs to be strengthened (Murthy 2001). This gap is due to lack of resources, lack of funding and will on parts of government and health agencies and stigma in addition to cultural beliefs associated with mental illness. Further, it has been pointed out that mental health services are centralized and concentrated near urban areas and mental health services have not integrated effectively into primary-health service (Saraceno et al. 2007). This is true by and large for all countries in South Asia. Unfortunately it has taken tragedies such as the Erawadi fire tragedy in which mentally ill patients that were kept chained up in a temple in South India were burnt alive because they could not escape from a fire to drive this lesson home. On the other hand, there are data to suggest that primary health care and mental health services can be integrated at minimal extra cost (Chisholm et al. 2000). However, subsequent experience has suggested that this is not always the case and a vast treatment gap persists (Saraceno et al. 2007).

ii. The following solutions have also been suggested. There needs to be political will to make more resources available to mental health services. This can be done by generating conditions for stronger national and international advocacy. Secondly, advocacy needs to be strengthened by factors such as education and media campaigns for destigmatizing mental illnesses. There is a need to strengthen primary and secondary level mental health resources and finally existing resources need to be used more efficiently by existing professionals (Saraceno et al. 2007).

In South Asia, these data have underlined the need to extend mental health services and to make them more comprehensive. This has been done in various ways. The most important of these is the institution of dedicated mental health programs. A recent review suggests that 80 % of countries in South East Asia (8/10) have a dedicated mental health plan (United Nations 2011). However, beds and other resources such as funding and professionals remains way below the levels seen in Europe or other developed regions of the world (United Nations 2011). All countries in South Asia have national level mental health programs (Jacob et al. 2007). However, all these countries have similar problems with the implementation of these programs (Gilani et al. 2005; WHO 2006). For instance, in India, because of lack of will and interest of state governments and the excessive burden on the already overburdened primary health workers, the national mental health program remained on paper and was not implemented to any significant degree for almost a decade and a half after its inception. It took judicial intervention, notably after the Erawadi fire tragedy alluded to above, that caused the strengthening of the mental health program. Funding has been increased and mental health has become a priority area. There is also a needed emphasis on manpower development and training in addition to development of treatment facilities in form of mental hospitals and other agencies. However, constant supervision and audit will be necessary to examine the feasibility of this new approach (Malik 2004; Jacob et al. 2007).

In addition to developing and strengthening the mental health programs, other methods have been tried to narrow the treatment gap. An interesting model is that of the outreach clinic for the homeless mentally ill run at the Institute of Human Behaviour and Allied Sciences (IHBAS) in New Delhi. A team of mental health professionals visits the Jama Masjid area of Delhi on a weekly basis and provides free mental health services and medications to the homeless mentally ill in the area (IHBAS 2011). Another innovative model that has been tried is the camp approach, especially in the treatment of substance abuse disorders (Chavan and Priti 1999). Given the paucity of trained personnel, the feasibilty of using lay counsellors in the treatment of common mental disorders has also been looked at with encouraging results (Patel et al. 2010).

An important role in the area of closing the treatment gap, advocacy and awareness related activities has been done by the nongovernmental organizations (Thara and Patel 2010). These agencies have been able to fill up gaps in some areas that are left behind by the organized sector such as in the field of mental health services for the homeless or poverty stricken. These agencies have also worked in the field of rehabilitation related activities.

iii. Another area where psychiatry has promoted the human right of health is in the field of disaster psychiatry. There is a gradual accumulation of data regarding epidemiological and interventional aspects following natural and man-made disasters (Chandra and WHO 2005; Chatterjee 2005).
iv. Judicial activism had tried to regulate the activities of magico-religious places of healing for the mentally ill persons. This was exemplified by the National Human Rights Commission of India's initiative in 1999 to investigate allegations of abuse at the Sultan Aayudeen Dargah in the Tamil Nadu state of India. However, the Erawadi tragedy took place shortly thereafter of which suo-moto note was taken by the Supreme Court of India. The Hon'ble court made observations with a view to strengthening the mental health program of India (Murthy and Nagaraja 2008).

1.5.3 The Influence of Psychiatry on the Human Rights of the Mentally Ill in the Organized Sector

There are a significant proportion of mentally ill people in South Asia who sooner or later get in touch with regular mental health services. By the organized sector, we mean the mentally ill who have been treated or are being treated by psychiatrist and mental health professionals in Mental hospitals, general hospital psychiatry units, or in private clinics. As we have seen, most human rights violations in the setting of psychiatric services have taken place in the government run hospitals (Natalie Drew et al. 2007). While deliberate repression of dissidents or political rivals has become rare, inadvertent human rights violations remain a distinct possibility. This is because the mentally ill are most vulnerable, often cannot speak for themselves; there are times when decisions have to be taken for them and there is always a possibility that people close to the patient may for a variety of reasons use mental illness as a pretext to get the state to take away the personal liberty, or take away the rights that a person has to manage and dispose off his or her property as deemed fit.

To regulate the activities of mental hospitals in particular and to safeguard the human rights of mentally ill patients, the most important step that has been taken is by the provision of legislative controls by the respective National governments (WHO 2005). In general mental health legislations lay down rules by which a person can get in touch with mental health services, the terms and conditions of admission and discharge. In general legislations try to achieve the following goals. They promote autonomy by ensuring mental health services are accessible for people who wish to use such services. They set clear, objective criteria for involuntary hospital admissions, and, as far as possible promote voluntary admissions. They also provide specific procedural protections for involuntarily committed persons, such as the right to review and appeal compulsory treatment or hospital admission decisions. In addition, they require that no person shall be subject to involuntary hospitalization when an alternative is feasible. They try to prevent inappropriate restrictions on autonomy and liberty within hospitals themselves (e.g. rights to freedom of association, confidentiality and having a say in treatment plans can be protected); and protect liberty and autonomy in civil and political life through, for example, entrenching in law the right to vote and the right to various freedoms that other citizens enjoy (WHO 2005).

In South Asia, all countries except Bhutan have dedicated mental health legislations for the regulation of mental health services. In Bhutan, specific sections of the penal code are dedicated to this issue (Jacob et al. 2007). India, Pakistan and Bangladesh inherited the British-era Indian Lunacy Law of 1912 which was clearly outdated. India replaced it with the Mental Health Act of 1987 (Sarkar 2004), Pakistan replaced it with the Pakistan mental Health Ordinance in 2001 (Government of Pakistan 2001), Bangladesh has submitted a draft of the Mental Health Act to the proper authority as of 2002 but it is yet to be implemented and the Indian Lunacy Act of 1912 remains in force (WHO 2006), Sri Lanka and Nepal have also enacted regulatory mental health legislations (WHO 2006; de Silva and Hanwella 2010).

Whereas these legislations mark an important step in the advancement of human rights of the mentally ill, human rights violations have been seen on a regular basis in all the South Asian countries (Jacob et al. 2007). A major reason for these human rights violations remains the non-implementation of these acts, poor awareness on part of the judiciary, police and psychiatrists and inadequacies of the acts per se (Trivedi 2002; Kala 2004; Math and Nagaraja 2008). While some have advocated a more meaningful implementation of the laws (Gilani et al. 2005), others have advocated changes in laws (Trivedi 2002) or changing the law altogether (Kala 2004). In India, as a result of the criticism of the Mental Health Act, the obvious human rights violations that took place under its jurisdiction, pressure from right-based groups and to better reflect India's position regarding the United Nations Convention on Rights of People with Disability (Nizar 2011), a new law called the Mental Health Care Act, 2010 has been drafted (Pathare and Sagade 2010). It is hoped that the new law would do away with the deficiencies that allowed human rights violations to occur previously.

Human rights violations of mentally ill persons have taken in state-run institutions in all South Asian countries. However, the reason for this has been the non-implementation of respective mental health legislations rather than a deliberate attempt on the part of the state to stifle dissent. In India, these issues have become a part of the public agenda and a subject of judicial activism (Murthy and Nagaraja 2008). The Supreme Court of India ordered a detailed inquiry into the state of specific mental hospitals (Dhanda 2000). The results were shocking at it was revealed that there was a significant proportion of mentally ill persons who were undergoing human rights abuses. In this scenario, in addition to legislative reforms, judicial activism has played an important role. The case of the judicial activism in the emancipation of the suffering mentally ill in India is particularly illuminating and can serve as a guideline to other South Asian countries that suffer from similar problems. The case of the observations of the Supreme Court of India regarding the activities of the magico-religious places of healing in India and in the aftermath of the Erawadi tragedy has already been mentioned. In addition, in a series of cases the Supreme Court also made observations and issued directives with respect to the human rights violations of mentally ill people languishing in mental hospitals, the mentally ill in prisons and in the case of mentally ill prisoners (Dhanda 2000). The process that was set in motion by the Supreme Court took the form of 3 phases. The first phase being that of the court intervention and the directive to the National Human Rights Commission (NHRC) to act as a watchdog for the state of human rights, the steps that were taken by the NHRC and finally the follow up of those efforts. It would not be amiss to describe these in some detail.

In the first phase the court responded to the Erawadi tragedy and to some public interest litigations by NGO, advocacy groups and individuals regarding lack of autonomy of patients, arbitrary detention, inhumane methods of treatments such as use of unmodified ECT and for greater say of families in decisions with respect of patients (Murthy and Nagaraja 2008). As a result, the court ordered the states of the union to make a mental health assessment, ordered setting up of new mental hospitals and to make the ones already existing more autonomous, increased financial outlay as a matter of right and not welfare for mentally ill patients, made the provision that legal aid should be freely available to every patient who may need it, and also directed the NHRC to monitor three mental hospitals who had been subject to litigation in the past.

In response to the directive, the NHRC took a series of steps. The NHRC in collaboration with a neuroscience institute undertook regular inspections of existing mental hospitals and rated them on indicators with relevance to the human right situation of the imamates. The effort was to find the situation on the ground, to sensitize the staff and to make suggestions for the future. It was found that most hospitals had deficiencies in structure and functioning with respect to human rights. This included architecture that was custodial rather than curative, poor living situation in the wards, provision of substandard food, and most importantly that in many cases the regulations made obligatory by the Mental Health Act, 1987 regarding admission, stay and discharge were not being followed. In many cases, there were no records available. There were also problems with regard to paucity of manpower, and lack of resources such as medicines. Most patients and staff were not aware of the provisos of the relevant legislations and were working in ignorance. Quite simply, the human rights of patients were not accorded the requisite attention. A report of these findings was prepared and recommendations were presented (Sekar and Murthy 2008). The recommendations included an emphasis on following the MHA, provision of a Board of Visitors (BOV), regular inspections by the NHRC among other things.

In the second stage, a series of workshops was organized to sensitize the staff and policy makers of the importance of human rights of the patients and to ensure that the law is followed in letter and spirit.

Finally, a follow-up was done on the basis of these recommendations in 2008 (Murthy and Sekar 2008). It revealed that almost all indicators were now better than before. This included infrastructure, construction of wards, facilities available to patients and staff. A major change was in the way admissions were being done. There were many more voluntary patients, court-ordered admissions had decreased somewhat. The institutions had also been able to reduce the number of long-stay patients despite the fact that there were many patients who did not have any place else to go to. Most hospitals now had regular BOV's and also legal rights were displayed prominently in the hospital premises. There was an increased emphasis on the sections of the MHA and record keeping. To sum up, on many indicators that had been identified, mental hospitals were doing better. However some problems have remained mostly due to lack of staff and too many patients (Murthy and Sekar 2008).

The other important fillips to human rights of the mentally ill have been the judicial interventions in the case of mentally ill locked up in jails and in the case of mentally ill prisoners. The Supreme Court ordered that the practice of keeping the mentally ill locked up in prisons as followed in some states of India is an infringement on their rights and ordered them to be transferred to mental hospitals (Murthy and Nagaraja 2008). The other pertains to the plight of mentally ill undertrials in India. It was observed that many mentally ill undertrials continued to languish in jails as they were never deemed to be fit to stand trial. Eventually their families deserted them and they would become the undertrials who had no place to go to and would not be released because of legal compulsions. It was ordered that no mentally ill undertrial remain in prison and should have the benefit of psychiatric care. It was also ordered that jail authorities should have psychiatrists and counselors to detect psychiatric problems and intervene earlier rather than later. Finally, it was ordered that if a prisoner who is undergoing treatment for a mental illness completed his term, he should complete the rest of his treatment as a free person (Murthy and Nagaraja 2008; Kallivayalil et al. 2009).

1.5.4 Survivors of Psychiatry Movement in South Asia

There are voices that are starting to be heard on the front of psychiatry survivor's groups in South Asia. Most of this has taken place in India (Mind Freedom 2011; MAD Pride India 2011). These voices have highlighted the plight of those in mental hospitals in India. Consumers of psychiatry services have also had their say in the proposed amendments to the Mental Health Act of India.

1.5.5 Psychiatry and Those Who have Undergone Human Rights Violations

South Asia has had a fair share of internal and external strife. However, there is little data that has been generated regarding the psychiatric aspects of human rights violations that have taken place in the setting of political violence or civil war. A few reports have come out on the psychological impact of civil war in Nepal (Tol et al. 2010) and Sri Lanka (de Jong et al. 2002; Siva 2010). These reports have confirmed a significant psychological impact of civil strife and human rights violations. These reports have also called for continued research and development of treatment strategies.

1.6 Rehabilitation of the Mentally Ill Disabled and the United Nations Convention on the Rights of People with Disability (UNCRPD)

There are laws for disability and rehabilitation in all South Asian countries (World Bank 2011). However, it is also true that these laws have been unevenly implemented and the realization that mental illness causes significant disability has been slowly accepted. In India, a major step that enabled this step was the development of an instrument that would allow quantification of disability due to mental illness (Mohan et al. 2003). Relevant laws were also drawn up to give legal sanction to the rights of disabled people with mental illness. Psychiatrists have been entrusted the responsibility of disability certification of the mentally ill disabled in India. However problems remain because of poor information and awareness regarding these laws, reluctance to implement these laws, and vague and ambiguous terminology used in the laws.

The UNCRPD was crafted in response to the fact that although pre-existing human rights conventions offer considerable potential to promote and protect the rights of persons with disabilities, this potential was not being tapped. Persons with disabilities were continued being denied their human rights and were kept on the margins of society in all parts of the world. UNCRPD reflects the most recent consensus of the United Nations General Assembly on the subject matter of the human rights of persons with disabilities and takes precedence over other previous related instruments. The basic purpose of the convention has been defined as "To promote, protect and ensure the full and equal enjoyment of all human rights and fundamental freedoms by all persons with disabilities, and to promote respect for their inherent dignity." This represents a paradigm shift in the understanding of disability from a medical to a social model that recognizes the limitations created by a disability not as a problem of the person but rather a problem of barriers in a non inclusive society (United Nations 2006).

The UN Declaration of Principles for the Protection of Persons with Mental Illness and for the Improvement of Mental Health Care states that it is permissible to deprive an individual of legal capacity by reason of mental illness, and authorize a personal representative to make decisions in his or her place. This is superseded by CRPD Article 12, which furthermore requires governments to provide access to support in exercising legal capacity and establish safeguards to prevent abuse of such measures, in particular to ensure respect for the rights, will and preferences of the person.

The UNCRPD may signal a change in the state of affairs and can have a major and profound impact on service delivery of mental health in this region. The UNCRPD makes it obligatory for the state governments to follow a right-based rather than a welfare approach with regard to disability and rehabilitation among other human rights of the disabled. It makes the governments open to audit and censure from international agencies if the human rights of the disabled are not being preserved. It also aims to look over interconnectedness of mental health service delivery system with the legal system. In other words, after the Universal declaration of Human Rights in 1948, a similar movement for the human rights of the disabled in particular is now taking place.

As of December 2011, Bangladesh, India and Pakistan have ratified the UNCRPD whereas Bhutan and Sri Lanka remain signatories to the same (United Nations 2011). The ratification of the UNCRPD paves the way for an obligatory rights-based approach towards the issue of disability and rehabilitation.

1.7 Conclusions and Future Directions

The above discussion underlines a few major themes. First and foremost is that mental illnesses stand in the way of the attainment of the right to health. To that extent, enlightened and developed psychiatry is a major force for the promotion and protection of human rights.

In South Asia, the major cause for human rights violations and the inability of the mentally ill to have access to the right to health has been a lack of interest and commitment from the governments. Quite simply, for a long time mental illness was not considered important enough. This was compunded by the chronic lack of resources and funding. As a result, legislations either continued as colonial, archaic and antiquated concepts or when these were changed, it was simply not implemented effectively enough. Similar was the case with mental health programs, either they did not and do not exist or were underperforming systems that did not attain the required goals. As a result, the treatment gap could not be bridged and the mentally ill continued to be abused and were left to their own devices.

But there are winds of change. The first is the realization that human rights are important and welfare is not a favor being done but an obligation that governments have to the weakest and most vulnerable. In South Asia, there is an increasing realization that there is no health without mental health and making psychiatry available to all those who need its services has become an urgent need. Over the past couple of decades, governments have been seen to be increasingly interested in reducing the burdens created by mental illness. Psychiatry and psychiatrists have shown that mental illnesses are treatable and have cleared a lot of misconceptions regarding the same. As a result, mental health programs have been strengthened everywhere, most notably in India. Increased funding is available for hospitals and clinics; manpower development has become a priority area and community outreach activities are increasing. There are some innovative programs for outreach programs for the homeless mentally ill. An important role is also being played by the Non-governmental organizations.

A telling fact is that it took judicial interventions in addition to change in mindsets of policy makers for this change to come about. In India, this has led to a tangible change in the state of mentally ill in mental hospitals and prisons and the mentally ill prisoners and undertrials. This has also led to the strengthening of the mental health service delivery in terms of infrastructure and manpower. It would be fair to say that in South Asia, human rights and psychiatry have become an important topic of discourse an there are tangible benefits, most notably in India.

There is a movement all over the world towards a rights-based approach for the vulnerable and weak. It is no longer alright to see the fit and healthy carrying the burden of the weak and disabled. Rather, disability is now a socio-economic construct and services and resources have to be made available to them as a matter of right and obligation so that they are able to fulfill their aspiration and reach their full potential.

The field of psychiatry has mostly been a positive force in the protection of human rights of the mentally ill. But, there have been major violations as well. However, in South Asia it would be safe to say that these violations have come about due to ignorance and paucity of resources rather than due to deliberation and malice. It is important to realize that by and large South Asian psychiatry has not been known to be a handmaiden of political repression despite widespread internal and external strife.

It is imperative that psychiatrists become aware of the winds of change, and become advocates of better human rights for the mentally ill. It is important because not only is it an obligation of the psychiatrists to do so, it will also allow many more people to benefit from the advances that science and other disciplines have made in the treatment of mental illnesses.

References

Adler, N., and S. Gluzman. 1993. Soviet special psychiatric hospital: Where the system was criminal and the inmates were sane. *The British Journal of Psychiatry* 163:713–720.

Beitz, R. 2009. *The idea of human rights*. Oxford: Oxford University Press.

Campbell, R. 2004. *Campbell's psychiatric dictionary*. Oxford: University press.

Chandra, V., and World Health Organization and Regional Office for South-East Asia. 2005. *Mental health and psychosocial support activities in response to the Tsunami disaster in Maldives: Detailed evaluation impact assessment and recommendations for disaster preparedness*. New Delhi: World Health Organization Regional Office for South-East Asia.

Chatterjee, P. 2005. Mental health care for India's tsunami survivors. *The Lancet* 365:833–834.

Chavan, B. S., and A. Priti. 1999. Treatment of alcohol and drug abuse in a camp setting. *Indian Journal of Psychiatry* 41:140–144.

Chisholm, D., K. Sekar, K. K. Kumar, K. Saeed, S. James, S. Mubbashar, and R. S. Murthy. 2000. Integration of mental health and care into primary care: Demonstration cost-outcome study in India and Pakistan. *British Journal of Psychiatry* 176:581–588.

Cohen, E. 1954. *The German concentration camp: A medical and psychological study*. Amsterdam: Paris.

Committee on Economic, and Social and Cultural Rights. 2000. The right to the highest attainable standard of health. International Covenant on Economic, Social and Cultural Rights. http://www2.ohchr.org/english/law/cescr.htm. Accessed 10 Dec 2011.

Dhanda, A. 2000. *Legal order and mental disorder*. New Delhi: Sage Publications India Pvt Limited.

Foucault, M. 2012. *Madness and civilization*. London: Routledge.

Gilani, A. I., U. L. Gilani, P.M. Kasi, and M. M. Khan. 2005. Psychiatric health laws in Pakistan: From lunacy to mental health. *PLoS Medicine* 2:e317.

Government of Pakistan. 2001. Mental Health Ordinance, 2001. http://www.emro.who.int/mnh/whd/Pakistan-Ordinance.pdf. Accessed 10 Dec 2011.

Gruskin, S. M. E., and D. Tarantola. 2007. History, principles, and practice of health and human rights. *The Lancet* 370:449–458.

Institute of Human Behaviour and Allied Sciences. 2011. Annual Report 2009–2010. Government of NCT of Delhi. http://delhi.gov.in/wps/wcm/connect/DoIT_IHBAS1/ihbas/annual+report/annual+report+2009-2010. Accessed 10 Dec 2011.

Jacob, K. S., P. Sharan, I. Mirza, M. Garrido-Cumbrera, S. Seedat, J. J. Mari, V. Sreenivas, and S. Saxena. 2007. Mental health systems in countries: Where are we now? *The Lancet* 370: 1061–1077.

de Jong, K., M. Mulhern, N. Ford, I. Simpson, A. Swan, and van der Kam. 2002. Psychological trauma of the civil war in Sri Lanka. *The Lancet* 359:1517–1518.

Kala, A. K. 2004. Why we should opt for a new mental health act, and not tinker with the old one? *Indian Journal of Psychiatry* 46 (2):96–98.

Kallivayalil, R. A., J. K. Trivedi, and A. Tripathi. 2009. Social factors and Psychiatry in India. *Current Opinion in Psychiatry* 22:482–487.

Locke, J. 1963. *The works of John Locke*. Aalen: Scientia Verlag.

MAD Pride India. 2011. https://www.facebook.com/madprideindia?sk=info. Accessed 10 Dec 2011.

Malik, S. 2004. Mental Health programme and legislation in India: Some observations and experiences. *Indian Journal of Psychiatry* 46:15–24.

Mann, J. M., L. Gostin, S. Gruskin, T. Brennan, and Z. Lazzarini. 1994. Health and human rights. *Health and Human Rights* 1:6–23.

Math, S. B., and D. Nagaraja. 2008. Mental health legislation: An Indian perspective. In *Mental health care and human rights,* ed. D. Nagaraja and P. Murthy. New Delhi: National Human Rights Commission.

Mind Freedom. 2011. Win human rights in mental health system. http://www.mindfreedom.org/photos/bhargavi-davar.jpg/view. Accessed 10 Dec 2011.

Mishra, L. 2008. Human rights in mental health care: An introduction. In *Mental health care and human rights,* ed. D. Nagaraja and P. Murthy. New Delhi: National Human Rights Commission.

Mohan, I., R. Tandon, H. Kalra, and J. K. Trivedi. 2003. Disability assessment in mental illnesses using Indian disability evaluation assessment scale (IDEAS). *Indian Journal of Medical Research* 121:759–763.

Murthy, R. 2001. Lessons from the erawadi tragedy for mental health care in India. *Indian Journal of Psychiatry* 43:362–377.

Murthy, P., and D. Nagaraja. 2008. Judicial interventions and NHRC initiatives in mental health care. In *Mental health care and human rights,* ed. D. Nagaraja and P. Murthy. New Delhi: National Human Rights Commission.

Murthy, P., and K. Sekar. 2008. A decade after the NHRC quality assurance initiative: Current status of government psychiatric hospitals in India. In *Mental health care and human rights,* ed. D. Nagaraja and P. Murthy. New Delhi: National Human Rights Commission.

Natalie Drew, M. F., S. Tang, J. Lamichhane, E. Chávez, S. Katontoka, S. Pathare, O. Lewis, L. Gostin, and B. Saraceno. 2007. Human rights violations of people with mental and psychosocial disabilities: An unresolved global crisis. *The Lancet* 368:1664–16675.

Nizar, S. 2011. Impact of UNCRPD on the status of persons with disabilities. *Indian Journal of Medical Ethics* 8:223–229.

Patel, V. 2007. Mental health in low- and middle-income countries. *British Medical Bulletin* 81: 81–96.

Patel, V., H. A. Weiss, N. Chowdhary, S. Naik, S. Pednekar, S. Chatterjee, De M. J. Silva, B. Bhat, R. Araya, M. King, G. Simon, H. Verdeli, and B. R. Kirkwood. 2010. Effectiveness of an intervention led by lay health counsellors for depressive and anxiety disorders in primary care in Goa, India (MANAS): A cluster randomised controlled trial. *The Lancet* 376:2086–2095.

Pathare, S., and J. Sagade. 2010. Amendments to the mental health act, 1987. Government of India, Ministry of Health & Family Welfare, New Delhi. http://www.wbhealth.gov.in/download/amendments.pdf. Accessed 10 Dec 2011.

Saraceno, B., M. van Ommeren, R. Batniji, A. Cohen, O. Gureje, J., Mahoney, D. Sridhar, and C. Underhill. 2007. Barriers to improvement of mental health services in low-income and middle-income countries. *The Lancet* 370:1164–1174.

Sarkar, J. 2004. A new mental health act for India: An ethics based approach. *Indian Journal of Psychiatry* 46 (2):104–114.

Sekar, K., and P. Murthy. 2008. NHRC/NIMHANS 2009 update on government psychiatric hospitals. In *Mental health care and human rights,* ed. D. Nagaraja and P. Murthy. New Delhi: National Human Rights Commission.

de Silva, V., R. Hanwella. 2010. Mental health in Sri Lanka. *The Lancet* 376:88–89.

Singh J. A., M. Govender, and E. J. Mills.2007. Do human rights matter to health? *The Lancet* 370:521–527.

Siva, N. 2010 Sri Lanka struggles with mental health burden. *The Lancet* 375:880–881.

Thara, R., and V. Patel. 2010. Role of non-governmental organizations in mental health in India. *Indian Journal of Psychiatry* 52:389–S395.

Thara R, and T. N. Srinivasan. 2000. How stigmatising is schizophrenia in India? *International Journal of Social Psychiatry* 46:135–141.

Thara, R., A. Islam, and R. Padmavati.1998. Beliefs about mental illness: A study of a rural South Indian community. *International Journal of Mental Health* 27:70–85.

Thara, R., R. Padmavati, and T. Srinivasan. 2004. Focus on psychiatry in India. *The British Journal of Psychiatry* 184:366–373.

Tol, W. A., B. A. Kohrt, M. J. Jordans, S. B. Thapa, J. Pettigrew, N. Upadhaya, and J. T. de Jong. 2010. Political violence and mental health: A multi-disciplinary review of the literature on Nepal. *Social Science & Medicine* 70:35–44.

Trivedi, J. K. 2002. The mental health legislation: An ongoing debate. *Indian Journal of Psychiatry* 44:95–96.

Tuckness, A. 2011. Locke's political philosophy. In The Stanford encyclopedia of philosophy, ed. E. Zalta. http://plato.stanford.edu/archives/win2003/entries/tuckness. Accessed 10 Dec 2011.

United Nations. 1966. Article 12, International covenant on economic, social and cultural rights, adopted and opened for signature, ratification and accession by United Nations General Assembly. United Nations. http://www2.ohchr.org/english/law/cescr.htm. Accessed 10 Dec 2011.

United Nations. 2006. Convention on rights of people with disability. http://www.un.org/disabilities/convention/conventionfull.shtml. Accessed 28 Aug 2013.

United Nations. 2011a. Convention on rights of people with disability. Convention and optional protocol. http://www.un.org/disabilities/default.asp?id=150. Accessed 10 Dec 2011.

United Nations. 2011b. *Mental health atlas 2011*. Geneva: World Health Organisation.

United Nations General Assembly. 1948. The Universal Declaration of Human Rights. United Nations. http://www.unhcr.org/refworld/docid/3ae6b3712c.html. Accessed 10 Dec 2011.

World Bank. 2011. Disability in South Asia region. http://web.worldbank.org/WBSITE/EXTERNAL/TOPICS/EXTSOCIALPROTECTION/EXTDISABILITY/0,,contentMDK:20183395~menuPK:417335~pagePK:148956~piPK:216618~theSitePK:282699,00.html. Accessed 10 Dec 2011.

World Health Organization. 1986. *The constitution of world health organization*. Geneva: World Health Organization.

World Health Organization. 2005. *WHO resource book on mental Health, human rights and legislation*. Geneva: World Health Organization.

World Health Organization. 2006a. WHO-AIMS report on mental health system in Nepal. Kathmandu: WHO and Ministry of Health.

World Health Organisation. 2006b. *AIMS report on mental health system in Bangladesh*. Dhaka: WHO and Ministry of Health & Family Welfare.

Zungu-Dirwayi, N., D. Kaminer, I. Mbanga, and D. J. Stein. 2004. The psychiatric sequelae of human rights violations: a challenge for primary health care. *The Journal of Nervous and Mental Disease* 192:255–259.

Chapter 2
Ethical Issues in Psychiatry in Southeast Asia: Research and Practice

Suresh Bada Math, Biju Viswanath and Ami Sebastian Maroky

2.1 Introduction

The ancient Greek words 'ethos' and its root 'ethica' are important for understanding the meaning of right and wrong (MacKenzie 2009). During the time of Aristotle, ethos came to mean a person's interior dwelling place, a reference to what a person carries within themselves: their attitudes, orientations, and disposition (Drane 1988). From the moral philosophers perspective, ethics is concerned not only with the question of whether an action is right or wrong but also covers the motives and consequences of the action in terms of whether they are good or bad (Bloch and Pargiter 2002). In simple words 'ethics' means application of values and moral rules to human activities.

Ethics, competence and autonomy of person with mental illness are dynamic in nature and vary significantly across time. Psychiatry as a medical science has been under constant scrutiny. Psychiatric disorders in general are still wrought with significant stigma, myths and biases. Attributions regarding the causation of illness are still colored with religious and supernatural fervour, and medical treatments are regarded with suspicion or considered ineffective or addictive. Considering the various dimensions of psychiatry, ethics plays a crucial role in safeguarding psychiatry as a profession. Ethics helps psychiatrists to be transparent and accountable in their practice. It also helps us to protect the rights of the persons with mental illness.

Ethical guidelines have been put forward by various national and international organizations for different groups of practitioners. In India, a code of ethics for psychiatrists was put forward by a committee which was approved by the Indian Psychiatric Society at its Annual National conference at Cuttack, Orissa (India) in 1989. The code was based on principles of responsibility, competence, benevolence, moral standard, patient welfare and confidentiality (Ahuja and Vyas 2008).

S. B. Math (✉) · B. Viswanath · A. S. Maroky
Department of Psychiatry, National Institute of Mental Health and Neuro Sciences, Bangalore 560029, India
e-mail: sbm@nimhans.kar.nic.in, nimhans@gmail.com

J. K. Trivedi, A. Tripathi (eds.), *Mental Health in South Asia: Ethics, Resources, Programs and Legislation,* International Library of Ethics, Law, and the New Medicine 58,
DOI 10.1007/978-94-017-9017-8_2

The ethical obligation of the psychiatrist includes, acting in the best interest of the patient, doing no harm, and observing justice (López-Ibor et al. 2010). In general, it can be seen that forums and discussions regarding ethical issues and guidelines mostly pertain to research rather than clinical practice. Teaching ethics in psychiatry during post-graduate training is generally lacking.

In the light of such disparities, this chapter discusses ethical issues in psychiatry in clinical practice and research separately. Though legal issues are part of ethics at large, the authors in this chapter have focused more on ethical issues.

2.2 Ethical Issues in Practice

2.2.1 Diagnosis

The field of psychiatric medicine has always grappled with accusations of being an inexact science. This is improving now with better diagnostic criteria and treatment guidelines. However, clinical observation and eliciting of symptomatology continues to be the scaffolding on which diagnostic criteria are applied and the relative lack of investigations and lab markers continue to place the practicing psychiatrist in ethical dilemmas of labelling people with mental disorders. Diagnosis of mental disorder is perceived with shame, blame, secrecy, exclusion, danger, discrimination and stigma. This leads to isolation and rejection of that person in all aspects of their lives and the label follows them everywhere till grave. Hence, psychiatrist should know the psychological and social consequences of his diagnosis in the patient's life (López-Ibor et al. 2010). There are instances where clinicians have labelled people's reactions to abnormal situations such as disasters, severe stress, grief and so forth as mental disorders. This issue of diagnosis needs to be kept in mind by the clinician before they label people with various diagnoses. Diagnosis in psychiatry calls for evidence based practice by the clinician.

2.2.2 Physician Patient Relationship

The innate ability of mental illnesses to affect the persons behaviour, emotions and cognitive functions, tilts the balance in the psychiatrist and patient relationship shifting more responsibility onto the psychiatrist. Consequently, exploitations and accusations of the same can occur on physical, emotional, financial and sexual issues. Issues of transference and counter transference and their infringement on the therapeutic alliance also pose difficulties for practicing psychiatrists. Clinician's behaviour that seeks to secure unfair or unlawful gains such as providing of false information (certificate) regarding illness, prescribing expensive branded drugs, unethical relationship with pharmaceutical companies, unnecessary investigations and physical examination, unwarranted admission to the hospital, fees splitting and so forth are becoming rampant in the present scenario.

In the contemporary world, the doctor-patient relationship has undergone a dramatic change in the last few decades. Enactment of the Consumer Protection Act 1986 (The Consumer Protection Act 1986.) and bringing the medical practise under the ambit of the act through a landmark judgement (Indian Medical Association vs. V.P. Shantha 1995) has created a major drift in the doctor-patient relationship. Patient is a consumer (service user) and doctor is a service provider (professional service). This definition of a commercial relationship has brought the medical practise to market. On one hand many of the doctors do consider this noble profession as a profitable business and on the other hand litigation against doctors is on the increase. Hence, medical professionals have resorted to 'defensive practice'. As a result, the doctor-patient relationship has deteriorated considerably.

This issue of doctor-patient relationship is very crucial in the healing process; hence there is an urgent need to educate the medical students in ethical practise at the earliest. Ethics needs to be part of the medical curriculum so that the erosion of ethics from clinical practise can be salvaged.

2.2.3 Involuntary Treatment

Many psychiatric patients require to be treated against their consent as they refuse to believe that they are ill and hence refuse treatment. This problem has been recognized by the governments of various countries and hence legal safeguards have been provided for the same. The Mental Health Act, 1987 of India (Mental Health Act 1987) provides for involuntary hospitalization with the consent of the relative taking care of the patient or through legal procedure. Such hospitalizations can be done in mental health facilities or general hospitals with psychiatric facility and are called supported admissions, indications for which include, the person having

(a) Recently threatened or attempted or is threatening or attempting to cause bodily harm to himself or herself and/or;
(b) Recently behaved or is behaving violently towards another person or has caused or is causing another person to fear bodily harm from him or her; and/or
(c) Recently shown or is showing a lack of competence to care for himself or herself to a degree that places the individual at risk of harm to himself or herself; (Draft of Mental Health Care Act 2010)

Majority of the mentally ill are treated as out-patients and clear guidelines and legal safeguards for the involuntary treatment in these cases is lacking (Agarwal 2001). Striking a balance between patients' autonomy and the need for involuntary treatment is often difficult in India. Although mandate requires that once the patient is better and capable of autonomy, the choice of hospitalization and treatment be made available to the patient, often the patient continues to stay in the hospital past requirements. This may be because the family members are not ready to accept them, they have no place to go, or the family wants them to stay in the hospital to avoid the inconvenience of keeping them at home. The lack of adequate social support to help the patients reassume their roles in the community often compel doctors to unnecessarily extend the hospitalization (Agarwal 2010). All this leads to doctors in these treatment facilities becoming overcautious and refusing to admit the patient.

In case of outpatient care, involuntary treatment may involve administration of the medicine surreptitiously by mixing in food or drink or by injections against the will of the patients with the consent of family members. This is being done keeping in mind the principle of beneficence and is ethically acceptable. However this practice runs the risk of misuse of drugs as in many cases the drugs are dispensed by proxy from the hospitals and pharmacies (Agarwal 2010). Relatives may also force doctors to prescribe treatment without actually seeing the patient citing reasons of unwillingness of the patient to come to the hospital. Proliferation of de-addiction centres has brought this issue into forefront. There are instances where patients with history of substance use are admitted against their will, tortured and at times beaten up by the de-addiction staff.

2.2.4 Electro Convulsive Therapy (ECT)

Available evidence clearly documents the efficacy of ECT in treating severe mental illness such as Schizophrenia (Tharyan and Adams 2005) and Depression (Pagnin et al. 2004). Hence, this treatment continues to exist in many developed and developing countries. The issue surrounding ECT are broadly two aspects, one regarding consent and the other modified vs unmodified ECT.

2.2.4.1 Consent for ECT

Some psychiatric hospitals continue to use ECT, without the consent of the patients or the legal guardian. This is common in situations when a patient is admitted to a mental hospital in a closed environment and family members are not available to give informed consent on behalf of the patient. Some hospitals have evolved standardized protocols for patients unable to provide consent for ECTs. One method has been to obtain the opinion of two independent psychiatrists and the consent of the hospital RMO or superintendent who acts as a surrogate guardian (Math and Nagaraja 2008).

2.2.4.2 Modified vs Unmodified ECT

The debate is whether to consider modified ECT (under anesthesia) or unmodified ECT (without anesthesia). While modified ECT is preferred over unmodified ECT, non-availability of anesthetists poses a practical difficulty as does the relatively greater cost of ECT. The issue of anaesthetist unavailability has led to some facilities continuing with unmodified ECT. The issue of cost should not be a factor for considering unmodified ECT treatment which, even if effective is inhumane. Consider the same issue in a different perspective: if it were proposed to conduct a surgical operation without anesthesia, how many would consent to it? Use of unmodified ECT has resulted in severe stigma attached to this potentially useful treatment. Considering the well established efficacy of modified ECTs, palatability of treatment and from a human rights perspective, modified ECT's should be mandated, unless specifically contraindicated. Hence, cost should not come in the way of ethical practise of psychiatry (Math and Nagaraja 2008).

2.2.5 Confidentiality

Confidentiality refers to therapist's responsibility of not disclosing information learned during treatment to any one without the patient's permission. The ethical challenge is the decision on whether to or how much is to be disclosed to family members considering that families are an important part of therapeutic activity. Persons with mental illness may not be in a position to understand the implication of treatment or refusal of treatment and give his consent. Family members play very crucial role in treatment choice and supervised medication. This issue needs to be addressed very carefully. Whenever possible, the patients need to be encouraged to share required information with the family. The treating doctor should keep in mind that confidentiality endures after death also; information should not be disclosed unless next of kin provide consent.

It is unethical to ask questions pertaining to confidential information to the patients in front of others including relatives. The case records of patients should not lie unguarded where they could be accessed by persons outside of the treating team. Publishing case records without hiding the identity of the patient or without his or her permission would also amount to breach of confidence (Agarwal 2010). Unfortunately, Right to information Act 2005 of India has raised a complex issue regarding confidentiality. There are instances when family members of the patients have applied for a copy of the patient's record under Right to Information Act. Confidential information if requested by family or employers should be provided only after explicit permission is obtained from the patient. However exceptions may be required in certain situations and can be practiced after ensuring that the patient is informed and consent is obtained wherever possible; only the relevant information should be disclosed; and the rationale for action should be documented. Following are few exceptions to breach confidentiality -

a) Tarasoff duty: When patient's acts are likely to harm others then it is the doctor's responsibility to protect others from harm (Walcott et al. 2001). This includes provision to protect a possible victim if plans of harm are disclosed to the doctor and also take steps to ensure that patients do not cause harm by negligence.
b) In life threatening emergencies make provision for care.
c) Disclosure of HIV status to spouse: In a landmark judgement, the Supreme Court of India stated that hospitals could not be charged with violating medical ethics

when they disclosed the HIV positive status of an infected individual to a person he or she intended to marry(Mr. X vs. Hospital Z 1998).

d) In forensic psychiatric assessment, patients should be informed about the purpose and nature of the examination. The psychiatrist should inform the patients about his/her obligations to provide information to specific agencies and an informed consent should be obtained. Refusal on the patient's side if present should be intimated to the court, continued directives of the court to reveal confidences should be carried out. This is commonly called 'double agentry'
e) Patient initiates litigation against the psychiatrist: The psychiatrist can reveal such confidences that are directly relevant to the case.

2.2.6 *Informed Consent*

The concept of informed consent has gained importance since 1950 and is still evolving. It is centred on three aspects:

a) Providing information- The patient should be informed about the diagnosis, nature of the illness and likely course without treatment and likely course with treatment through the various treatment options available. Details on the treatment recommended, reasons for the same, side effects, duration of treatment and cost should be provided. Therein arises an important question; How much is to be informed? This becomes highly relevant in developing countries where majority of the population is uneducated.
b) Competence-can the patient understand the information and make rational decisions?
c) Autonomy-can the patient takes autonomous decisions without being influenced by the disease process cultural factors, or other extraneous factors?

The Supreme Court of India in its judgment (Samira Kohli v Dr. Prabha Manchanda 2008) on January 16, 2008, held that a doctor has to seek and secure the consent of the patient before commencing a 'treatment'. Giving the judgment, the three judge bench said that "the consent so obtained should be real and valid; the patient should have the capacity and competence to consent; his/her consent should be voluntary; and his/her consent should be on the basis of adequate information concerning the nature of the treatment procedure, so that he/she knows what he/she is consenting to". This judgment comprehensively sums up the consent process.

Consent in psychiatry is complicated because, persons suffering from mental illness are unable to process the provided information adequately, and the possibility that they would refuse treatment due to lack of insight into the fact that they are ill puts them at a disadvantage. In such cases, consent is usually taken from the caregiver. Where possible, the family should be encouraged to seek a second opinion from another competent psychiatrist of their choice. However, exceptions for consent are, life threatening emergencies, therapeutic privilege and incompetence due to severe mental illness or in case of minors.

2.2.7 *Boundary violations*

The therapeutic relationship between a doctor and the patient is established solely with the purpose of therapy and whenever this relationship deviates from its basic goal of treatment it is called boundary violation and becomes non therapeutic.

In psychiatry, as the therapeutic relationship is prolonged and becomes more personal as many confidential matters are discussed, there is likelihood of developing strong emotional bonds, and greater possibilities of boundary violations. It is the doctor's duty to preserve the boundary. Boundary violations include:

a) Sexual activity with a patient, ex-patient or with the patient's family member.
b) Business relationship with a current patient.
c) Permitting the psychiatrists ideology to influence clinical decisions.
d) Consultations held at social places and outside office hours (excluding emergencies).
e) Financial: fees charged should be reasonable; fee reduction policies should be transparent. Avoid accepting gifts and favours from patients or relatives.
f) Dress and language that is not formal, provocative or abusive

Unfortunately, "the curtain of culture" thinking continues to pervade and justify many actions that clearly need to be condemned and controlled. There is no effective ethical or legal framework to address such issues (Bhan 2010) in developing countries like India.

2.2.8 *Conflict of Interest*

Conflict of interest is "a set of conditions in which professional judgment concerning a primary interest (such as patients' welfare or validity of research) tends to be unduly influenced by a secondary interest (such as financial gain)".The relationship between physician and pharmaceutical industry, research on patients and physicians relationship with health providers are examples of such conflict of interest.

It is believed that the pharmaceutical lobby influences medical publishing, as well as doctors prescribing habits. There is good evidence that pharmaceutical industry is a main sponsor of research (Healy 2004). It is therefore important that the profession remains vigilant about research reports funded by pharmaceutical companies. It is now accepted by almost all ethical organizations including MCI that only small gifts could be accepted by the doctors from the pharmaceutical industry. Pharmaceutical sponsored Continuing Medical Education (CME)s are continually under debate. The health service providers also at times influence clinical decisions, on the basis of availability of medicines in the hospital pharmacy or provision for reimbursement or insurance.

2.2.9 *Ethical Issues in Day- to-Day Clinical Work*

Ethical behaviour needs to be a part and parcel of day to day clinical activity. Psychiatric history taking and examination should be need based and unnecessary details of intensely personal matters like sex or other emotional relationship should only be obtained when needed. Patient should be informed about the diagnosis, treatment and prognosis of the illness. Patient and families would have many queries regarding marriage, likelihood of inheriting the disease or the effect of disease on education, career, work or family life. These questions need to be answered after weighing available evidence. Well intentioned half lies or evasions are unethical and need to be avoided. Ethical considerations should influence even minor clinical decisions.

2.3 Ethics in Research

Research is defined as, "Any systematic investigation, designed to develop or contribute to generalizable knowledge" (Armstrong and Sperry 1994). Human health related research needs participation from human subjects. The relationship between the investigator and research subjects should be based on honesty, trust and respect (NIH 2008).

Historically, ethical research had its origins following the Nazi medical war crimes. This led to the development of the Nuremberg Code in 1947, which was the first international standard for ethical conduct of research (Weindling 2001). Later on, the infamous Syphilis study at Tuskegee was exposed in 1972 (Katz et al. 2006). Following this, the National Commission for the Protection of Human Subjects of Biomedical and Behavioural Research was founded in 1974. In 1979, the commission drafted the Belmont Report—Ethical Principles and Guidelines for the protection of Human Subjects of Research (NIH 2008).The three basic ethical principles enlisted in this report are: (a) Respect for persons, (b) Beneficence, and (c) Justice.

The principle of, "Respect for persons" states that individuals should be treated as autonomous agents and that persons with diminished autonomy are entitled to additional protection (NIH 2008). These include pregnant women, prisoners and children. In applying this principle, the researcher should ensure that the participant comprehends the risks and potential benefits of participation and that an informed consent is taken. In addition, the participant has to be free from any undue influence/coercion which might affect his/her decision to participate (Shah and Basu 2011). The principle of "Beneficence" deals with maximization of benefits and minimization of harm (also called "Non-Maleficence"). The principle of "Justice" ensures that individuals and groups are treated fairly and equitably in terms of bearing the burdens and receiving the benefits of research. For further reading on the ethical principles in research, readers may access the website http://phrp.nihtraining.com, hoisted by the National Institute of Health for training on research ethics.

These ethical issues are all the more pertinent in Southeast Asia, which consists of third world countries with great cultural and linguistic diversity. India, which houses one-seventh of the worlds' total population, is rapidly becoming a research hub for human research. Following globalization and industrialization, multinationals have an emerging interest for research in this area. There has been growing concern that research in developing countries like India will lead to exploitation and injustice (Chaturvedi 2008). This concern is all the more pronounced in case of clinical trials and biological research funded by foreign agencies. However, there is also concern that by imposing unnecessary and expensive regulatory burdens, scientific research may be hampered. The equation is a delicate one, and there is increasing concern internationally that renewed vigilance (or hyper-vigilance in some cases) is out of balance. In pursuit of the goal to protect human subjects from exploitation, we have been left with various regulations, which are either too broadly or too narrowly applied/implemented without serving any purpose (Math 2004).

2.3.1 Clinical Drug Trials and Ethics

"Clinical trial" refers to a systematic study of new drug(s) in human subject(s) to generate data for discovering and/or verifying the clinical, pharmacological (including pharmacodynamic and pharmacokinetic) and/or adverse effects with the objective of determining safety and/or efficacy of the new drug (Schedule Y 2005). Majority of the investigational new drugs are discovered in developed countries by multinational pharmaceutical companies. These multinational pharmaceutical companies are the sponsors for investigational new drugs clinical research. Sponsors prepare and submit the protocol to the regulating authorities (like FDA, DCGI) for initiating the study as per the rules of the countries where it plans to carry out the research.

As per the Schedule Y of the Drugs and Cosmetic Act, 1940, Phase-I of the investigational new drug discovered in other countries are not allowed. However for investigational new drug discovered in India, clinical trials are required to be carried out in India right from Phase I to Phase III. The Schedule (Schedule Y 2005) allows investigational new drug Phase II and Phase III trials to be carried in India for both drug discovered in India or abroad. Multinational pharmaceutical companies are outsourcing their phase II and phase III trial to developing countries like India. Main reasons are as follows:

a) Cost effective (Cheaper by 40 %),
b) Easy availability of skilled labor,
c) Easy availability of participants,
d) Cutting down on duration,
e) No rigorous monitoring in developing countries,
f) Certain randomized placebo controlled clinical trials are ethically unacceptable in developed countries,
g) Golden opportunity to launch the molecule in developing countries at the earliest once it gets approved.

For the above reasons multinational pharmaceutical companies are finding developing countries as the new labs for their investigational new drug clinical trials (Math 2004).

2.3.2 *Regulations*

All clinical trials in India require prior permission from the Drug Controller General, India (DCGI) and approval by the concerned hospital's Ethics Committees. Indian Council for Medical Research has published a detailed set of guidelines in 2000 titled "Ethical Guidelines for Biomedical Research" on Human Subjects, which seeks to update the Policy Statement on "Ethical Considerations Involved in Research on Human Subjects" that was brought out by the Indian Council of Medical Research (ICMR 2000). These form the regulatory network within which clinical trials ought to be conducted in India.

The ethical and legal codes that govern medical practice also apply to clinical trials. The trial follows a carefully controlled protocol, a study plan which details what researchers will do in the study. Before carrying out the trials, the protocol for the same should be reviewed and approved by the Ethical Committee of the institution in which the volunteers are being subjected to the trial (for example, patients in a hospital). The Ethical Committee would be a team of 5 to 12 persons, comprising of doctors from the institution, independent observers, ethicists and lawyers (ICMR 2000).

2.3.3 *Ethical Issues*

2.3.3.1 Recruitment Fees

Recruitment fees are offers of money to physicians, nurses, or other health professionals in reward for their referral of patients eligible for research participation (Lemmens and Miller 2003). In simple words recruitment fees (finder's fees) and recruitment incentives are those paid/offered by the corporate research sponsors to the investigators for recruiting subjects into the trial. They are, rather, most often integrated into the budgets of clinical trials, usually described only as payment of "administrative costs." Since existing guidelines generally allow researchers and research personnel to be compensated for extra time spent on research, finder's fees can easily be hidden among bona fide expenses. Offers of finder's fees ranging between $ 2000 and $ 5000 per subject are now regarded as "common" in the United State (Goldner 2000). The payment of finder's fees for subject recruitment can be seen as a classical example of an increasingly commercial research environment.

Indian researchers are also not behind in the race. Clinical trials have been conducted illegally, predominantly at private clinics not recognized as research centres (Sims and Kuhnlein 2003). It is by design that the sponsors recruit private practitioners in the clinical drug trials as investigators so that they can manipulate them easily by giving money and receiving the required results. This undoubtedly explains the increasing number of clinical trials and researchers in developing countries (Martin and Kasper 2000). At the same time the conduct of clinical trials is shifting from the teaching hospitals of academic institutions and research institutions to the offices of private physicians and CROs (Bodenheimer 2000).

Recruitment fees give rise to another set of concerns often overlooked, that is when financial interests in patient recruitment are significant; physicians may give priority to this type of activity over patient care. Ethical Guidelines for Biomedical Research on Human Subjects, (ICMR 2000), is silent regarding finders' fees and conflict of interest. To overcome these problems there is a strong need for modification in current ethical guide lines with regard to finders' fees, financial disclosure and conflict of interest.

2.3.3.2 Informed Consent Related Issues

The ICMR ethical guidelines (2000) for conducting research in human subjects makes it mandatory for investigators in India to seek written informed consent from subjects volunteering for medical research. In a recent report that examined ethical aspects of studies published in the Indian Journal of Psychiatry, informed consent was mentioned in 51 % of the studies in 2000, which rose to 82 % in 2007(Chaturvedi and Somashekar 2009). Consent was reported to be written in 40 %, while language of the consent form was noted in only 3 %. In developing countries like Bangladesh, it has been shown that even when care has been taken to get proper informed consent, participants may have a very limited understanding of their basic rights concerning a clinical trial (Lynoe et al. 2001).

A study conducted in JIPMER, Pondicherry (Gitanjali et al. 2003), India in a clinical mock trial demonstrated that the educational status did not play any role in patients who consented to participate in the study. However gender played a very important role in which a female refused to participate in the study indicating that were not in a position to make independent choices. Another important finding of this study was that understanding of informed consent was poor in those who consented. This raises a very serious question whether all subjects participating in studies being conducted in India are "truly informed". If this is the status of a mentally sound individual participating in a clinical trial, the understanding of informed consent in clinical trials involving mentally ill patients is highly questionable.

With regard to research in mentally ill patients, Mental Health Act 1987 (Sec 81) has clearly stated that no mentally ill person under treatment shall be used for purposes of research, unless- (i) Such research is of direct benefit to him/her for purposes of diagnosis or treatment; or (ii) Such person, being a voluntary patient, has given consent in writing or where such person (whether or not a voluntary patient) is incompetent by reason of being a minor or otherwise, to give valid consent, the guardian or other person competent to give consent on his/her behalf, has consented in writing for such research. ICMR ethical guide lines states that it is mandatory to obtain the written informed consent of the prospective subject or in the case of an individual who is not capable of giving informed consent, the consent of a legal guardian (ICMR 2000).

However these guide lines have failed to express what constitutes “not capable” or “incapable”? How to assess “not capable” or “incapable”? Who is going to assess this? Determining the proper standards and procedures to assess capacity poses a major challenge in formulating policy on research involving subjects with mental disorders that may affect decision making capacity. Persons with such disorders vary widely in their ability to engage in independent decision making. They may retain such capacity, or possess it intermittently, or be permanently unable to make decisions for themselves. Decisional impairment in mentally ill patients is not a static phenomenon; in fact it is a dynamic phenomenon. This decisional impairment depends upon various factors like diagnosis, severity of the illness, availability and response to treatment. A capacity assessment process must adequately protect the interests of individuals with conditions that increase the risk of decisional impairment (Math 2004). The consent process should be continuous and not a onetime procedure. Study designs must, therefore, provide for this contingency by regular assessment of capacity and take informed consent if the capacity for decision changes. In cases where the investigator had taken the consent of a legal guardian since patients’ decisional capacity was impaired, but during the study if the patient improves substantially and gains the decisional capacity, then it is the responsibility of the investigator to take informed consent from the participants (Math 2004).

2.3.3.3 Institutional Review Board (IRB) Related Issues

In a recent study that examined ethical aspects in published research articles in the Indian Journal of Psychiatry, it was found that ethics committee approvals were mentioned in only 2% of the studies in 2000, which increased to 25% in 2007 (Chaturvedi and Somashekar 2009). This raises a question whether research conducted in developing countries should be at different standards from those applied in the developed countries (Angell 1997; Varmus and Satcher 1997).

Increase in clinical trials since the past one decade, has increased IRBs workloads. Inadequate resources, poor infrastructure and poor knowledge about the ethical issues are haunting the current IRBs of developing countries. This is because local research entities do not design the protocols, and sponsors will remove protocols from institutions whose IRBs demand modifications. In developing countries like India, where private practitioners and private hospitals are available in plenty, sponsors will be able get their work done in another site where IRBs are ready to take up these kinds of opportunities for money. Investigators will vocally resist changes requested by a local IRB on the ground that other IRBs have already approved the study (Math 2004). There are instances where clinicians working in government institutes approach private IRBs for getting the required approval.

2.3.3.4 Should IRBs be made Responsible for not Executing Their Duties

Primary responsibility of the IRB is to safeguard the welfare and the rights of the participants. The IRBs are entrusted not only with the initial review of the proposed research protocols prior to initiation of the projects but also have a continuing responsibility of regular monitoring of the approved programmes to foresee the compliance with ethical guidelines during the entire period of the project (ICMR 2000). If an IRB fails to monitor research or halt a study that is unethical, it violates its duty of care. Other breaches include approving inadequate informed-consent documents and permitting conflicts of interest on part of investigators or even IRB members themselves. Unfortunately there are no mechanisms to bring these erring IRBs within the purview of the justice system.

2.3.4 Transfer of Biological Material from Developing Countries

A lot of biological research in India is funded by the international community. Human material with potential to be used in biological psychiatry research includes blood, stem cell lines, brain tissue, and cerebrospinal fluid. The ICMR has established guidelines for transfer of biological material abroad for research purposes. Where exchange of material is envisaged as part of a collaborative research project, the project proposal as a whole must be routed through the appropriate authorities for evaluation and clearance. The exchange of human materials should be an integral part of a collaborative project, which should have been approved by the Institutional Review and Ethics committees, and not be a separate activity.

2.4 Recommendations for Medical Ethics Guidelines for Studies Conducted in Developing Countries Sponsored by Developed Countries

The ethics of research related to healthcare in developing countries (Nuffield Council on Bioethics 2002) and The Council for International Organizations of Medical Sciences (WHO 2002) have emphasized the need for guidelines so that exploitation can be minimized. The report makes a number of recommendations on key issues in developing world research such as

a) The standard of care,
b) The issue of participant access to study interventions post research.
c) For research that is to be conducted in a developing country but is funded by a body in a different country the proposed should be subject to independent ethical review in the sponsors' country(ies) in addition to the country(ies) in which the research is to be conducted,
d) Ensure that the research conducted in a developing country meets the research needs of that country

e) Individual informed consent must be obtained from all participants.
f) All participants must receive equal consideration and care.

Research performed in less-developed countries by scientists from developed countries, where the objective is to develop and test therapies that will never be available or affordable for populations in less-developed countries, may not be ethically appropriate (McMillan and Conlon 2004). Wendler and colleagues proposed a model to avoid exploitation of developing countries by scientific researchers from developed countries (Wendler et al. 2004). Specifically, institutional review boards should demand that by default the best methods are applied in all cases and approve research using less than the worldwide best methods only when it satisfies the following 4 conditions:

(a) *Scientific necessity:* investigators must use less than the worldwide best methods to answer the scientific question posed by the trial;
(b) *Relevance for the host community:* answering the scientific question posed by the trial will help address an important health need of the host community;
(c) *Sufficient host community benefit:* the trial will produce a fair level of benefit for the host community; and
(d) *Subject and host community non-maleficence:* subjects and the host community will not be made prospectively worse off than they would be in the absence of the trial.

Those who are involved in research should be required to have some understanding of, and be sensitive to, the social, economic, and political milieu that frames the context in which their research is taking place and that greatly influences the health of their participants (Benatar and Singer 2000).

In conclusion, the ethical obligation of the psychiatrist includes acting in the best interest of the patient, doing no harm, and observing justice. The psychiatrist can perform best if he is sensitive on the one hand to the consequences of his diagnosis, and, on the other hand, if he applies evidence based practise in day-to-day clinical practice. Though the ethics in clinical practice is deteriorating in contemporary world, however there is little being done to safeguard. This calls for urgent measures such as inculcating ethics in medical curriculum and strong monitoring mechanisms. Regarding monitoring of ethical practices in research, IRBs routinely fail to provide adequate protection to the participants. This issue needs to be addressed with the enactment of legislation to safeguard the participants and also research.

References

Agarwal, A. K. 2001. Ethical issues in the practice of psychiatry. *Indian Journal Psychiatry 2001* 43 (1):16–21.

Agarwal, A. K. 2010. A review of Indian psychiatry research and ethics. *Indian Journal Psychiatry* 52:297–305.

Ahuja, N., and J. N. Vyas. 2008. *Textbook of post graduate psychiatry*. New Delhi: Jaypee Brothers Medical.

Angell, M. 1997. The ethics of clinical research in the third world. *New England Journal of Medicine* 337:847–849.

Armstrong, J. S., and T. Sperry. 1994. The ombudsman: Business school prestige—Research versus teaching. Interfaces 24(2):13-43.

Benatar, S. R., and P. A. Singer. 2000. A new look at international research ethics. *British Medical Journal* 321:824–826.

Bhan, A. 2010. Boundary violations in patient care: Need for evolving professional practice standards and training. *Indian Journal Med Ethics* 7:192.

Bloch, S., and R. Pargiter. 2002. A history of psychiatric ethics. *Psychiatric Clinics of North America* 25:509–524.

Bodenheimer, T. 2000. Uneasy alliance—clinical investigators and the pharmaceutical industry. *New England Journal of Medicine* 342:1539–1544.

Chaturvedi, S. K. 2008. A review of Indian publications on ethical issues regarding capacity, informed consent, and placebo controlled trials. *The Internet Journal of Mental Health* 5: 1531–2941.

Chaturvedi, S. K., and B. S. Somashekar. 2009. Reporting ethical aspects in published research articles in the Indian Journal of Psychiatry. *Indian Journal Psychiatry* 51:34–37.

Draft of Mental Health Care Act. 2010. Draft Mental Health Care Act, Ministry of Health and Family Welfare, Government of India. http://www.prsindia.org/uploads/media/draft/Draft%20 Mental%20Health%20Care%20Act,%202010.pdf. Accessed 21 Dec 2011. (India., M. o. H. F. W. G. o. (ed.). New Delhi).

Drane, J. F. 1988. *Becoming a good doctor: The place of virtue and character in medical ethics*. Kansas City: Sheed and Wood.

Gitanjali, B., R. Raveendran, D. G. Pandian, and S. Sujindra. 2003. Recruitment of subjects for clinical trials after informed consent: Does gender and educational status make a difference? *Journal of Postgraduate Medicine* 49:109–113.

Goldner, J. A. 2000. Dealing with conflicts of interest in biomedical research: IRB oversight as the next best solution to the abolitionist approach. *Journal of Law, Medicine & Ethics* 28:379–404.

Healy, D. T. 2004. Transparency and trust: Figures for ghost written articles was misquoted. *British Medical Journal* 329:1345.

ICMR. 2000. *Ethical guidelines for biomedical research on human subjects*. New Delhi: Indian Council of Medical Research.

Indian Medical Association vs. V. P. Shantha. 1995. Indian Medical Association vs. V.P. Shantha 1995 (6):SCC 651.

Katz, R. V., S. S. Kegeles, N. R. Kressin, B. L. Green, M. Q. Wang, S. A. James, S. L. Russell, and C. Claudio. 2006. The tuskegee legacy project: Willingness of minorities to participate in biomedical research. *Journal Health Care Poor Underserved* 17:698–715.

Lemmens, T., and P. B. Miller. 2003. The human subjects trade: Ethical and legal issues surrounding recruitment incentives. *Journal of Law, Medicine & Ethics* 31:398–418.

López-Ibor, J. J., M. I. López-Ibor., and H. Helmchen. 2010. Ethics of diagnosis and classification in psychiatry, ethics in psychiatry. In: *Ethics in Psychiatry* (pp. 199-207). Springer Netherlands.

Lynoe, N., Z. Hyder, M. Chowdhury, and L. Ekstrom. 2001. Obtaining informed consent in Bangladesh. *New England Journal of Medicine* 344:460–461.

MacKenzie, C. R. 2009. What would a good doctor do? Reflections on the ethics of medicine. *Journal of Hospital for Special Surgery* 5:196–199.

Martin, J. B., and D. L. Kasper. 2000. In whose best interest? Breaching the academic-industrial wall. *New England Journal of Medicine* 343:1646–1649.

Math, S. B. 2004. *Clinical drug trials sponsored by foreign investigators in India*. PGDMLE: National Law School of India University.

Math, S. B., and D. Nagaraja. 2008. Mental health legislations: An Indian perspective. In *Mental Health; Human Rights,* eds. Prof Pratima Murthy and Prof D. Nagaraja. New Delhi; Bangalore: Publisher; National Institute of Mental Health and Neuro Sciences (Deemed University); India and National Human Rights Commission. http://nhrc.nic.in/Publications/Mental_Health_ Care_and_Human_Rights.pdf.

McMillan, J. R., and C. Conlon. 2004. The ethics of research related to health care in developing countries. *Journal of Medical Ethics* 30:204–206.

Mental Health Act. 1987. Mental Health Act. Government of India. In India G. o. (ed.).

Mr. X vs. Hospital Z. 1998. Mr. X. vs. Hospital Z. 1998. 8 SCC 296.

NIH. 2008. Protecting human reseach participants. http://phrp.nihtraining.com. Accessed 10 Dec 2011.

Nuffield Council on Bioethics. 2002. The ethics of research related to health care in developing countries. http://www.nuffieldbioethics.org/sites/default/files/Ethics%20of%20research%20related%20to%20healthcare%20in%20developing%20countries%20I.pdf. Accessed 9 Dec 2011 (London).

Pagnin, D., de. V. Queiroz, S. Pini, and G. B. Cassano. 2004. Efficacy of ECT in depression: A meta-analytic review. *Electroconvulsive therapy* 20:13–20.

Samira Kohli v Dr. Prabha Manchanda. 2008. Samira Kohli v Dr. Prabha Manchanda (2008). CPJ 56 (SC).

Schedule, Y. 2005. Schedule Y to the Drugs and Cosmetics Act, 1940. Amended in January 2005. http://dbtbiosafety.nic.in/act/Schedule_Y.pdf. also visit http://cdsco.nic.in/html/schedule-y%20(amended%20version-2005)%20original.htm. Accessed 15 Dec 2011.

Shah, R., and D. Basu. 2011. Coercion in psychiatric care: Global and Indian perspective. *Indian Journal Psychiatry* 52:203–206.

Sims, J., and H. V. Kuhnlein. 2003. *Indigenous peoples and participatory health research: Planning and management/ preparing research agreements. Centre for Indigenous Peoples' Nutrition and Environment(CINE) and World Health Organization.* Geneva: World Health Organization.

Tharyan, P., and C. E. Adams. 2005. Electroconvulsive therapy for schizophrenia. Cochrane Database of Systematic Reviews CD000076.

The Consumer Protection Act. 1986. http://ncdrc.nic.in/1_1.html. Accessed 27 Dec 2011.

Varmus, H., and D. Satcher. 1997. Ethical complexities of conducting research in developing countries. *The New England Journal of Medicine* 337:1003–1005.

Walcott, D. M., P. Cerundolo, and J. C. Beck. 2001. Current analysis of the Tarasoff duty: An evolution towards the limitation of the duty to protect. *Behavioral Sciences & the Law* 19:325–343.

Weindling, P. 2001. The origins of informed consent: The international scientific commission on medical war crimes, and the nuremberg code. *Bulletin of the History of Medicine* 75:37–71.

Wendler, D., E. J. Emanuel., and R. K. Lie. 2004. The standard of care debate: Can research in developing countries be both ethical and responsive to those countries' health needs? *American Journal Of Public Health* 94:923–928.

WHO. 2002. Council for International Organizations of Medical Sciences (CIOMS)/World Health Organization International Ethical Guidelines for Biomedical Research involving Human Subjects. http://www.cioms.ch/. Accessed 10 Dec 2011. (Geneva, World Health Organization).

Chapter 3
Ethical Handling of Religious and Spiritual Issues: South Asian Perspective

Gurvinder S. Kalra, Nilesh Shah, Nicholas Deakin and Dinesh Bhugra

3.1 Introduction

In countries and societies where religion and spirituality play a big role, it is inevitable that these values may well carry more weight than legal frameworks. Furthermore, when societies are kinship based and socio-centric as they are in South Asia, the beliefs about formal laws and structures may well be different from those societies where formal law has a bigger role. In the former kinship based looser structures may be seen as more important than rigid imposed legal expectations. Each society has a different contract with psychiatry and thus may have varying expectations related to religious and spiritual aspects of both the individual patient and their families and carers.

Gough (1936/1957), in the context of the development of the social contract, has provided an overview of both the historical contract and the context within which the social contract emerges. Gough notes that society was formed first and that the governing regulations—including ethical values—emerged later; and that within such a context there was an understanding that the monarchs had a divine right to rule and also had essential powers, especially related to healing and supporting of the ruled. He talks of ancient Greece where the context of democracy developed and the changes that occurred in the Western societies between the Fifth and Eleventh centuries. Regrettably, he does not talk about civilizations in the East such as those of India and China which may have had different patterns of growth. In the religious context he describes the role of the Catholic Church in the development of the social contract. However, embedded within the religious and social contract is the ethical framework. In South Asia however, even though organised religions have existed for millennia, their role is more philosophical rather than simply confined to organised activities such as rituals and taboos.

D. Bhugra (✉) · G. S. Kalra · N. Shah · N. Deakin
Institute of Psychiatry, London, England
dinesh@dineshbhugra.net

J. K. Trivedi, A. Tripathi (eds.), *Mental Health in South Asia: Ethics, Resources, Programs and Legislation,* International Library of Ethics, Law, and the New Medicine 58,
DOI 10.1007/978-94-017-9017-8_3

Religious beliefs and practices are extremely common throughout different parts of the world (Rosmarin et al. 2011) and form a core of the human existence. Increasingly in medicine and mental health, spirituality is becoming an important aspect of clinical practice with academic interest by clinicians and researchers alike. Spirituality refers to the transcendent relationship between an individual and a higher being (Peterson and Nelson 1989). It is a broader concept that should be seen as different from organised religious boundaries. It is a part of being human; every individual has a spiritual aspect apart from physical and emotional aspects. It is often thought of as being present as a quiet force in the background, acting as an inspiration and an orientation in our life (Waaijman 2002). On the other hand, religion, though often confused with spirituality, is an institution that is composed of rituals, customs and dogmas that survive on the basis of fear and blind faith, established in the name of God. Obviously religion also is an organised activity with very clear rituals and taboos. McSherry et al. (2009) caution that despite two terms (spirituality and religion) being confused often, they are not meant to be used interchangeably.

As noted above, interest in spirituality and spiritual issues in the context of clinical practice is high (Richardson 2002) and continuing to increase with these issues gaining prime importance in various areas of life. In psychiatric consultations; spiritual care is now recognised as an essential component of healthcare practice (Milligan 2011). Although spirituality is the most basic aspect of our being, it is also the least understood and likely to be confused with other aspects of an individual's functioning. Though psychiatric patients may hide their religious and spiritual feelings for fear of lack of understanding and sympathy from therapists (Bhugra and Osbourne 2004), they are often open to discuss these issues with their therapists; at the same time if they are encouraged. It is also likely that therapists and consultants themselves often feel poorly prepared to help their clients with these issues for the simple reason of having received no training in dealing with them or even exploring them.

The word ethics has been derived from the Greek term ethikos, meaning "rules of conduct that govern natural disposition in human beings" (see Oxford English Dictionary). In simpler terms ethics means principles of right conduct between different groups of people or individuals. Professional ethics thus refers to appropriate ways to act professionally. In the profession of psychiatry, ethics is a complex and often controversial issue.

Through successive editions of the 'Principles of Biomedical Ethics' (1979–2009), Beauchamp and Childress have proposed and refined a system for medical ethics based on the four principles of respect for autonomy, non-maleficence, beneficence and justice. This is worthy of note here. They claim that principlism has emerged following engagement with 'considered moral judgments… the way moral beliefs cohere… and professional values' (Beauchamp and Childress 2009). Briefly, autonomy involves respecting an individual's status as a self-governing moral agent; nonmaleficience with avoiding causing harm; beneficence with contributing positively to others' welfare; and justice with ensuring that everyone is treated fairly (Beauchamp and Childress 2009). Each principle in this normative ethical theory is intended to guide, and offer justification for, specific actions and general rules about them (Beauchamp and Childress 2009). Each principle has been described as 'prima facie' so they are all equally valuable in medical practice, and hold unless the situation leads to inter-principle conflicts. Principlism combines traditional and more contemporary values in medical ethics, ensures that decisions consider a range of factors and may help health professionals reach a set of universal moral values. Whilst principlism may appeal in theory as 'basic moral content' (Beauchamp and Childress 2009), in reality there is often conflict between the four principles and some critics argue that it should only be used to highlight ethical issues rather than serve as a decisive theory to be utilised in every moral dilemma. This becomes particularly important when we consider its use by those inexperienced in the realm of medical ethics who may find the management of conflicting principles particularly difficult. This said, the wide potential application of the four principles within psychiatric and medical practice leads some to argue that they 'constitute moral DNA' (Gillon 2003). Whilst there exists no universally accepted approach to such ethical dilemmas, principlism reaches similar results to many other ethical theories, including the practical approach proposed by the British Medical Association (Sommerville 2003). This success is inextricably linked to the fact that the four principles fuse the historical, Hippocratic principles of beneficence and non-maleficence with more contemporary patient considerations including justice and respect for autonomy (Beauchamp and Childress 2009). Indeed, as a testament to the content of their approach, even critics have conceded that the four principles are almost always involved in any real discussion about medical ethics (Harris 2003). In reality, it is very difficult to think of a case where psychiatrists should ignore that patients are independent moral agents (with some exceptions in paediatrics and where the law commands it in relation to mental health), not strive to do good, not minimise harm and where it is appropriate to distribute limited resources in a way that discriminates unfairly against certain groups of patients. The fusion of historical principles and more modern considerations make principlism an important theory that psychiatrists should use when faced with practical ethical dilemmas. Honest discussion with patients and their families in relation to these can foster confidence in the management plan being proposed. Where there is inner conflict between the principles, the relative weights of them should be considered and other theories and professional opinions should be sought.

Patients who have strong religious values and those who are more spiritually oriented have additional mental health related issues which must be taken into account in both assessments and in management (King 1978; Worthington 1986). Knowing ethical ways of dealing with various religious and spiritual issues of patients will help avoid ethical conflicts between psychiatrists and their patients and will improve the therapeutic relationship by increasing therapeutic adherence and alliance. The ethical conflicts and dilemmas that arise in psychiatric practice are somewhat different from those that arise in general medical or surgical practice. Not only in most societies psychiatrists are given the option of treating patients against their will and taking away their liberty whereas in other branches of medicine there is a clear assumption that they (the patients) want to be treated and may have higher levels of motivation related to levels of insight. An additional factor which may often play a role but is ignored is the 'Explanatory Model'—the explanations that the patients have for their experiences. In psychiatric practice these explanatory models will be significant in both help-seeking and utilising services. Psychiatric consultations involve a close relationship between the patient and therapist that can often lead to intense transference and counter-transference, which raise further ethical dilemmas and occasionally may be maliciously utilised by both the patient and the therapist. Moreover the distinction between normal and abnormal behaviors in the psychiatric context is hazy, which can easily lead one to question psychiatric diagnoses and treatments, most of which are aimed to change these behaviors and may be utilised for certain vested interests.

3.2 South Asian Religions

3.2.1 Hinduism

Hinduism is one of the oldest religions in existence and ranks as the world's third largest religion after Christianity and Islam (Wangu et al. 2009). To many, Hinduism is a religion, a culture or a way of life in itself. About eighty percent of India's population belongs to this religion (Narayanan 2010). Although thought of as a unified religion, it is divided into hundreds of internal divisions by caste, community, language and geography (Narayanan 2010). This religion comprises of multiple philosophical schools and beliefs coming together and recognises several sacred writings which contribute to these basic schools.

Two great epics, the Mahabharata and the Ramayana, have had a great influence on the principles and philosophy of Hinduism. *Karma*, *Samsara* and *Moksha* are some of the basic tenets of Hinduism. *Karma* is the idea that one's deeds will have an effect either in this life or in the subsequent life. *Samsara* means the cycle of a soul's birth, death, and rebirth. *Moksha* refers to the release of a soul from the cycles of *samsara* (Wangu et al. 2009). The idea of *Karma* governs all Hindu actions, since they believe that all their actions and deeds will have an effect either in this life or the next life.

3.2.2 Islam

Islam is said to be the world's fastest growing religion (Gregorian 2004) and the followers of Islam are called Muslims. The word Islam comes from an Arabic word that means 'surrender' or 'submission' (Farah 2003) and refers to the complete submission of individuals to the will of God that is supposed to be the only way in which an individual can achieve peace of mind (Maqsood 1995).

The basic principles of Islam encourage a belief in *Tawhid* (one-ness or unity of God, who is considered to be the absolute), *Nowbowat* (belief in the prophetic mission of Muhammad), *Ma'ad* (belief in the final day of judgment after death) and *Barzakh* (a state of waiting before the final day of judgment) (Maqsood 1995). There are also references to life after death in Islam, referred to as *Akhirah* (Maqsood 1995).

Islam lays particular significance on prayers that have to be offered five times a day as a reminder of the commitment of Muslims to Islam and God (Farah 2003; Gregorian 2004), thus inculcating God in the trivia of daily life (Farah 2003).

3.2.3 Sikhism

Often described as the newest and smallest of the world's religions (Nesbitt 2005), Sikhism began with the birth of the first guru of Sikhs, Guru Nanak Dev, in 1469 in Northern India (Kaur 2011). Unlike Hinduism, this religion rejects the caste system and believes in the equality of human beings in the sight of the guru (Duggal 1988).

Seva (selfless voluntary service), *langar* (cooking and eating irrespective of caste, religion or sex), and *sangat* (congregation) are an important part of the Sikhism doctrine and practice (Kaur 2011). The general approach of Sikhism is *Sahaja* (the easy way), and focuses on practical *sadhana* (worship of the God), that transcends the paths of *Jnana* (knowledge), *Bhakti* and *Karma* and rejects ritualism that makes man excessively dependent on various rituals (Duggal 1988) and hence may distract man from God. The basic teachings of Sikhism defer karmas done under influences of ego as they lead one to a vicious cycle of birth and rebirths while karmas of *kirtan, bhakti* and *seva* lead to union with the Lord.

3.2.4 Jainism

Jainism is thought to have originated in Northern India from where it later on spread to other parts of the country (von Glasenapp and Shrotri 1999). Jainas are the followers of *Jinas*, which means conqueror or the victorious (von Glasenapp and Shrotri 1999), the one who has conquered the worldly passions by one's strenuous efforts, the one who has attained supreme knowledge and is free from all sorts of worldly attachments. Often Jainism has been referred to as a set of principles that is preached by *Jinas* and is purely of human origin, emanating from a *Jina*, rather than some non-human or divine figure (Titze and Bruhn 1998).

Some of the basic tenets of Jainism include that of "*Asarah Samsarah*," and "*ahimsa*" (Titze and Bruhn, 1998; von Glasenapp and Shrotri 1999). While the former means "vain is the life in the world," the latter refers to the principle of non-violence towards living beings. The doctrines of Jainism understand that all *Jivas* or living beings, from the lowest to the highest level, are subject to suffering, pain and death. Jainism also believes in the doctrine of karma and that karma flows into an individual's soul by activities of body, speech and our mind. A good karma leads to influx of reward (*Punya*), while bad karma leads to influx of sin (*Papa*) (Shah

1998). Jainism describes nine types of *Punya*(donating food, drink, clothing, shelter and bed; doing good by thoughts, words and works, showing reverence to elders) and 18 types of *papa* (killing of living beings, untruthfulness, theft, un-chastity, excessive attachment to possession, anger, pride, greed, inclination, hatred, pugnacity, slander, passing on information secretly, captiousness, pleasure or displeasure, hypocrisy and heterodoxy) (von Glasenapp and Shrotri 1999). Jain texts describe five vows (*Anuvratas*) to every individual: *ahimsa*, *satya* (to always speak the truth), *asteya* (refrain from stealing), *Brahmacharya* (living in chastity), *Parigraha-tyaga* (not to desire new things greedily) (von Glasenapp and Shrotri 1999).

3.2.5 Buddhism

Growing out of the teachings of Buddha and an offshoot of Jainism (Thomas 1877), Buddhism has more often been considered a philosophy rather than a religion since it does not believe in a Supreme Being but on the contrary denies the existence of God (Keown 2009; Harvey 1990). However this tends to restrict the definition of religion to a belief in God.

Buddhism is supposed to have evolved alongside Hinduism in India and hence some of its concepts such as that of samsaras and karma seem to have been inherited from Hinduism (Cantwell 2010).

Buddhist practices have been increasingly influencing psychotherapy. Heravadan Buddhist mindfulness meditation and cognitive therapy practices have been used in the treatment of a client with depressive disorder with a degree of success.

3.3 Religious and Spiritual Issues

In the past, psychiatry as a profession sometimes has had a negative view of religious and spiritual issues, when such beliefs and practices were thought to have a pathological basis and were considered symptoms of mental illness (Verghese 2008). Psychiatrists may come across various spiritual issues in their clinical practice, for instance, possession states, mystical experiences, and near death experiences. They may also come across various other symptoms such as loss of interest in religious activities including prayers or meditation in major depressive disorder, religious delusions, or hearing voices of God or other religious figures in schizophrenia. Koenig (2009) argues that while religious beliefs and practices can serve as psychological resources for coping with stress and lead to improvement in overall health and well being of an individual (Maselko and Kubzansky 2006), they are often so intricately entangled with various disorders that it becomes difficult to determine whether they are a resource or a liability. For example, sometimes it is difficult to differentiate religious beliefs and rituals from delusions and compulsions, especially if these are rigid in nature or are followed rigidly by the individual. Hence, it is important that various normative spiritual and religious experiences be differentiated from psychotic episodes which are usually more intense and terrifying, often containing special messages from God or other religious figures and are associated with socio-occupational dysfunction (Greenberg and Witzum 1991).

A number of studies highlight the relation between religion/spirituality and mental health and psychiatric disorders. Huquelet et al. (2009) reported an inverse relation between religious involvement and substance use/abuse, with religion playing a protective role toward substance misuse in 14 % of their total sample. This may be explained by the promotion of qualities such as honesty and responsibility by religion and spirituality, which in turn may lead to abstinence (Tonigan 2003). Similar findings were also observed by Haber et al. (2012). They examined a sample of 4002 female adolescents/young adults and their families, and found an inverse association between risk factors for alcohol use and religious dimensions in their functioning. Furthermore they also noted that milestones of alcohol use such as initial drink, first intoxication, regular use, heavy consumption and alcohol dependence were also linked with religion/spirituality dimensions.

Propst et al. (1992) reported significantly lower post-treatment depression in patients with non-psychotic and non-bipolar depression, who received religious cognitive-behavior therapy (RCT). Patients were given RCT in 18–20 one-hour sessions over a period of three months. While religion can protect an individual against depression (Brown and Prudo 1981), loss of faith and religion can precipitate depression and in the long run may also make the prognosis of depression worse (Dew et al. 2010).

Religion may also have a protective role in suicide by condemning such acts (Huquelet et al. 2007; Stompe and Ritter 2011), for example death by suicide is considered haram (a sin) in Islam (Stack and Kposowa, 2011). Stack and Kposowa (2011) report that persons who live in nations with relatively high levels of religiosity and religious commitment, and who are engaged with a religious network, generally find suicide less acceptable. In this context, mosaic religions (that are based on the idea of man as the image of God, such as Christianity, Islam and Judaism) refuse suicide more strictly compared with the Eastern religions of reincarnation (which are based on the concept of transmigration and rebirth, such as the Eastern religions) (Ritter et al. 2011).

A collaborative study at three centers in India (Vellore, Chennai, Lucknow) reported better prognosis in patients of schizophrenia who spent more time in religious activities at two- and five-year follow up (Verghese et al. 1989, 1990).

Religious and spiritual beliefs and practices also help an individual in phase of life problems such as dealing with loss of spouse (Michael et al. 2003).

Sansone et al. (2011) used a cross-sectional sample of outpatients and reported statistically significant lower religion/spirituality (RS) status in individuals with borderline personality symptoms than those without this type of psychopathology. These researchers examined RS status using the 12-item Functional Assessment of Chronic Illness Therapy-Spiritual Well-Being Scale (FACIT-Sp −12), while borderline symptomatology was assessed using two self-report measures, the borderline personality scale of the Personality Diagnostic Questionnaire −4 and the Self-Harm Inventory.

Regular religious activities such as involvement in Church activities improves the quality of relationships and also parenting skills of parents which in turn leads to enhanced school performance and psychosocial adjustment in their children (Brody 2003).

Several studies point out that religious interventions such as prayers may also improve success rates of in-vitro fertilisation, decrease hospitalisations in septic patients, increase immune functions, improve rheumatoid arthritis and produce reduction in anxiety levels (Coruh et al. 2005). Post and Wade (2009) reviewed recent empirical research on religion and spirituality in psychotherapy and found it as an effective adjunct to traditional therapy interventions. These are in fact powerful medium that help in the healing process (Chaudhry 2008).

Mohr and Huquelet (2004) caution that in some instances in patients, religion itself may become part of the problem. In some cases, in managing various conditions patients may be supported by their community or family by using spiritual activities. For example, speaking in tongues may be seen as clinically pathological but socially this is not an issue if it is being carried out in the Church or in the religious context. In some cases pathology may be ignored using spiritual activities, and beliefs and rites and rituals whereas in others it may lead to being extruded from the kinship.

3.4 Ethical Principles and Dilemmas Ethics: Using Bhagvadgita as an Example

It has been argued that psychiatry has ignored religion and spirituality as possible sources of strength and well-being (Bhugra and Osbourne 2004) and psychiatrists have been seen as less religious than other specialists (Galanter et al. 1991; Curlin et al. 2005) though as noted above this is beginning to change. Religion and spirituality are arguably neglected areas in psychiatric mental health assessment and intervention (Mohr 2006) but clinicians need to know that these are significant aspects in their patients' social functioning. Therefore, they may benefit from an awareness of religious and spiritual issues in the lives of their clients and may in fact be able to provide more accurate assessments and effective treatment plans to them (Yarhouse and VanOrman 1999). For a successful psychiatric practice, it is important to establish a harmonious relationship between religion, spirituality and psychiatry (Turbott 1996). Integrating spirituality into healthcare systems requires incorporation of knowledge about spirituality into skillful assessment of patients, and using the existing psychiatric knowledge for patient care.

Psychiatrists need to follow the first principle of ethics (Ross (1939), respect for autonomy of the patient, through respecting the decision-making capacities of psychiatric patients and helping them make informed choices regarding treatments. Respecting a patient's autonomy will also include respecting their socio-cultural and religious values and beliefs. A psychiatrist's ideologies should play as little role as possible in patient's decision-making. Professionalism has at its core the four principles proposed by Beauchamp and Childress (2009). Across cultures the notion of patient autonomy will differ especially in socio-centric cultures hence clinicians must look out for these. The treating psychiatrist has to weigh the risk-benefit ratio of different choices that patients make and act in a way that benefits the patient (beneficence), thus avoiding causing any harm to patients (non-maleficence) whilst considering autonomy and wider societal issues of justice (Beauchamp and Childress 2009, pp. 12–13). The latter becomes particularly important where resources are scarce and rationing may well mean that treatment choices are restricted by financial constraints.

It is inevitable that in conjunction with the patients' explanatory beliefs their religious beliefs will influence the way they access and agree for various treatment modalities. Psychiatric patients often feel the need that their therapists should be aware of their spiritual beliefs and needs (D'Souza 2002). On the other hand, clinicians and therapists may be influenced by their personal attitudes towards religion. Transference and counter-transference related to religious values must be taken into account.

Religious psychotherapy (RPT) recognizes and utilises religious beliefs of clients in treatment in order to manage mental health problems. Berry (2002) reviewed randomized controlled research on use of RPT and found that RPT was as effective as standard treatment and that these results were both statistically and clinically significant. Similar findings of significantly faster improvements and response in patients of anxiety and depression have also been reported in other studies, when they received religious-sociocultural psychotherapy than those who received standard treatment (Razali et al. 1998; Azhar and Varma 1995).

Greenberg and Witztum (1991) recommend that therapists may need to have at least some basic knowledge of doctrines and rituals of various religions; they should assess patients using its terminology and if possible should approach these patients through the social organisation of their religious group. Clinicians need to maintain a non-judgmental attitude towards the patient's religious and spiritual leanings.

For example, Molzahn and Sheilds (2008) enumerate various factors that contribute to the reluctance seen in nurses to discuss spirituality with their clients: not having the right words, lack of education, viewing spirituality as somebody else's forte, influences of secularism and diversity in society, and the current health-care context. A number of these factors may in fact be similar and applicable to psychiatrists too.

Koening (2008) suggests the use of bio-psycho-socio-spiritual approach to psychiatric patients and stresses the importance of taking a history of religious and spiritual issues, the way patients experience these areas of life, and the meaning that they have in their lives. It is advisable to respect and support patients' religious beliefs if these help them to cope better, for example if praying everyday reduces the patient's stress, then it should be encouraged. However, if such beliefs affect patient's mental health negatively, then they can be challenged. Koening (2008) encourages teaming up of psychiatrists with religious workers; however this comes with a caveat of prejudices or myths that could arise in either of the team members.

Many beliefs and rituals can act as anti-anxiety processes. Religious practices, such as going to church and temple, are socially acceptable ways to deal with problems. Similarly, little bits of superstitions can also act as a good stress buster by projecting our anxieties. For instance, eating sweet yogurt before going for exams may actually reduce the student's anxiety.

Religious rituals have different backgrounds to their origins. For example, the practice of Indian Brahmans of sprinkling water around their food plates before sitting down to eat food began at a time when food was usually eaten sitting on the ground, which would attract different insects and ants with the smell of food. So to prevent them from reaching the plates, sprinkling water helped. However, with time, eating food on dining tables became customary; yet the sprinkling behavior which was earlier a necessity now became more of a ritual. Looking at this and other rituals, psychiatrists may find them unacceptable or irrational. But we also need to understand that they have their own importance, that individuals may have a certain amount of attachment to these rituals and find solace in them perhaps in the way they are able to control these.

Condemning or ridiculing religious practices of individuals will only hamper the therapeutic alliance and relationship. Instead, the pragmatic way is to go along with them, simultaneously introducing them to the biomedical model of disease and the need for treatment. This is important since religious gurus, places and other traditional healers may be their first contact in cases of emergencies and thus cannot be completely ignored. However, we as psychiatrists need to iterate the fact that practices such as chaining or flogging of psychiatric patients may harm them and actually aggravate the illness. Suggesting to them that a combination of both medicine and prayers (*dawaa aur duaa*) is likely to give better results than either of these alone may help them understand that, although they may go to the religious places and seek blessings, they have to continue taking their medications.

3.5 Do's and Don'ts in Clinical Practice

- Always respect his/her religious or spiritual belief.
- Change your frame of reference.
- Don't condemn or ridicule their practices.
- Introduce them to the biomedical system of medicine.
- Consider core ethical principles.

3.6 Conclusion

Spiritual and religious approaches may enhance contemporary psychiatric practice particularly as there will be a number of patients who will see religious and spiritual aspects as very important parts of their lives and functioning. Their ethical dilemmas need to be identified, acknowledged and worked with using established frameworks in order to improve therapeutic alliance, adherence and outcomes. Under these circumstances the psychiatrist must be aware of ethical dilemmas that the society may pose and therefore will need to take these into account in formulating any management plans. Socio-centric communities may feel that the kinship based laws are more appropriate for them which may also add another dimension of potential tension.

References

Azhar, M. Z., and S. L. Varma. 1995. Religious psychotherapy in depressive patients. *Psychother Psychosom* 63 (3–4): 165–168.

Beauchamp, T., and J. Childress. 2009. *Principles of biomedical ethics*, 6th ed. New York: Oxford University Press.

Berry, D. 2002. Does religious psychotherapy improve anxiety and depression in religious adults? A review of randomized controlled studies. *International Journal of Psychiatric Nursing Research* 8 (1): 875–890.

Bhugra, D., and T. R. Osbourne. 2004. Spirituality and psychiatry. *Indian Journal Psychiatry* 46 (1): 5–6.

Brody, G. H. 2003. Religiosity and family relationships. *Journal of Marriage and Family* 56 (4).

Brown, G. W., and R. Prudo. 1981. Psychiatric disorders in a rural and an urban population: etiology of depression. *Psychological Medicine* 11: 58.

Cantwell, C. 2010. *Buddhism: The basics* (Illustrated edition). New York: Taylor and Francis.

Chaudhry, H. R. 2008. Psychiatric care in Asia: Spirituality and religious connotations. *Intertional Review Psychiatry* 20 (5): 477–483.

Coruh, B., H. Ayele, M. Pugh, and T. Mulligan. 2005. Does religious activity improve health outcomes? A critical review of the recent literature. *Explore* New York 1 (3): 186–191.

Curlin, F. A., J. D. Lantos, C. J. Roach, S. A. Sellergren, M. H. Chin. 2005. Religious characteristics of US physicians: A national survey. *Journal of General Internal Medicine* 20: 629–634.

Dew, R. E., S. S. Daniel, D. B. Goldston, W. V. McCall, M. Kuchibhatla, C. Schleifer, M. F. Triplett, and H. G. Koenig. 2010. A prospective study of religion/spirituality and depressive symptoms among adolescent psychiatric patients. *Journal of Affective Disorders* 120 (1–3): 149–157.

D'Souza, R. F. 2002. Do patients expect psychiatrists to be interested in spiritual values? *Australas Psychiatry* 10: 44–47.

Duggal, K. S. 1988. *Philosophy and faith of Sikhism* (revised edition). Pennsylvania: Himalayan Institute Press.

Farah, C. E. 2003. *Islam: Beliefs and observances* (7th edition). New York: Barron's Educational Series.

Galanter, M., D. Larson, and E. Rubenstone. 1991. Christian psychiatry: the impact of evangelical belief on clinical practice. *American Journal Psychiatry* 148: 90–95.

Gillon, R. 2003. Ethics needs principles—four can encompass the rest—and respect for autonomy should be "first among equals". *Journal Medicine Ethics* 29: 307–312.

Gough, J. W. 1936/1957. *The social contract*. London: Oxford University Press.

Greenberg, D., and E. Witztum. 1991. Problems in the treatment of religious patients. *American Journal Psychother* 45 (4): 554–565.

Gregorian, V. 2004. *Islam: A mosaic, not a Monolith*. Washington, DC: Brookings Institution Press.

Haber, J. R., J. D. Grant, T. Jacob, L. B. Koenig, and A. Heath. 2012. Alcohol milestones, risk factors, and religion/spirituality in young adult women. *Journal Stud Alcohol Drugs* 73 (1): 34–43.

Harris, J. 2003. In praise of unprincipled ethics. *Journal Medicine Ethics* 29: 303–306.
Harvey, B. P. 1990. *An introduction Introduction to buddhism: teachings, history and practices* (Illustrated edition). Cambridge: Cambridge University Press.
Huquelet, P., L. Borras, C. Gillieron, P. Y. Brandt, and S. Mohr. 2009. Influence of spirituality and religiousness on substance misuse in patients with schizophrenia or schizo-affective disorder. *Substance Use and Misuse* 44 (4): 502–513.
Huquelet, P., R. S. Moh, V. Jung, C. Gillieron, P. Y. Brandt, and L. Borras. 2007. Effect of religion on suicide attempts in outpatients with schizophrenia or schizo-affective disorders compared with inpatients with non-psychotic disorders. *European Psychiatry* 22 (3): 188–194.
Kaur, N. G. 2011. *Sikhism: an introduction* (Illustrated edition). New York: I.B.Tauris.
Keown, D. 2009. *Buddhism: A brief insight* (Illustrated edition). New York: Sterling Publishing Company Inc.
King, R. 1978. Evangelical Christians and professional counseling: A conflict of values? *Journal of Psychology and Theology* 6 (4): 276– 281.
Koenig, H. G. 2009. Research on religion, spirituality, and mental health: a review. *Canadian Journal Psychiatry* 54 (5):283–291.
Koening, H. G. 2008. Religion and mental health: what shall psychiatrists do? *Psychiatry Bulletin* 32: 2001–2003.
McSherry, W., R. McNidder., and H. Wordsworth. 2009. Delivering spiritual care. Interview by Lynne Pearce. *Nursing Standards* 23 (28): 22–23.
Maqsood, R. W. 1995. *Islam: Examining Religions* 2nd ed. Oxford: Heinemann.
Maselko, J., and L. D. Kubzansky. 2006. Gender differences in religious practices, spiritual experiences and health: results from the US General Social Survey. *Social Science Medicine* 62 (11): 2848–2860.
Michael, S. T., M. R. Crowther, B. Schmid, and Allen, R. S. 2003. Widowhood and spirituality: coping responses to bereavement. *Journal of Women and Aging* 15 (2–3): 145–165.
Milligan, S. 2011. Addressing the spiritual care needs of people near the end of life. *Nursing Standard* 26 (4): 47–56.
Mohr, W. K. 2006. Spiritual issues in psychiatric care. *Perspectives in Psychiatric Care* 42 (3): 174–183.
Mohr, S., and P. Huquelet. 2004. The relationship between schizophrenia and religion and its implications for care. *Swiss Medical Weekly* 134 (25–26): 369–376.
Molzahn, A E., and L. Sheilds. 2008. Why is it so hard to talk about spirituality? *Canadian Nurse* 104 (1): 25–29.
Narayanan, V. 2009. *Hinduism: understanding religions* (Illustrated edition). New York: The Rosen Publishing Group.
Nesbitt, E. M. 2005. *Sikhism: A very short introduction* (Illustrated edition). Oxford: Oxford University Press.
Peterson, E., and K. Nelson. 1987. How to meet your clients' spiritual needs? *Journal of Psychosocial Nursing* 25: 34.
Post, B. C., and N. G. Wade. 2009. Religion and spirituality in psychotherapy: a practice-friendly review of research. *Journal of Clinical Psychology* 65 (2): 31–46.
Propst, L. R., R. Ostrom, P. Watkins, T. Dean., and D. Mashburn. 1992. Comparative efficacy of religious and nonreligious cognitive-behavioral therapy for the treatment of clinical depression in religious individuals. *Journal of Consulting and Clinical Psychology* 60 (1): 94–103.
Razali, S. M., C. I. Hasanah, K. Aminah, and M. Subramaniam. 1998. Religious-sociocultural psychotherapy in patients with anxiety and depression. *Australian and New Zealand Journal of Psychiatry* 32 (6): 867–872.
Richardson, R. 2002. *Spirituality: what What does it mean to be spiritual?* (Illustrated edition). Downers Grove: InterVarsity Press.
Ritter, K., W. Zitterl, and T. Stompe. 2011. Religion and suicide—part 2: confessions, religiousness, secularization and national suicide rates. Neuropsychiatry 25 (3): 127–134.
Rosmarin, D. H., S. Pirutinsky, and K. I. Pargament. 2011. A brief measure of core religious beliefs for use in psychiatric settings. *International Journal of Psychiatry in Medicine* 41 (3): 253–261.

Ross, W. 1939. *The foundation of ethics*. Oxford: Clarendon Press.
Sansone, R. A., A. R. Kelley, and J. S. Forbis. 2011. Religion/spirituality status and borderline personality symptomatology among outpatients in an internal medicine clinic. *International Journal of Psychiatry in Clinical Practice.* [Epub ahead of print].
Shah, N. 1998. *Jainism: The world of conquerors* Vol. 2. Brighton: Sussex Academic Press.
Sommerville, A. 2003. Juggling law, ethics and intuition: practical answers to awkward answers. *Journal of Medicine Ethics* 29: 281–287
Stack, S., and A. J. Kposowa. 2011. Religion and suicide acceptability: a cross-national analysis. *Journal of Science Study Religion* 50 (2): 289–306.
Stompe, T., and K. Ritter. 2011. Religion and suicide—part 1: the attitudes of religions towards suicide. *Neuropsychiatry* 25(3): 118–126.
Thomas, E. 1877. *Jainism: The early faith of Asoka.* London: Trubner and Co. Forgotten Books.
Titze, K., and K. Bruhn. 1998. *Jainism: A Pictorial Guide to the Religion of Non-violence.* 2nd ed. Delhi: Motilal Banarsidass Publishers.
Tonigan, J. S. 2003. Project match treatment participation and outcome by self-reported ethnicity. *Alcoholism Clinical and Experimental Research* 27: 1347.
Turbott, J. 1996. Religion, spirituality and psychiatry: conceptual, cultural and personal challenges. *Australian And New Zealand Journal Of Psychiatry* 30: 720–27.
Verghese, A., J. K. John, S. Rajkumar, J. Richard, B. B. Sethi, and J. K. Trivedi. 1989. Factors associated with the course and outcome of schizophrenia: results of a two year follow up study. *British Journal of Psychiatry* 154: 499–503.
Verghese A., J. K. John, S. Rajkumar, J. Richard, B. B. Sethi, and J. K. Trivedi. 1990. Factors associated with the course and outcome of schizophrenia: results of a five year follow up study. *British Journal of Psychiatry* 32: 211–216.
Verghese, A. 2008. Spirituality and mental health. *Indian Journal of Psychiatry* 50: 233–237.
Von Glasenapp, H., and Shrotri, S. B. 1999. *Jainism: Anm Indian Religion of Salvation* (Illustrated edition). Delhi: Motilal Banarsidass Publishers.
Waaijman, K. 2002. *Spirituality: Forms, Foundations, Methods* (Illustrated edition). Leuven: Peeters Publishers.
Wangu, M. B., J. O'Brien, and M. Palmer. 2009. *Hinduism: World religions* (4th ed). New York: Infobase Publishing.
Worthington, E. 1986. Religious counseling: A review of published empirical research. *Journal of Counseling and Development* 64: 421–430.
Yarhouse, M. A., and B. T. Van Orman. 1999. When psychologists work with religious clients: applications of the general principles of ethical conduct. *Professional Psychology Research and Prctice* 30 (6): 557–562.

Chapter 4
Ethics and Psychiatry: East Versus West

T. Maniam and Fifa Rahman

4.1 Introduction

Psychiatric patients are more vulnerable than most other patient groups. Their illness stigmatizes and renders them mostly powerless in society. In terms of social status they occupy the lower rungs of society. Apart from a few wealthier and/or more enlightened Asian countries, in many other countries many severely ill psychiatric patients are indigent, and with poorly funded public social services, if they have no caring family to help, the mentally ill are left to wander the streets, sleep in five-foot ways or under bridges. Many are abused; others are incarcerated in unhygienic and stultifying nursing homes or psychiatric institutions. In such circumstances, because they are voiceless and powerless, no other group of patients needs more a thoroughly ethical approach to caring. Ethics protects the helpless and upholds the dignity of the person.

At the outset a curious reader might wonder on what basis ethical rules of practice are drawn up. If East is different from West, do they have different guiding principles? This, as the reader would be aware, is not as simple a question as it appears at first. What is the basis of ethics? Is there a moral law underlying ethics? If so, whose law? Is it the belief in God that gives weight to this law? We neither have the space here, nor the expertise of a moral philosopher or ethicist, to discuss these questions apart from making a brief comment. We think it is reasonable to say that there are some differences between East and West partly due to the increasing secularization of western society. David Berlinsky (2009), a secular mathematician-philosopher, quotes Richard Rorty thus: "The West has cobbled together, in the

T. Maniam (✉)
National University of Malaysia (UKM) Medical Centre, JalanYaacob Latif, 56000
Kuala Lumpur, Malaysia
e-mail: tmaniam@yahoo.com

F. Rahman
Malaysian AIDS Council, Kuala Lumpur, Malaysia

J. K. Trivedi, A. Tripathi (eds.), *Mental Health in South Asia: Ethics, Resources, Programs and Legislation,* International Library of Ethics, Law, and the New Medicine 58,
DOI 10.1007/978-94-017-9017-8_4

course of the last two hundred years, a specifically secularist moral tradition—one that regards the free consensus of the citizens of a democratic society, rather than the Divine Will, as the source of moral imperatives." It is to be expected that in such a moral climate practices like physician assisted suicide would take root, since in a democracy what the majority supports (or at least what the leaders think the majority support) would be codified as law. One remembers Feodor Dostoevsky's warning in his book *The Brothers Karamazov* more than a century ago that if God does not exist, everything is permitted. Though globalization and the near-ubiquitous presence of electronic media have influenced the East and led to a dilution of many eastern values, we think it is fair to say that much of Asia holds, at least in name, to ancient beliefs and value systems. To take an example, for adherents of theistic religions, the dignity of the individual is grounded in their being created by God and therefore a person's life is of value and his rights must be respected.

The question of ethics in the practice of medicine in general and psychiatry in particular, cannot be studied in a vacuum, but within the cultural milieu in which the patient and the psychiatrist find themselves. In the wake of World War II, many Asian nations obtained independence from their colonial masters and began their journey towards modern nationhood. However the social structures of these countries have been slow to change to match the rapid socioeconomic changes and thus have impeded the transformation of these countries into egalitarian societies. Even where progressive laws have been put in place, their implementation has been uneven. Consequently, the treatment of socially disadvantaged groups may leave much to be desired. Caste systems, the politics of ethnic dominance, the economic and social disadvantages of belonging to minority ethnic groups or migrant labor all may affect the ethics of medical practice. However, there is a danger of overemphasizing the differences between East and West. Asia is becoming an increasingly wired continent, and its values are ever changing with global influence of the electronic media. In many ways the younger generation of Asians is not very dissimilar from its western peers. At the same time we need to be careful about drawing conclusions about the 'average Asian', just as there is no such thing as an average patient. It is a fallacy ably pointed out by Inkeles and Levinson (1969) that modal statistical characteristics influence our perception of national character.

The medical profession in Asia, as elsewhere, is generally held in high regard in society. It is seen as being honest and ethical in its practice (Gallup 2011). However, this high regard is in danger of being eroded. As the cost of medical care escalates the increasing utilization of complicated technology in diagnosis and treatment, gives rise, in the public mind, to suspicions that the decision to use such procedures in some cases is to pay for the mortgage on costly equipment. It is not uncommon to hear of a private medical clinic being referred to as a 'business' rather a medical practice. This carries the danger that doctors may be viewed as belonging to a 'money-grubbing' profession rather than a caring one. One already hears these comments and that not uncommonly.

An additional threat to the respect of patients is the not infrequent reports of unethical relationships between medical professionals and the pharmaceutical industry. While in the West there are increasing calls for the regulation of this relationship, in Asia such calls to regulate are few and far between. In many developing Asian countries pharmaceutical firms feed the unbridled greed of practitioners, and the profession runs the risk of gradually losing the respect of patients and of society in general. What happened to the status of the clergy in the United States in the last few decades, which, in the aforementioned Gallup poll (Gallup 2011), went from being among the top most highly respected to 5th in the ranks, is a warning to the medical profession.

4.2 Paternalism Versus Autonomy

Many Asian countries are just beginning to emerge from their feudal past. The ruling classes and others in the upper classes held sway over lesser mortals. Doctors used to come from these upper classes until wider access to education began to level matters somewhat. In many Asian nations the medical profession is paternalistic especially towards the psychiatrically ill, deciding unilaterally what is good for the patient. Often discussions with patients on matters of ethics such as informed consent, for example, take the form of a monologue, with the doctor telling rather than listening. This is changing slowly, and is expected to gain momentum with increasing literacy rates and the advent of electronic media. It is not uncommon for patients to read up on their illnesses and their treatment and side-effects and raise their concerns with their doctors. The paternalistic professional would see this as challenging his exalted status as the learned doctor. Questions by patients and their relatives are sometimes seen as nuisance, or worse, as challenging the superior knowledge of the doctor. Offence may be taken. One not uncommonly hears of patients' complaint that a particular doctor had dismissed their queries and concerns by a fiat statement, "Are you the doctor or am I?" Such paternalistic attitudes are gradually disappearing.

Sometimes treatment procedures may be conducted without the patient fully understanding the consequences.

> An example, albeit an extreme one, seen by the first author many years ago, was the case of a woman who was being treated for depressive disorder; she gave a past history of a surgical procedure when she was in her 40s, which resulted in her uterus being removed. She had signed consent for surgery but had never been told it had anything to do with her uterus leave alone that it might result in her losing it. There was in reality no informed consent. It is possible that this happened because she was seen as a psychiatric patient.

Such extreme cases do not happen anymore, but less invasive procedures, such as intramuscular injections and the use of physical restraints, are often carried out without due regard for the patient's dignity and autonomy. A couple of decades back the first author has observed some hospitals take blanket consent for procedures from relatives of psychiatric patients at the time of the patients' admission. This was consent for giving electroconvulsive treatment under general anesthesia or transfer to a psychiatric institution hundreds of miles away. No explanations were given. Fortunately such practices have stopped. 'Consent' is not valid if it is not informed consent. It must not be presumed that psychiatric patients, even those with psychotic illnesses, are unable to provide informed consent. The information normally provided to non-psychiatric patients when their consent is obtained should not be truncated in the case of the mentally ill on the erroneous belief that they cannot understand such information. In general most of them do. In the event that they are, in legal language, of unsound mind, then responsible relatives may give consent on their behalf. In some countries who may give consent on the patient's behalf is spelt out by law (for example, the Malaysian Mental Health Act 2001 specifies these). The information provided for informed consent procedures should include a description of the procedure, the likely side-effects, what alternative treatments are available, why this particular treatment or procedure is preferred, the patient's right to refuse or withdraw consent, and that he/she would continue to receive the usual treatment after such refusal. This applies to informed consent for research as well. If consent has been taken from a legal representative because the patient was of "unsound mind", then fresh consent must be obtained from the patient at the earliest opportunity when he has regained competence.

It must be noted, however, that in Asian settings, personal autonomy may not be the highest good. Non-western societies are not as individualistic as their Western counterparts. Asians are more family oriented and the welfare of the family and society at large figures large in their decisions about ethics. Psychiatrists must take this into account. Most ethical guidelines are based on the Western model and the individual rights of the patient take precedence over the rights of the family, and the community. While individual rights rightly form the backbone of any discussion of ethics, and therefore must be carefully safeguarded, these must be seen in the context of the social background. In much of Asia the family is the main caregiver for the patient. Social welfare aid and properly run nursing homes are often non-existent. Legal aid may be available on paper but often does not reach the patient. When patients are in trouble, legally or financially or in any other way, it is to the family they turn to and therefore it is in the individual patient's interest that the rights and welfare of caregivers are also nurtured. Safeguarding the patient's autonomy while taking care of the legitimate concerns of the family may present thorny problems. Another case illustrates this:

> A young woman in her 20s was admitted to the psychiatric ward by her family. She had adjustment disorder with complicated relationship problems. After a few days the patient asked to be discharged but requested that her family not be informed since they would insist that she remain warded. The doctor was in a dilemma, since it was the family who were taking care of her, providing shelter and financially supporting her. However she was an adult who had the right to request for discharge. The patient was discharged and went to live with her friend. Later the doctor had to face the wrath of visiting family members.

In a Western setting the above-cited case is a no-brainer, but not so in the Asian environment. Collective interest often takes precedence over individual ones. Finding a balance between the two often requires wider consultation with peers. It must be noted that respecting the rights of psychiatric patients would help to reduce the stigma attached to mental illness. Stigma may also be reduced by improvements in the settings in which psychiatric treatment is provided. All too often psychiatric facilities get meager funding and such facilities compare poorly with medical and surgical facilities. It is obvious that this will increase the stigma, which, as Norman Sartorius frequently says, is attached to not only to psychiatric illness and patients but also to psychiatric medications, clinics, and wards and to anyone who works in psychiatry including psychiatrists.

4.3 Boundary Issues in Asian Setting

A second area of interest is the concept of therapeutic boundary. Boundaries exist in all relationships. The doctor-patient relationship is a therapeutic relationship, but it is an unequal relationship, in that the physician is in a position of power over his patient. The unethical physician, who is motivated by self-interest, may impose decisions upon the patient which may not be in the latter's best interest. Therefore therapeutic boundaries have been formulated primarily for the protection and well-being of patients. These are particularly necessary in psychiatry because of the special vulnerabilities of psychiatric patients mentioned earlier.

But in the ways how therapeutic boundaries are adhered to may vary between East and West. As has been noted before: "In Western practice therapeutic boundaries are more rigidly drawn, no doubt for good reasons. Boundary violations adversely affect the course of therapy and may harm patients in very serious ways and may even permanently scar them. So there are very cogent reasons to be circumspect about therapeutic boundary. But in the East patients tend to see the doctor/therapist as more than a fee-for-service practitioner. The therapist, like the traditional healer, is part of the community, and is seen as being equally accessible. Her presence is sought at community functions such as weddings of patients' children; her advice is sought on areas outside of the immediate concerns of therapy. She is plied with gifts during festivals, and declining to receive them is deemed to be insulting. Some patients of course, might abuse this, and use the 'friendship' of the therapist as a reason to avoid bringing up difficult issues, or to seek favors that might impinge on the therapist's time etc. So the Asian therapist needs to walk a tight rope—she would have to modify the more rigid practices in relation to boundary issues, at the same time protecting patients from harmful boundary violations" (Maniam 2009).

Boundaries encompass many aspects of the therapeutic setting—verbal and non-verbal. The latter include behavioral aspects as well as components of non-verbal communication such as spatial (distance), vocalics and oculics.

Boundary issues may be regarded as falling into two distinct, yet related, groups—boundary crossings and boundary violations. Firstly, boundary violations, which constitute professional misconduct inimical to the doctor-patient relationship, would, one would expect, be similar cross-nationally. They are generally harmful, unethical and may be subject to disciplinary or legal proceedings. Boundary violations such as inappropriate physical contact with patients and business relationships would be similarly regarded as violations in both Eastern and Western contexts. Boundary crossings, on the other hand, are much more complicated and are sometimes unavoidable. These are situations which are not as serious as violations and perhaps at times might be helpful for patients; some authors consider them to be therapeutic at times. Examples include receiving gifts from patients, sharing personal information with patients and meeting patients outside of the therapy situation. What constitutes a boundary crossing is to a great extent contextually determined, and may differ between cultures, therapies, therapists as well as the clients involved, besides the setting in which the therapy is conducted. The concepts of boundaries in therapy have evolved over time, and should not be seen as a set of rigid unchanging rules. It is instructive to remember that Freud himself appears to have had a rather flexible view of boundary crossings: he is known to have had a meal with his client, and had conducted therapy while walking his dog. Therapists need to be sensitive to the differing contexts to avoid becoming excessively rigid, yet alert not to become exploitative in the guise of flexibility. What may begin as boundary crossings may become violations.

How then shall we protect our patients, maintain reasonable boundaries and still manage not to appear stand-offish? The therapist must wisely decline to advice patients on areas that are beyond their expertise, even when such advice is sought. We must refuse the seduction of the inherent flattery when our advice is sought on subjects far and wide. Patients must be redirected to relevant experts. We must also anticipate situations that may lead to the crossing of proper boundaries.

There are Ethical Codes for medical practitioners in each country. Those who belong to the medical profession are subject to the Code of Conduct of the profession which is legally enforceable. In many countries counselors and non-medical psychotherapists have governing bodies to provide and enforce guidelines. Therapists who work with government agencies are subject to the rules and regulations of service. We note, of course, that rules and regulations notwithstanding, abuse of patients' trust does take place even in countries where ethical codes are strictly enforced. The first author has a patient now who, while staying overseas in a developed country where psychiatry is highly regulated, was abused by a senior therapist, who was a recognized leader in his field.

4.4 Special Situations

4.4.1 Ethical Treatment of Psychiatric Patients with Medical Illness

Psychiatric patients as a group experience high medical morbidity. As many as 60 % of psychiatric patients may have some medical disorder. Diabetes mellitus, hypertension, metabolic syndrome and thyroid disorders are not uncommon afflictions. These are among some of the reasons why psychiatric patients have shorter life spans. For a number of reasons adduced below these may not be adequately treated:

1. The illness may remain undetected
2. Patients do not wish to undergo more investigations and receive more drug treatments
3. Patients do not understand the serious health implications of these disorders
4. Patients are reluctant to visit an endocrine or general medical clinic; most do not have a family doctor or general practitioner
5. Medical and endocrine clinic personnel may not be sympathetic to the difficulties psychiatric patients experience in following dietary restrictions
6. The long waiting lists and clinic waiting times outside medical clinics discourage patients
7. Patients' lack of insight

In such situations it is ethically incumbent upon the psychiatric doctor to institute treatment for these medical conditions and refer to their medical colleagues if there are special problems. In other words psychiatrists need to be physicians for their patients.

4.4.2 Ethics of Non-Treatment

Is it ethical to let a patient remain untreated just to protect his autonomy? Would it protect his dignity if someone who is severely depressed, exercises his right to refuse treatment, but becomes, more psychotic, attempts suicide, takes up abusing alcohol, loses his job and is turned out of his billet? It is quite apparent that the best interest of the patient is served by some form of treatment which might initially be against the patient's personal wishes, but which serves his wider welfare. Some human rights activists, but not all, defend the right of the patient to refuse treatment, even if that means the patient might die by committing suicide. This would not go down well in most Asian societies where family ties are strong, where one does not belong to oneself alone, but to a family and by extension to one's community. Extreme individualism fails to see that human rights are not an end in themselves but a means to promote both the wellbeing of the individual and the community one belongs to.

4.4.3 Physician Assisted Suicide (PAS)

There have been increasing calls in some Western countries to legalize physician assisted suicides. PAS differs from euthanasia in that the doctor enables a patient to commit suicide by placing in his (the patient's) hands the means of causing his own death; it is the patient who carries out the final act of administering the lethal procedure. Patients who suffer from terminal and disabling illnesses sometimes seek recourse to this means of ending their suffering. A number of countries like the Netherlands have enacted laws permitting doctors to assist in patients' suicide, as has the state of Oregon in the United States. The argument goes that when a person is seriously ill, when he has lost control over his bodily functions and becomes dependent on another person, then such a person has lost his human dignity and therefore may choose to end his life. Doctors therefore must assist him to take this ostensibly dignified exit. This stance raises three, among others, very thorny problems.

Firstly, it presents a great dilemma to the families of ill people. When an elderly parent is very ill and has become totally dependent should they seek to terminate their parent's life? In the East respect for one's parents is ingrained from the early years, and filial attachment/piety demands that sick or disabled parents are to be cared for and not put to death. At the very worst disabled and dependent parents might be sent to a nursing home.

Secondly, PAS places the doctor in an untenable position in terms of his role. From the role of a care-provider he has to switch the role of a terminator of life. In such a situation it is conceivable that patients would become confused.

Thirdly, where would it all end? Would life be terminated for economic reasons, or to free up much needed beds? Would the depressed suicidal patient be granted his wish, when his depression might be treatable? Jochensen and Keown (1999) reporting a survey, with comparison with an earlier study, raise disturbing questions about the practice of euthanasia and physician assisted suicide in Netherlands. They show that among psychiatrically ill patients who requested to have their lives ended, and whose request was denied, only 16 % went on to commit suicide without doctors' assistance. More significantly 35 % no longer wished to die, and among a further 10 % there was a diminishing of the wish to die.

More ominously Jochensen and Keown noted that in large numbers of cases life was terminated without the express wish of the patient. Citing this and other reasons they came to the dismal conclusion that in spite of attempts to tighten this practice "the practice of euthanasia remains beyond effective control." While "slippery slope" arguments are generally logically suspect, in this case there appears to be evidence that over time the situation might worsen.

4.5 Relationship of Psychiatrists with Pharmaceutical Companies

There has been increasing concern about the nature of the relationship of doctors with pharmaceutical firms and the influence this relationship exerts on clinical decision making. Where many western professional organizations have begun limiting the role that 'Big Pharma' plays in medical affairs it is the authors' opinion that in Asia there is too much uncritical acceptance of involvement and influence of pharmaceutical firms. While it is widely accepted that drug companies have an important role in continuing medical education (CME), it is less well-recognized that gifts and sponsorships do skew medical decision-making. Gifts produce a sense of obligation to reciprocate, and this is usually an unconscious thing, and the main way a doctor can reciprocate is by prescribing the drug marketed by the sponsoring firm. We can easily persuade ourselves that we are acting in the best interest of the patient, when we are mostly acting to satisfy our unconscious need to reciprocate. There needs to be more checks and balances regarding this. Apart from physician education there need to be guidelines. Many have proposed that drug companies be required to place their CME budgets with professional organizations who would organize the CME activity and determine its scientific content. Alternatively, there have been suggestions that pharmaceutical firms pay a special tax which would be set aside for physician education, thus distancing the doctor from immediate pecuniary contact with the firm.

4.6 Research

Good Clinical Practice (GCP) guidelines on the ethical issues in psychiatric research have been formulated in many Asian countries. By far and large they are based on the Helsinki Declaration—which grounds its ethical approach on the basic principles of patient autonomy, beneficence and justice. GCP guidelines across the world are similar, since the results of research are published internationally, and regulatory bodies evaluate the ethical practices of all the countries that contributed data. Therefore in the area of research there is unlikely to be material differences of significance between the East and West.

What is possible, however, is that the rigorousness of implementation of the guidelines may vary across countries. For example there may be subtle pressures on less educated or poor patients to consent to participate in a clinical trial, though the Ethics Committees and GCP guidelines have laid down clear rules on these.

4.7 Legal Aspects

In a fairly conservative Asian country where the concept of autonomy is beginning its nascent rise, the question of ethics in psychiatry is a delicate balance of legal principles, newly enacted but inadequate mental health law, and therapeutic privilege.

Ideally, legislation should be geared towards principles that the autonomy of the patient is protected, that he or she is protected from inhumane and degrading treatment, the rights of those caring for the person is taken into account, and communication of risks and benefits to the public is effective. The task of the legislator is difficult; he must draft rules that balance rights effectively, yet at the same time remain practical. The concern exists of legislators overstepping boundaries and impeding psychiatric practice by unreasonably restricting decision-making ability. Gardner notes that there is: '… an uneasy relationship between medicine and law. Conflicts can and do emerge between competing rights and interests…'

(Gardner 2000) There is an idea that excessive intermingling of medicine and law can adversely affect therapeutic relationships (Sage 2001) as the doctor must have a certain degree of decision-making power, and should not fear the court in exercising his discretion for the benefit of the patient. The conflict is adequately summed up in the words of William M Sage: 'Substituting autonomy for beneficence is not always comfortable for physicians but intuitively appeals to lawyers' (Sage 2001).

Mental health law currently applicable in Malaysia is contained in the Mental Health Act 2001 (hereinafter MHA 2001). Despite its name, there was a delay of 9 years before the Act actually came into force, despite receiving Royal Assent (as required in Malaysian lawmaking) on 6 September 2001 and being published in the government Gazette on 27 September 2001. The delay was caused by a multitude of reasons, the foremost of which the coming-into-force of provisions related to private healthcare.

That aside, when the MHA 2001 came into force, it was recognized as a definite improvement from the previous statutes, with more detailed provisions on voluntary and involuntary treatment. In the 2001 Act, involuntary patients can be admitted to a psychiatric hospital by way of application by a relative (Mental Health Act 2001 (Act 615), Laws of Malaysia, s10(1)(a)) or upon recommendation of a registered medical practitioner or medical officer. (Mental Health Act 2001 (Act 615), Laws of Malaysia, s10(1)(b)) Subsequent to that, within 24 hours of the involuntary patient to the psychiatric hospital, a medical officer or registered medical practitioner shall make an examination of the patient to ascertain whether his detention is further necessary. (Mental Health Act 2001 (Act 615), Laws of Malaysia, s10(3)) If such detention is no longer necessary, the Medical Director may discharge him. (Mental Health Act 2001 (Act 615), Laws of Malaysia, s10 (5)(a))

The Indian Mental Health Act (Mental Health Act 1987, India) regulates involuntary admissions under Section 19 of the Act, albeit not explicitly stating the words 'involuntary patient'—it instead states 'any mentally ill person who does not, or is unable to, express his willingness for admission as a voluntary patient, may be admitted and kept as an inpatient in a psychiatric nursing hospital or psychiatric nursing home on an application made in that behalf by a relative or friend of the mentally ill persons if the medical officers-in-charge is satisfied that in the interest of the mentally ill persons it is necessary to do so' (Mental Health Act 1987, India, s19).

JK Trivedi suggested in 2009 that the provision could be improved by including a set criteria for involuntary admission, and that two accredited medical practitioners—one of them a psychiatrist—certify mental disorder prior to involuntary admission. In addition to that, the author criticized the procedures in the Act for further detention—in that the requirement to obtain a reception order from a Magistrate was 'overly legal, cumbersome and complicated for patients and their family' (Trivedi 2009).

This may be contrasted with involuntary patient admission provisions contained in the New South Wales, Australia Act which contains general provisions in addition to specific provisions on involuntary admission, which state that 'a person must not be involuntarily admitted unless an authorized medical officer is of the opinion that the person is a mentally ill person or a mentally disordered person and no other care of a less restrictive kind, that is consistent with safe and effective care, is appropriate and reasonably available to the person.' (Mental Health Act 2007 (NSW), s12(1)).

In regard to ECT, Section 77(1) of the Malaysian Mental Health Act (Mental Health Act 2001 (Act 615), Laws of Malaysia) states: 'Where a mentally disordered person is required to undergo surgery, electroconvulsive therapy or clinical trials, consent for any of them may be given by (a) by the patient himself if he is capable of giving consent as assessed by a psychiatrist (b) by his guardian in the case of a minor or a relative in the case of an adult, if the patient is incapable of giving consent and (c) by two psychiatrists, one of whom shall be the attending psychiatrist, if there is no guardian or relative of the patient available or traceable and the patient himself is incapable of giving consent.

In our view it is interesting that in subsection (b) the Act uses the word 'relative' in relation to adult incapable of giving consent. In the jurisdiction of New South Wales, Australia, the use of the word 'relative' alone is expressly not used in isolation to avoid situations where the patient is subjected to treatments consented to by relatives who simply view him or her as a burden and do not really take into account his or her views or concerns. The words used are 'primary carer, relative, or friend'. (Mental Health Act 2007 (NSW), s26) As it is, public perception of mental illness in Malaysia still operates along the lines of ostracizing from the society in general, and laypersons still refer to the mentally ill as 'crazy people'. Thus there may be an increased burden on the law to legislate thoroughly on this matter not only to regulate the administering of ECT, but also to influence public perception of the mentally ill. The law should operate to ensure that the patient is treated as humanely as possible, and one way of ensuring that is to make sure that the person making decisions on the patient actually cares for his or her welfare.

Section 77 of the Malaysian Mental Health Act does not detail what should be done in a situation where the patient refuses ECT and therefore is an involuntary patient. It only details those who are 'incapable' of giving consent but not those who are 'capable' but merely refuse consent. This opens avenues for arbitrary action towards patients. The establishment of a Mental Health Review Tribunal similar to that in New South Wales, Australia would remedy these defects.

In determining what steps the law needs to take, it is necessary to examine what current generally-accepted clinical practice is. In Malaysia, current clinical practice in obtaining informed consent for ECT involves the following:

1. Both the patient and a responsible adult caregiver must give written consent (in case of emergencies or if the patient is not able to give consent, then a caregiver's consent is sufficient)
2. Prior to obtaining consent, both the patient and the caregiver must be informed of:

   a. The alternative treatments available
   b. The reason for choosing to administer ECT (benefits of ECT and likely consequences if ECT is not given)

c. The ECT procedure including the administering of general anaesthesia
d. The risks and side-effects of ECT
e. Likely number of ECT treatments that may be administered
f. Their right to withdraw consent at any time before completion of course of ECT treatment

3. Both the patient and the caregiver must be given adequate time to ask questions and receive satisfactory answers.

It should be noted, however that none of these steps are mandated by law, but instead are steps that are mutually recognized and understood by competent practitioners of psychiatric practice in Malaysia.

Informed consent forms in Malaysia for ECT are very general. To some, arguing over the definition of informed consent in the context of ECT is an exercise in futility, rooted in the claws of the overzealous lawyer or legislator. But we must remind ourselves of the patient pool that ECT is centred around. The patients may have impaired thinking processes and are more vulnerable to bias than the average person, and so the question of consent is doubly delicate.

A more detailed consent form, however, does not necessarily mean increased patient understanding. Cassileth et al. noted that even with competent patients, an elaborate informed consent form does not necessarily mean that they comprehend and appreciate fully the nature and risks involved in receiving ECT (Cassileth et al. 1980). Richard et al. state: '[To] assume that the patient's decision is based firmly on a complete understanding of the issues and without outside influence is neither realistic nor achievable and ultimately underestimates significantly the finitude of human beings' (Richard et al. 2010). It may be possible to construe from these statements that consent forms should be *appropriately* detailed as opposed to more heavily detailed, and that an improvement of consent forms alone is insufficient to protect the rights of patients. This point in particular is especially pertinent to mental health patients who may require more 'careful' and/or repeated provision of information, both via the consent forms and via doctor-patient interactions.

It may, therefore be up to the law to set down legal framework to help medical practitioners communicate to patients concise information on ECT. An amendment to the Malaysian Mental Health Act 2001 to include the above 'mutually recognized' steps would help to increase transparency and enable uniformity in the obtaining of informed consent in the administering of ECT.

As a whole, ethics and law relating to electroconvulsive therapy should seek to ensure that mentally ill persons are protected from harm and inhumane and degrading treatment, to ensure that services are accessible and to ensure that patient's rights to information are protected, whilst maintaining doctor's rights in regard to the doctrine of therapeutic privilege.

Mental health litigation or litigation involving mental health issues are on the rise in Malaysia, and show the tendency of the courts to increasingly recognize the rights of persons with mental illness or mental disorder(s). For example, in the 2003 case of *TenagaNasionalBhd v. SugumaranJohansonSundram*, the claimant (S) had been absent from work without authorization for due days due to a mental condition and was hospitalized during this time. The defendant employer (TNB) dismissed the claimant due to this absence, and it was held that the actions of the defendant were unjust, and that the defendant had failed to consider the special circumstances of the claimant.

However, the opposite occurred in the case of a government officer who was dismissed—according to the claimant—due to his mental condition. In *Dr Meer Ahmad Mydin Meera v Public Services Commission Malaysia & Ors* (2010), the claimant was a civil servant employed at the Ministry of Health. His employment was terminated in the wake of several incidents. The government was held not liable because of a clause in government contracts that they do not have to give reasons for termination.

However, there seem to be a paucity of cases debating particular ethical dilemma—such as cases initiated by hospitals, seeking to carry out electroconvulsive therapy on non-consenting or involuntary patients, for example. There is also a lack of judicial discussion on causative links between particular traumatic head injuries and psychiatric illnesses or disorders diagnosed post-injury.

This is contrasted with discussions in the West, where medico-legal and ethics questions are complex, and are discussed in open court. In the South Australia case of *Hawker & Ors v Miller*, for example, discussed whether the schizoaffective disorder suffered by the plaintiff was a result of the head injury suffered in a vehicle accident, or whether the disorder was independent of the accident. Equally interesting is another discussion in the New South Wales District Court in *Nuhic v Orahim*, where as a result of preexisting injuries, evidence showed that the plaintiff had vulnerabilities to psychiatric illness prior to the motor vehicle accident, and hence required a detailed consideration of monetary damages.

Neither in Asia has the concept of Mental Health Review Tribunals evolved. These tribunals are sometimes referred to as 'problem-solving courts' (Nolan 2009), although this term may be misleading as to the tribunal structures. These tribunals are intended to act as 'therapeutic agents' (Donnelly et al. 2011), and usually consist of lawyers, psychiatrists, psychologists, and social workers working together to make ethical decisions in regard to individual mental health dilemmas.

The efficacy of these tribunals is still being debated—some studies have found that 'negatively perceived hearings can have adverse effects on therapeutic relationships.' (Donnelly et al. 2011) Others say that these tribunals only deal with the issue at hand—such as committal proceedings or involuntary treatment, but do not take into account patient issues such as 'objections against particular medicines or side-effects and desires for psychosocial counseling' (Carney 2011).

4.8 Conclusion

Asian countries have come a long way in the last half a century in their ethical treatment of psychiatric patients. However, there are pockets of Asian society where there are areas for improvement. Newer ethical challenges have arisen and these include the need to balance the right of the patient to privacy and confidentiality versus the right of the caregiver to know and to be involved in the management of the patient. Even when new more enlightened mental health legislation has been introduced, its implementation is often hampered by social and economic factors (Kallivayalil et al. 2009). Another area that poses a challenge is the close involvement of the pharmaceutical companies with the psychiatric profession. A fresh approach is needed to balance their role as providers of CME with their marketing activities that may unduly influence the clinical decisions of psychiatrists.

References

Berlinsky, D. 2009. *The Devil's Delusion*, 39. New York: Basic Books.

Carney, T. 2011. Mental health tribunals as governance: lessons from an Australian study? Dublin University Law Journal, Forthcoming; Sydney Law School Research paper No. 11/12. SSRN: http://ssrn.com/abstract=1753334.

Cassileth, B. R., R. Zuphis, K. Sutton-Smith, and V. March. 1980. Informed consent—why are its goals imperfectly realized? *The New England Journal of Medicine* 302:k896–900.

Donnelly, V., A. Lynch, D. Mohan, and H. G. Kennedy. 2011. Working alliance, interpersonal trust and perceived coercion in mental health review hearings. *International Journal of Mental Health Systems* 5:29–45.

Dr Meer Ahmad MydinMeera v Public Services Commission Malaysia & Ors. 2010. 5 CLJ 675.

Gallup. 2011. Ethics: Record 64% rate honesty, ethics of members of congress low. http://www.gallup.com/tag/Ethics.aspx. Accessed 8 Feb 2012.

Gardner, J. 2000. Mental illness—freedom and treatment. *Law in Context* 17 (2): 120.

Hawker & Ors v Miller. 2009. SADC 50.

Inkeles, A., and S. J. Levinson. 1969. National character: The study of modal personality and social cultural systems. In *The handbook of social psychology. reading,* eds. G. Lindzey and E. Aronson. MA: Addison-Wesley.

Jochensen, H., and J. Keown. 1999. Voluntary euthanasia under control? The latest empirical evidence on euthanasia in the Netherlands. *Journal of Medical Ethics* 25:16–21.

Kallivayalil, R. A., J. K. Trivedi, and A. Tripathi. 2009. Social factors and forensic psychiatry in India. *Current Opinion in Psychiatry* 22:482–487.

Maniam, T. 2009. The assimilation of western psychotherapeutic practices in Asia. Proceedings of the 3rd Congress of ASEAN Association of Psychotherapy. Jakarta.

Mental Health Act. 1987. (India).

Mental Health Act. 2001. (Act 615), Laws of Malaysia.

Mental Health Act. 2007. (NSW).

Nolan, J. L. Jr. 2009. *Legal accents, legal borrowing*. New Jersey: Princeton University Press.

Nuhic v Orahim. 2011. NSWDC 200.

Richard, C., Y. Lajeunesse, and M.–T. Lussier. 2010. Therapeutic privilege: Between the ethics of lying and the practice of truth. *Journal of Medical Ethics* 36:353–357.

Sage, W. M. 2001. The lawyerization of medicine. *Journal of Health Politics, Policy and Law* 26 (5):1179–1195.

TenagaNasionalBhd v. SugumaranJohansonSundram. 2003. 3 ILR 1041.

Trivedi, J. K. 2009. Mental health act, salient features, objectives, critiques and future directions. In *Forensic psychiatry: Clinical practice guidelines for psychiatrist in India,* eds. S. Gautam and S. A. Awasthi, 11–19. India: Indian Psychiatric Society.

Chapter 5
Ethics in Counselling and Psycho-Therapy: What We Need to Learn—South Asian Perspective

Arun Kumar Gupta and Anuradha Menon

5.1 Introduction

'Counselling and psychotherapy are umbrella terms that cover a range of talking therapies. They are delivered by trained practitioners who work with people over a short or long term to help them bring about effective change or enhance their well-being.' (BACP)

The real revolution in modern psychiatry has been the concept of 'moral treatment' and a deep understanding of the dynamics within the therapeutic relationship, be it one of prescribing or of a particular talking therapy. However, nowhere is the relationship between the treating clinician and the patient more examined than in the context of a talking therapy.

The idea of a Hippocratic oath laying the foundation for what is considered as 'good practice' in medicine is understood as a type of professionalism expected of medical practitioners; in other words, the actual clinician. In various countries, these principles are tailored to fit the contemporary demands of professional practice. The GMC guidelines in the UK are an example, as is the ethical guidance recommended by the Indian medical council.

This chapter will try to address the importance of being able to consider ethical aspects of being a practitioner in psychotherapy and counselling. This is somewhat different from following a professional code of practice. The main reason for this distinction is that practitioners of psychotherapy often hail from varying backgrounds, with varying professional origins. Talking therapies are even more varied, with constant reinterpretation in response to society's changing profile. Ethical

A. K. Gupta (✉)
Northumberland, Tyne & Wear NHS Foundation Trust, Cherry Knowle Hospital, Ryhope, Sunderland SR2 0NB, UK
e-mail: ArunKumar.Gupta@ntw.nhs.uk

A. Menon
Psychoanalytic Psychotherapy & Adult Psychiatry, Leeds and York NHS Partnerships Foundation Trust, Southfield House, 40 Clarendon Road, Leeds LS2 9PJ, UK

J. K. Trivedi, A. Tripathi (eds.), *Mental Health in South Asia: Ethics, Resources, Programs and Legislation,* International Library of Ethics, Law, and the New Medicine 58,
DOI 10.1007/978-94-017-9017-8_5

practice in psychotherapy and counselling hinges on the actual process of therapy, and not the skill set. In a climate of change, codes of practice need to hold true to important basic constructs so as to protect a fundamental tenet of all treatments: 'above all, do no harm.'

Jeremy Holmes, an eminent medical psychotherapist in the UK (Holmes 2001) writes eloquently about the difficult task of a therapist: 'This discrepancy between the ethical standards expected of a profession and the realities of psychotherapeutic work is a crucial theme for therapists. It does not mean that in order to practise psychotherapy one must have the ethics of a saint and the wisdom of a sage. Therapists are no less flawed than anyone else; indeed most have to a greater or lesser extent suffered and sinned, and the process of recovery has played its part in choosing their profession. There are however three interrelated qualities in therapists which go some way to ensuring ethical practice. First is the capacity for self-reflection: the ability to see thoughts, feelings and actions as they arise, and to think about them. Second, the capacity to put those thoughts, feelings and potential actions into words rather than being drawn into enactment in the powerful interpersonal field which comprises the therapeutic situation. Third is the ability to attend closely to boundaries, and so to beware of the dangers and possibilities created by therapeutic intimacy and its limitations. They must focus intensely on the patient, yet keep in mind the borderland between the ethical and the professional, the political and the personal, the private and the public.'

We live and work in extraordinarily diverse settings. As therapists, our role needs to be tailored to the setting, and this is often a dynamic process over time. The provisions of contextually sensitive and appropriate services are also a fundamental ethical concern.

The British association of counselling and psychotherapy defines a 'Practitioner' as including anyone undertaking the role(s) of counsellor, psychotherapist, trainers and educators for these roles, providers of coaching and mentoring in association with counselling and psychotherapy, supervisors, and practitioner researchers. The term 'client' is used as a generic term to refer to the recipient of any of these services. The client may be an individual, couple, family, group, organisation or other specifiable social unit. Alternative names may be substituted for 'practitioner' and 'client' in the practice setting, according to custom and context (BACP 2010).

5.2 Ethics: Definitions and Basic Concepts

The field of ethics (or moral philosophy) involves systematizing, defending, and recommending concepts of right and wrong behaviour. One could think of ethics as having six intertwined dimensions: professional ethics, theoretical ethics (philosophical ethics/moral philosophy), clinical ethics, social ethics (including regulatory and policy ethics), virtue ethics and cultural ethics. All of these pertain to psychotherapy.

Herlihy and Corey (1996) distinguish between *mandatory ethics*, which means functioning according to minimum legal standards, and *aspirational ethics*, which means to function at a higher standard in accordance with the spirit behind the literal meaning of the code. To say that ethical values are different for different cultures would be inaccurate; just as most basic human needs, ethical needs cannot be but universal. However, a sensitive interpretation of universal ethical values in the context of a culture is crucial for effective practice as a therapist. In this way, fundamental values are identified while recognizing that different cultures may express those values through their own different culturally learned behaviours.

Jordan and Meara (1990) distinguished between principle ethics—which focus on rational, objective, universal and impartial principles mandating actions and choices—and virtue ethics, which focus on the counsellor's motives, intentions, character, and ethical consciousness that recognize the need to interpret principles differently in each cultural context.

Houser et al. (2006) present a hermeneutic framework to demonstrate the importance of contextual issues in ethical decision making. Ford (2006) likewise believes that ethical issues in counselling should be more grounded in the context of philosophical approaches to thinking about ethics as an alternative to abstract, code-based legalistic discussions about ethical issues.

One of the characteristics of contemporary society is the coexistence of different approaches to ethics. This 'ethical diversity' raises complex questions about practitioners needing to be responsive to differences in client abilities, needs and culture and taking into account variations between settings and service specialisations.

For a practitioner, a good way to begin thinking about ethical practice is to understand overarching concepts that would be essential in any type of therapy or counselling; and would, in a sense, bridge diversity in approaches. To do this, it is important to define certain values and principles.

5.3 Values of Counselling and Psychotherapy

The fundamental values of counselling and psychotherapy include a commitment to:

1. Respecting human rights and dignity
2. Protecting the safety of clients
3. Ensuring the integrity of practitioner-client relationships
4. Enhancing the quality of professional knowledge and its application
5. Alleviating personal distress and suffering
6. Fostering a sense of self that is meaningful to the person(s) concerned
7. Increasing personal effectiveness
8. Enhancing the quality of relationships between people
9. Appreciating the variety of human experience and culture
10. Striving for the fair and adequate provision of counselling and psychotherapy services

5.4 Principles of Psychotherapy and Counselling (Adapted from BACP 2010)

Traditionally there are certain principles that are defined as being central to practice. This is often linked to the idea of 'responsibility' as an ethical practitioner. A fair and just code of ethics needs to do more than reflect the cultural values of those who wrote the code. Kitchner (1984) described four of the basic moral principles that provide a foundation for the ethical code of counsellors as autonomy, beneficence, non-maleficence, and fairness. These four principles are presumed to be universally valued regardless of the cultural context.

In practice, these could be elaborated as follows:

5.4.1 Fidelity: Honouring the Trust Placed in the Practitioner

The practitioner's first duty is to the client. Thus confidentiality is regarded as an obligation arising from the client's trust. Being trustworthy is regarded as fundamental to understanding and resolving ethical issues.

The link between confidentiality and trustworthiness is a complex one. In clinical situations, therapists may become privy to information that involves third parties in a way that raises issues of risk. Deciding how to respond to such a situation requires a constant awareness of the interplay between the internal world of the client, and thus the therapy, and the other, 'external' world of the therapist and client. It would be impossible to be prescriptive on such a matter. Confidentiality, therefore, is a given in most situations, but with some important exceptions. For a practitioner in some south Asian regions, dealing with anxieties that arise within a patient's family can be fraught with more problems than compared to the western context. It is sometimes impossible to confine discussions about the patient to a one to one relationship with an individual; indeed patients often take for granted family members' access to their treatment records. The idea of fidelity in these situations would have to be reinterpreted in a cultural context, taking into consideration the client's situation.

5.4.2 Autonomy: Respect for the Client's Right to be Self-governing

The principle of autonomy speaks for the rights of the client in choosing to engage in a therapeutic process. Respecting this means that the practitioner is sensitive to and opposes the manipulation of clients against their will, even for beneficial social ends. This not only protects best interests, it also puts the client at the heart of the process in being empowered to exercise an informed choice at all times.

Ideally, the process of psychotherapy should be self directed. However, as a therapist in India, for example, one would often be approached by friends and family who wish to access treatment for a family member who seems to have little insight into their condition. In such cases, the beginning of a therapeutic process may be far from ideal, and contrary to the central value of autonomy. These scenarios require careful case by case consideration.

In a study, Varma and Ghosh (1976) found that the Indian psychotherapist led a relatively more active role than that the Western counterpart, suggesting, sympathising, manipulating the environment, teaching and reassuring. The Indian psychotherapists, in suggesting departures from the Western model, pleaded for greater flexibility, greater activity on the part of psychotherapist and greater use of suggestions and reassurance (Varma 1982).

Case Example A therapist based in a suburban area in northern India, Dr X, was approached by the wife and 22 year old son of a patient, Mr. Y. They described a difficult situation at home in which Mr Y would become angry over trivial issues, refuse to communicate with any friends and family and thus isolating himself.They were concerned that he might be experiencing a problem at work which he found difficult to talk about. The therapist encouraged the family to persuade Mr Y to meet him in a neutral setting. This was arranged and after introductions Mr Y was ambivalent about the meeting. At this point the therapist frankly admitted to his role in the situation and the family concerns. Mr Y was told about his options in a clear and forthright manner. He eventually made the choice to confide his problems to the practitioner, which turned out to be health concerns. Further discussions led to a gradual involvement of concerned family members and thus an effective support system was established around the 'problem'. The therapists' role in this scenario was clearly that of a facilitator, but this stance proved valuable in the long term.

5.4.3 Beneficence: A Commitment to Promoting the Client's Well-being

The principle of beneficence puts the onus on the practitioner to act in the client's best interests in a particular way. This involves good training, understanding limits of competence, adequate monitoring and supervision, good research techniques and above all, a commitment to constant efforts to improve quality of the therapeutic experience on offer.

In the absence of a strict system of revalidation and constant evaluation of competence in practitioners, in south Asian countries one sometimes comes across an 'assumption of competence'. This is often linked to a lack of supervision, peer or otherwise, and lack of proper evidence based practice. In therapeutic settings which involve engaging in a client's inner world, one is then left with a reliance on the therapist's moral code and personal belief systems. This, while sometimes useful, can become at the very least, an anti-therapeutic activity and at its worst, a focus of dangerous enactments into which the therapist enters. In psychotherapy, supervision is crucial; it is of the greatest importance so as to avoid an imposition on an individual and a family by a biased therapist who may be acting, often unconsciously, under the influence of his or her own internal standards.

Although south Asian practitioners engage in activities that contribute to a process of continuing professional development, it is doubtful whether most institutions would offer a consistent supervisory experience for practising therapists and trainees in psychotherapy.

5.4.4 Non-maleficence: A Commitment to Avoiding Harm to the Client

The principle of 'above all, do no harm' resonates when one thinks about the enormous responsibility carried by the therapist/counsellor who becomes the repository of the client's trust. Non-maleficence involves: avoiding sexual, financial, and emotional or any other form of client exploitation; avoiding incompetence or malpractice; not providing services when unfit to do so due to illness, personal circumstances or intoxication. The ethical responsibility also extends to a broader role in mitigating harm caused by other professionals in questioning malpractice, aiding investigations into unethical practice by colleagues, and so on.

The issue of Boundary violations has preoccupied many western researchers. This is hardly helped by often unrealistic and frankly unhelpful depiction of therapists by a visual media as glamorised, all powerful figures that may enter into inappropriate relationships with clients who place trust on them. While popular media is fast catching up with a more mature understanding of the sensitive nature of a therapist-patient relationship, it is worthwhile trying to understand just why this issue is such a serious one. Sarkar (2004) refers to Boundary violations as representing an attack on the security of the relationship between the patient and the doctor. It is obvious that there is a power imbalance between doctor and patient, which can lead to exploitation. In sexual relationships, the harm comes because of the parallels with incest (Kardener 1974) and the danger to the patient of making transferential fantasies real. Simon (1995) describes the 'slippery slope' of sexual misconduct and how it is often a gradual erosion of professional identity rather than a single event.

Unfortunately the popular media, for e.g.: cinema does occasionally portray scenarios where boundary violations are clearly seen: this does pose problems in managing the public perception of the role of a therapist.

5.4.5 Justice: The Fair and Impartial Treatment of All Clients and the Provision of Adequate Services

Practitioners have a duty to strive to ensure a fair provision of counselling and psychotherapy services, accessible and appropriate to the needs of potential clients. A commitment to fairness requires the ability to appreciate differences between people and to be committed to equality of opportunity, and avoiding discrimination against people or groups contrary to their legitimate personal or social characteristics.

In the South Asian context, clients come from diverse backgrounds that are diverse in aspects such as cultural, economic, religious. Being non judgemental as well as keeping an open mind in the face of values that may be drastically different from one's own is a challenge.

5.4.6 *Self-Respect: Fostering the Practitioner's Self-Knowledge and Care for Self*

There is an ethical responsibility to use supervision for appropriate personal and professional support and development, and to seek training and other opportunities for continuing professional development. It is thus important for the practitioner to engage in a lifelong commitment to reflective practice. In a western psychoanalytic training personal therapy 3–5 times a week is compulsory; this is not yet universally acknowledged in South Asian counties that do not offer such training.

Reflective practice has been advocated in various Indian myths as well as in religious texts. The idea of moving away from a didactic and rigid 'hierarchy' between the therapist and patient, towards a more interactive and iterative relationship, is the hallmark of most therapies. In a South Asian context the more favoured mode of reflective practice seems to be a reliance on peer supervision, as well as informal supervision from more experienced colleagues. However, particularly in the training period one needs to be able to access regular external supervision where relevant issues could be explored.

These principles are neither comprehensive nor discrete. Many ethical questions require the practitioner to combine these principles and make complex decisions for individual cases. In certain conditions, it may seem that a principle is being contradicted. Sometimes what needs to happen is that the patients' benefits need to be weighed up against other considerations.

The following scenarios are some examples of how a therapist would need to think flexibly around ethical principles.

1. *Disclosure of clinical information*

E.g. this could become an issue when risky clinical situations come up, and at these times the therapist may then have to break confidentiality agreements. This is a difficult decision that needs to be carefully thought through on a case by case basis.

2. *Getting consent for research purposes*

E.g.: In therapeutic work obtaining information for research might have an unplanned effect on the therapeutic process and relationship. Again, the issue of hierarchy in a more traditional setting may mean that the client does not feel empowered to have an opinion regarding consent for research.

3. *The idea of 'harm' inflicted by a therapy that seeks to do no harm*

E.g. A client may end up negatively re-evaluating a previously 'stable' and accepted personal relationship as a result of introspection in therapy. In this case, what is considered as healthy from an intrapsychic point of view may not fit in with societal expectations of an individual.

5.5 Personal Moral Principles

Reliance on principles alone may detract from the importance of the practitioner's personal qualities and their ethical significance in the counselling or therapeutic relationship. This reflects the inherent tension between a rather generic view of client needs and the needs of a particular individual.

All therapies, without exception, will stress on the value of the therapeutic relationship. The key idea here is the quality of the actual clinician and what he or she is able to bring to the client by virtue of training, personal values, effective practice, competence and so on. No two practitioners are alike. However, one would expect a good practitioner to be empathic, sincere, respectful and fair. The process of therapy is a gruelling although potentially fulfilling one for both the client and the practitioner; the demands it places on both are immense. A good practitioner needs to be resilient for the client. Wisdom and courage are prerequisites if we think about how the therapeutic relationship needs to engender an atmosphere of safety and security.

5.6 Culture, Therapy and Ethics

As therapists, one grapples daily with ethical dilemmas that centre on the type of therapy, cultural influences, individual differences, past experiences, and so on. There are several hundred diverse models of therapies each of which is founded on different assumptions, requires different skills, calls for different training programmes, accepts different forms of accreditation (or not, as the case may be), uses different techniques and predicts different outcomes. The field of psychotherapy and counselling (psychoanalysis, cognitive behaviour therapy, gestalt therapy, transactional analysis, person centred therapy, to name a few) is a disparate, non-unified discipline (Laungani 2004).

Culture controls our lives and defines reality for each of us, with or without our permission and/or intentional awareness. A "culture-centred" approach to counselling and psychotherapy recognizes culture as central and not marginal, fundamental and not exotic, for all appropriate counselling interventions. While mental health problems are similar across cultures, the complex classification of the appropriate helping responses across cultures has given rise to a global variety of counselling styles that are complex and ever changing.

Anthropologists have tended to take a relativist position when classifying and interpreting behaviours across cultures (Geertz 1973). Psychologists, by contrast, have linked social characteristics and psychological phenomena with minimum attention to cultural differences (Bernal et al. 2002).When counsellors have applied the same interpretation to the same behaviour regardless of the cultural context, cultural bias has been the consequence.

Adopting cross cultural perspectives makes clarifying, evaluating and justifying the values tied to therapeutic goals even more daunting, because some argue that traditional therapy goals reflect western cultural values (Tjeltveit 1999).Thus these goals may or may not be transferable transculturally.

Desai (1982) listed several cultural issues that need to be addressed for the practice of psychotherapy in India. The basic unit of Hindu society is the family, rather than the individual. Thus, individual growth is subordinated to family integrity. The matter of collaboration versus individuation needs to be handled carefully. The Hindu family strives to maintain integrity by prohibiting expressions of anger and hostility. The balance of suppression versus expression of emotion thus requires delicate management in therapeutic situations.

Pande (1968) pointed out that in eastern, relationship-oriented societies, no formal agenda is necessary for the cultivation of a relationship; however, in western work and activity-oriented societies, the absence of an agenda would be disturbing.

Let us consider, for example, the idea of 'autonomy', a concept regarded by Strupp (1980) as an essential therapeutic value and endorsed by 96 % of US therapists (Jensen and Bergin 1998) as 'important for a positive, mentally healthy lifestyle'. Researchers in a traditional setting in India interpret the centrality of this concept in a different way.

Varma (1988) noted that in India "there is a greater degree of mutual interdependence that in the west." Thus he argues:

"It is questionable how far western psychotherapy with its high emphasis on autonomy and individual responsibility can be prescribed for such a society; and accordingly what modifications are required in the rules and practice of conventional psychotherapy."

Practitioners in such settings will be familiar with client attitudes that rely heavily on the knowledge, attitude and credibility of the therapist. The idea of 'the doctor knows best' is a difficult one to work with as a therapist when one is trying to take up a therapeutic, non-judgemental stance.

Roland (1991) writes about how when patients come for therapy in an eastern setting, they may relate to the therapist as a hierarchical superior, usually modelled after a family elder, with expectations that the therapist shall take care of them completely. Therapy as such seemed to be practiced in a *guru-chela relationship*, a paradigm of preceptorship (Varma 2008).

In such cases, the idea of autonomy needs to be interpreted flexibly. The authors feel that psychotherapy does not impose any aims or goals on a client; the aim of therapy is agreed jointly by the client and therapist. The therapist is often a guide or catalyst. Therefore all forms of psychological treatments are generic in nature and can be applied to any culture if the therapist remains sensitive to individual needs. The experience at teaching CBT in India by the author (Gupta and Ogilvie 2008, 2009) demonstrates that same principles of CBT are effective in India as well.

Finally, it is worthwhile noting that although cultural issues tend to be noticed only when cultural differences between patient and therapist are clearly evident, all psychotherapy is cross-cultural in that no two people have internalized identical constructions of their cultural worlds. Even if the therapist is treating someone from a similar cultural background, there are always communication problems to overcome and differences in values to be negotiated.

5.7 Does the Type of Therapy Matter?

Comparative study of indigenous healing practices and modern psychotherapy has revealed the existence of universal elements of the healing process that are probably important factors whatever the form of therapy: the cultivation of hope, the activation of social support and the enhancement of culturally sanctioned coping (Frank 1961; Kleinman and Sung 1976; Torrey 1986).

Corey et al. (2007) point out that all of the contemporary therapeutic models need to recognize the cultural contexts in which behaviours are learned and displayed. Each therapy and each ethical code will reflect the values of its cultural context. This statement seems to imply that the essential difference between various therapies lies not in the modality, but in the cultural interpretation of the 'essence' of each style.

It would follow that a Western-based code of ethics is based on a preference for individualism rather than collectivism as the preferred worldview. Individualism applies to societies in which everyone is expected to look after themselves, whereas collectivism applies to societies in which people are integrated into cohesive groups and/or relationships that protect the members of the group in exchange for their loyalty. A comprehensive code of ethics needs to respect the values of both individualistic and collectivistic cultural contexts.

However, the fundamental therapeutic orientation may be supernatural, natural, biomedical, socio-philosophical or psychological. In folk therapy, healing practices, ceremonies or health-promoting exercises may be applied to resolve the problems, without being perceived as 'psychological therapy.' In contrast, in other practices, particularly professional psychotherapy, the therapist and the patient both recognize that the procedure is primarily for 'treating or resolving a psychological problem' and that they are engaged in an activity for that perceived purpose. Thus, there exists a broad spectrum of 'psychotherapy,' in terms of basic orientations, methods and goals to be achieved (Tseng 1999).

That is why it is important that ethical codes are developed and agreed locally, in accordance with accepted modes of practice.

5.8 Existing Regulatory Bodies in South and South-East Asia

The Singapore Psychological society was established in 1979 and lays down a code of professional ethics for all practitioners, as well as providing registration. The Psychological association of the Philippines, established in 1962, recently set out a code of ethics via their scientific and professional ethics committee in 2008–2009. From a detailed search, we were unable to identify any other specific professional or regulatory bodies aimed at psychologists/psychotherapists in the other countries within south Asia. All of them, excluding Bhutan, which did not have a medical council as a regulatory body, had well established medical, dental and nursing councils which regulated and provided registration for professionals in these areas. In the Indian context, although the registration for psychologists (specifically clinical and rehabilitation) was offered by the Rehabilitation council of India (RCI), there seemed to be no codified ethical practice guideline that was used nationally. The RCI was established as a statutory body in 1992. The yoga and psychotherapy association of India is an exception, being a non-profit organisation that is affiliated to the Asian federation for psychotherapy. The federation follows a code of ethics, most recently updated in October 2008 in Beijing.

This presents a problem in terms of the quality and credibility of treatments offered currently. With no unifying ethical code of practice that holds the practitioner to their model of therapy, it becomes difficult to clearly establish the effectiveness of therapy as well as to compare national results in a systematic manner.

5.9 Literature on Contemporary Views on Ethics in India

Unlike in the UK and the US, a lack of a specific regulatory body in the south Asian countries forces us to consider broader ethical values and principles which govern doctors and other professionals offering counselling and psychotherapy.

The Medical council of India (2002) published its revised code of ethics. However, these are general principles for doctors nationally and do not always address the particular context of the psychotherapist.

An attempt at creating an ethical framework for clinical application was made by the Indian Psychiatric Society when it adopted a code of Ethics for Psychiatrists in 1992, outlining the major principles of (i) responsibility, (ii) competence, (iii) benevolence, (iv) moral standards, (v) patient welfare, and (vi) confidentiality. A number of Presidential Addresses at Indian national conferences have focused on various aspects of the subject (Agarwal 1994; Anthony 1995; Kumar 2001; Kuruvilla 1998). However these suggested guidelines are yet to be codified within a legal framework.

A review article by Agarwal (2010) examines the concerns relating to ethics in psychiatric practice today. He concludes that there seems to be a lack of clear ethical guidelines for psychiatrists who practice psychotherapy in a clinical context.

Desai (2006) writes about the Indian context, and how concerns have been expressed by individual professionals and the Indian Psychiatric Society about the basic issues of ethics in psychiatric practice as well as the need to identify responsibilities of psychiatrists. There are many difficulties and obstacles in appropriate recognition of ethical responsibility of all health professionals, and particularly mental health professionals in developing countries. These relate to the absence of clear ethical guidelines, lack of consistent application, the attitude of professionals involved, and the still limited assertion of the ethical rights by the consumers. The responsibility of being a competent psychiatrist with adequate skills seems to be undermined by varying standards of education and indifferent monitoring mechanisms. This is an important editorial which emphasizes newer dimensions to ethical behaviour: like advocacy, social responsibility and certain unconventional areas are included.

There are very few studies on psychotherapy process or outcome research in India (Manickam 2010). Psychotherapy is included in the curriculum of postgraduate training in psychiatry. However, there are no guidelines available at the national level on how and what skills need to be imparted. Introducing psychotherapy case submissions with adequate supervision as part of the curriculum is likely to generate the interest and improve the skills of the trainee psychiatrist (Shamsunder et al. 1993).

Verma and Kaur (2000) point out that psychotherapist face several ethical dilemmas in the course of therapy, especially since there is no single right way or treatment of choice. This would call for some degree of introspection on the part of the therapist at every stage of therapy from intake to termination. They highlight that this is an important area that has received very little attention.

5.10 Developing an Ethical Code for Psychotherapy and Counselling in India: A Proposal

Paul Chodoff (1991) identified four major responsibilities of psychiatrists, viz. (i) the responsibility of competence or the need to master their task; (ii) the responsibility of ethical behaviour or to police their ranks; (iii) the responsibility of accountability or to be accountable to the public; and (iv) the responsibility of advocacy or to be advocates for the mentally ill persons.

Lazarus (1999) concurred with the views of Chodoff and, while saying that 'it is likely that the twenty-first century psychiatrist will occupy new roles but continue in the clinical treatment of the mentally ill', suggested several guidelines that address the various roles taken up by psychiatrists in providing a service for the mentally ill patient. These are important but broad based ideas.

Professional ethics can only be evaluated in the context of the value system currently prevalent in society. Needless to say, a physician is also part of the society and its prevailing moral and ethical standards have to influence him. With respect to psychotherapy the field of clinical application is wide, as is the background of a potential therapist. Practitioners would by definition hail from different professional backgrounds, the common denominator being the work within the mental health field. Needless to add, a codified ethical framework is an essential aspect of a well-established regulatory body. It has several functions:

1. Clarifying the role of a psychotherapist
2. Standardisation of interventions
3. Establishing a code of conduct for practitioners and by definition, providing a framework for disciplinary proceedings in case of unethical conduct
4. Improving quality of training and revalidation
5. Supporting practitioners in their professional development and thus encouraging better outcomes in therapeutic practice
6. Responding to the evolving needs of society by redefining and updating standards regularly

However, regulatory bodies and the ethical frameworks that they espouse contribute only to certain aspects of ethical dilemmas. This is because codes of professional ethics are generated from within the profession; therefore they are more narrowly construed. They also apply to the profession's obligations and the ideal behaviour to which the therapist is to aspire. Codes of professional ethics do not pertain to the ethical analysis of clients. They also do not generally include a process of arriving at a standard. All these factors limit the application of a professional ethical code to psychotherapy and counselling. By necessity, the other 5 ethical dimensions (theoretical, social, cultural, virtue and clinical) would need to be considered.

Psychotherapists, in a sense, have always answered ethical questions: by drawing upon consensus, training, experts, experience, intuition, rational arguments, science and so on (Tjeltveit 1999). The challenge is to determine the best, or the best possible combination of these approaches to address a particular ethical question.

5.11 Gaps in the Indian Context

The field of Psychology in India is both a thriving academic discipline and a vital professional practice. A number of universities offer courses which contribute to undergraduate, professional and research degrees in psychology. The Rehabilitation council of India (RCI) monitors the registration and regulation of clinical psychologists and clinical psychology programs in India. A qualified psychologist is often in a position to offer psychotherapeutic interventions within hospital and clinic-based populations. There is scope to work alongside and to contribute to psychiatric interventions, as well as to offer complex investigations such as psychological testing. However the lack of a regulatory body limits quality control and standardisation of treatments. This is a significant deterrent to an attempt to improve practice and modulate treatments to changing needs of society. Furthermore, it limits the current credibility of training offered. A standard ethical code that unifies practitioners would address this gap in the future.

5.12 Conclusions

There are many reasons why modifications and adjustments in psychotherapy and counselling may be needed, including: the basic personality or ethnic character of the population concerned (Doi 1964); commonly shared belief and value systems (Hoch 1990); basic philosophical attitudes (Rhee 1990; Varma 1982); and orientations toward, and expectations of psychotherapy (Neki et al. 1985; Ng 1985; Tseng 1995). Beyond this, it is also obvious that the practice of psychotherapy is profoundly influenced by socio-cultural factors and political ideology (Tseng 1999). In this context, the understanding and application of ethical mores to the therapeutic field in a multicultural society remains a challenging and dynamic area of study.

References

Agarwal, A. K. 1994. Ethics in psychiatry. *Indian Journal of Psychiatry* 36:5–11.

Agarwal, A. K. 2010. A review of Indian psychiatry research and ethics. *Indian Journal of Psychiatry* 52 (Suppl1): S297–S305.

Antony, J. T. 1995. Contemporary society, health care scenario and the psychiatrist. *Indian Journal of Psychiatry* 37:53–8.

Bernal, G., J. Trimble, K. Burlew, and F. Leong. 2002. *Handbook of racial and ethnic minority psychology*. Thousand Oaks: Sage.

British Association of Counselling and Psychotherapy. 2010. Ethical framework for good practice in counselling and psychotherapy. London: BACP.

Chodoff, P. 1991. The responsibility of the psychiatrist to his society. In *Psychiatric ethics,* ed. S. Bloch and P. Chodoff. Oxford: Oxford University Press.

Corey, G., M. S. Corey, and P. Callanan. 2007. *Issues and ethics in the helping professions*. Belmont: Thomson Brooks/Cole.

Desai, P. N. 1982. Learning psychotherapy: A cultural perspective. *Journal of Operational Psychiatry* 13 (2): 82–87.

Desai, N. G. 2006. Responsibilities of psychiatrists: Need for pragmatic Idealism. *Indian Journal of Psychiatry* 48:211–4.

Doi, T. 1964. Psychoanalytic therapy and 'Western man': A Japanese view. *International Journal of Social Psychiatry* 1: 13–18.

Frank, J. D. 1961. *Persuasion and healing: A comparative study of psychotherapy*. New York: Schocken.

Ford, G. G. 2006. *Ethical reasoning for mental health professionals*. Thousand Oaks: Sage.

Geertz, C. 1973. *The interpretation of cultures*. New York: Basic Books.

Gupta, A. K. and A. E. Ogilvie. 2008. Pilot study evaluating qualitatively the outcome of an introduction to cognitive behaviour therapy course in India. Poster presentation at BABCP annual conference, Edinburgh.

Gupta, A. K. and A. E. Ogilvie. 2009. Qualitative evaluation of the impact of basic CBT training amongst mental health professionals in India. Poster presentation at BABCP annual conference, Exeter.

Herlihy, B. and G. Corey. 1996. ACA ethical standards casebook 5th ed. Alexandria: American Counselling.

Hoch, E. M. 1990. Experiences with psychotherapy training in India. *Psychother Psychosom* 53: 14–20.

Holmes, J. 2001. *Values and ethics in the practice of psychotherapy and counselling,* USA: Open University press.

Houser, R., F. L. Wilczenski, and M. A. Ham. 2006. *Culturally relevant ethical decision-making in counseling.*

Jensen, J. P. and A. E. Bergin. 1988. Mental health values of professional therapists: A national interdisciplinary survey. *Professional Psychology: Research and Practice* 19 (3): 290–297.

Jordan, A.E. and N. M. Meara. 1990. Ethics and the professional practice of psychologists: The role of virtues and principles. *Professional Psychology: Research and Practice* 21:107–114.

Kardener, S. H. 1974. Sex and the physician–patient relationship. *American Journal of Psychiatry* 131:1134–1136.

Kitchner, K. S. 1984. Intuition, critical evaluation and ethical principles: The foundation for ethical decisions in counselling psychology. *The Counselling Psychologist* 12:43–55.

Kleinman, A and L. Sung. 1976. Why do indigenous practitioners successfully heal? *Social Science and Medicine* 13 (B): 7–26.

Kumar, K. A. 2001. The role and road ahead. Indian *Journal of Psychiatry* 43:5–11.

Kuruvilla, K. 1998. The future of psychiatry: The need to return to the field of medicine. *Indian Journal of Psychiatry* 40:3–12.

Laungani, P. 2004. *Asian perspectives in counselling and psychotherapy.* Brunner: Routledge Association.

Lazarus, J. A. 1999. Ethical conduct of the psychiatrist. In *Psychiatry in the new millennium*, ed. S. Weissman, M. Sabstein, and H. Eist Washington DC: APA.
Manickam, L. S. S. 2010 Psychotherapy in India. *Indian Journal of Psychiatry* 52 (Suppl1): 366–S370. doi:10.4103/0019–5545.69270
Medical Council of India (MCI). 2002. *The Indian Council* (Professional Conduct and Ethics) *Regulations*.
Neki, J. S., B. Joinet, M. Hogan, J. G. Hauli and G. Kilonzo. 1985. The cultural perspective of therapeutic relationship—A viewpoint from Africa. *Acta Psychiatrica Scandinavica* 71:543–550.
Ng, M. L. 1985. Psychoanalysis for the Chinese—Applicable or not applicable? *International Review of Psychoanalysis* 12 (4): 449–460.
Pande, S. K. 1968. The mystique of "Western" psychotherapy: An eastern interpretation. *The Journal of Nervous and Mental Disease* Â146 (6): 425–432.
Rhee, D. S. 1990. The Tao, psychoanalysis and existential thought. *Psychotherapy and Psychosomatics* 53: 1–27.
Roland, A. 1991. Psychoanalysis in India and Japan: Toward a comparative psychoanalysis. *American Journal of Psychoanalysis*. 51(1): 1–10.
Sarkar, S. P. 2004. Boundary violation and sexual exploitation in psychiatry and psychotherapy: a review. *Advances in Psychiatric Treatment* 10:312–320.
Shamasunder, C, M. Verghese, R. R. Raguram, S. Jain, S. Girimaji, and S. Seshadri, et al. 1993. Psychotherapy programme for psychiatry residents at NIMHANS: A descriptive account. *Indian Jounal Psychiatry* 35: 215–7.
Simon, R. I. 1995. The natural history of therapist sexual misconduct: identification and prevention. *Psychiatric Annals* 25: 90–94.
Strupp, H. H. 1980. Psychotherapy: assessing methods. *Science* 207(4431):590.
Tjeltveit, A.J. 1999. Ethics and values of psychotherapy. London: Routledge.
Torrey, E. F. 1986. Witchdoctors and psychiatrists: The common roots of psychotherapy and its future, 73–74. New York: Harper and Row.
Tseng, W. S. 1995. Psychotherapy for the Chinese: Cultural considerations. In *Chinese societies and mental health*, ed. T.Y. Lin, W. S. Tseng, and E. K. Yeh . Hong Kong: Oxford University Press.
Tseng, W. S. 1999. Culture and Psychotherapy: Review and Practical Guidelines. *Transcult Psychiatry* 36:131.
Varma, V. K. and A. Ghosh. 1976. Psychotherapy as practised by the Indian psychiatrists. *Indian Journal Psychiatry* 18: 177–86.
Varma, V. K. 1982. Present state of psychotherapy in India. *Indian Journal Psychiatry* 24:209–226. (Reviewed in: Transcultural Psychiatric Research Review 21(4), 291–291, 1984.)
Varma, V. K. 1988. Culture, personality and psychotherapy. *International Journal of Social Psychiatry* 34(2):142–9.
Varma, V. K., and N. Gupta. 2008. *Psychotherapy in a traditional society: context, concept and practice*. New Delhi: Jaypee Brothers Medical Publishers.
Verma, S. K., and H. Kaur. 2000. Ethics in health psychology. *Psychological studies* 45:136–138.

Part II
Resources and Opportunities/Mental Health Programs and Policies (their current status and future prospects)

Chapter 6
Programs of Mental Health and Policies in South Asia: Origin and Current Status

M. A. Javed

6.1 Introduction

Asia, the largest continent in the world both in area as well as population faces a number of problems in various social domains. With a population exceeding 4 billion, it comprises of a number of cultures, ethnic and religious groups who speak different languages and live in a varying climate ranging from very cold to hot and tropical weather. The changing economic conditions add to the diversity and one can find very rich countries in the continent along with a number of poor and low income countries that struggle even for the basic needs. When it comes to health care systems, Asian countries again show a wide variation that is often based on historical, cultural and at times colonial heritages. Mental illnesses that unfortunately affect over 450 million persons in the world who, as per World Health Organisation's reports (WHO 2011), suffer from a range of mental and neurological disorders also show a significant presence in Asia. Asia therefore takes a major burden of care for the mentally ill people and many countries in the continent struggle in their efforts to provide better mental health services to their people (Patel et al. 2007; Jacob et al. 2007).

Salient points related to extent of the mental health problems in South Asian countries are mentioned below.

- 450 million people in the world suffer from various types of mental disorders.
- 1 out of 4 persons seeking primary health care suffers from mental disorders.
- 10% of the global burden from non communicable diseases is accounted for by neuropsychiatric conditions in low and middle-income countries.
- 5 of the 10 leading causes of disability worldwide are mental disorders.
- 70% of cases of depressive disorders all over the world are either not properly diagnosed or treated.
- High rates of suicide and alcohol/drug abuse are of special concern to the South-East Asia Region.

M. A. Javed (✉)
Fountain House, 37- Lower Mall, Lahore, Pakistan
e-mail: afzal@afzaljaved.co.uk

J. K. Trivedi, A. Tripathi (eds.), *Mental Health in South Asia: Ethics, Resources, Programs and Legislation,* International Library of Ethics, Law, and the New Medicine 58,
DOI 10.1007/978-94-017-9017-8_6

Mental health by no means is standardised and is extremely varied in perception and practices. Many norms of psychiatric care practices in economically developed countries may only be inapplicable but, may also be highly impractical to mental health care systems in many parts of Asia (WHO 1973). This is particularly true of some aspects of mental health care that are a part of routine care in the developed countries, such as community care for the mentally ill, social security for the disabled, hostels for discharged mental patients, and free treatment for the mentally ill. These and many other aspects of mental health care are not available in vast areas of Asia and often substituted by a remarkably resilient, but not always highly successful, family care alternative and a strong heritage of traditional medical care for majority of the mentally ill that has its own limitations.

The available data suggest that about 23 % of the world's population lives in the South Asian region, and based on the general epidemiological reports about incidence and prevalence of mental disorders, it can be assumed that one fifth of psychiatrically ill patients from all over the world live in these countries. It is worth noting that despite vast geographical, economical, political, cultural and religious diversities, the factors influencing mental health generally remain the same throughout this wide region. The situation becomes even worse in most of these countries where lack of mental health policy and programs, limited budgets for mental health care, and limited number of professionals make the practice of mental health care more difficult. There is huge scarcity of resources to address the mental health needs of the population along with the negative social attitudes towards mental health, massive underestimation of the suffering of mentally ill people, lack of political empathy, and the lack of mental health leadership which in turn increases the real challenges in providing appropriate care (Saraceno et al. 2007; Desjarlais et al. 1995).

Major highlights of South Asian countries related to mental health are as follows.

- Multi ethnic, multi religious & multi cultural influences
- Hindu, Muslim, Buddhist & Christian influence
- Ancient history dating back to at least 2000 years
- Meeting place of East Asian & South East Asian civilisation
- History of invasion and influences from all other regions
- Very rich in natural resources
- Experience worst natural & other disasters
- Growing economy but still low income countries

It is also worth noting that available mental health services are usually urban-centered and confined to hospital-based facilities, with the result that 80–90 % of populations have no access to treatment. Mentally ill are the worst victims of social, emotional and institutional abuse and the mainstream society fails to acknowledge their suffering as human beings who require more attention and support. Once people are labeled as mentally ill, their civil and human rights are taken away and may even be suspended for ever. They are also exposed to discrimination that results in further damages to their self esteem. Human rights violations of mentally ill people are not only evident in many countries but also constitute a major obstacle in getting mental health as a priority (Trivedi et al. 2007; Javed 2007; Lancet Global Mental Health Group 2007).

Mental health issues in South Asia can be discussed from different perspectives. The situation may vary from country to country and the topics may range from national policies to actual practices, ethics to stigma, discrimination to involvement of patients in their care, gender biases to domestic violence, management of disasters to practice of preventive psychiatry and capacity building to efforts aimed at generating more interest among the general public. It thus makes a long list of desirable topics which need a continuous and constant reminder for further deliberations (Sharma 2006; Niaz 2003; Lolas 2006; Crisp et al. 2000; Rothman et al. 2009).

Mental health is generally neglected in most of the South Asian countries (WHO 2011). These countries spend much less of their GDP on health than many other low-income countries of the world. In most countries programs on mental health development generally do not get priority from the government authorities and unfortunately in many countries, the policies are difficult to get approved and even after the approval the implementation is still hard to achieve. This chapter will look at the current situation of mental health programs in some of the South Asian countries and would discuss the areas that need urgent attention both by the governments as well as by other mental health services providers.

India, the largest country in terms of area and population in SAARC region and the second most populated country of the world with a population of more than one billion, continue facing a number of problems in the provision of mental health services. In India nearly three fourths of the total source of health expenditure comes mostly from the households whereas the public funds account for less than one quarter. While the most common method of financing mental health care is tax-based or social insurance in developed countries, in India, like many other countries in the South-East Asia regions 'out-of-pocket payments' is used as the primary method of financing health care (Khandelwal et al. 2004; Chatterjee 2009; Saxena et al. 2003). Although the private sector is growing very fast and that of NGOs are also appearing as a main stake holder in the provision of mental health services in many parts of the country but their share is very minimal at this stage.

India was one of the first countries in the region to formulate a national mental health Program. As early as 1980, the Central Council of Health and Family Welfare (CCHFW) recommended for adaption of a National Mental Health Program for India (NMHP) (DGHS 1982). The need for community care for people with mental disorders and setting up of model programs was a salient feature of the Indian National Mental Health Program right from the beginning and has changed the mental health scene significantly after many states in the country initiated implementation of this Program. While the adoption of the national mental health program was a major step forward in the development of mental health, numerous problems including budget allocations, lack of manpower and preferences of professionals and policy makers in its implementation did delay the process at many stages of its implementation. There were also practical difficulties of its application in larger population especially in rural settings and in the periphery where health resources were already limited (Haq et al. 2008). The availability of trained manpower have further limited its application as the number of professionals never reached to the figure that would have been enough to keeping up with the needs and requirements. Similarly most of these professionals, where and if present, have been mostly based in urban areas.

However despite these limitations National Mental Health Program in India has witnessed an encouraging response at different levels of its implementation. Several publications and research reports have looked at various aspects of this program and have discussed the factors for its success as well as its failure. As the NMHP primarily focuses on community involvement, the need for decentralizing the services from the urban areas and its impact have been specifically highlighted in these reports. While funding may be a problem in different areas of the country, delayed receipt of funds, inappropriate and interrupted dispersal of funds, administrative hurdles in the full utilization of available funds and a variety of organisation and bureaucratic obstacles have been flagged as major issues in the real practice. The salient features of NMHP like community mental health programs, district mental health programs and setting priorities for primary care mental health services have also been evaluated by many researchers (Patel 2010) and the issues whether patients receive minimum and adequate treatment at this level is still debated.

Looking at the current scene, Indian NHMP has no doubt played an important role in setting up new directions and formulating priorities for mental health in the country. Considering the current developments in many areas of mental health, it is generally felt that a lot has still to be done towards further development of mental health care in India including continuous and regular reviews of the mental health Program, enhancing the capacity to train mental health professionals, organising higher training programs, exploring further research opportunities in different areas of mental health and provision of affordable and accessible clinical services at all levels of society (Srinivasa 2007; Isaac 2011).

Pakistan with a population of over 160 million also struggles with the mental health programs due to a number of reasons (Shafique 1995; Gadit 2006a). Mental illnesses are stigmatized and widely perceived to have supernatural causes. Ironically, Pakistan has a dearth of prevalence and incidence studies as research has not been an important and strong component among the health professionals (Gadit 2006b). There are very few organized national studies on mental health morbidity and a proper and reliable information management system is not in place. There is no compulsion on hospitals to send vital information to the authorities and hence figures about mental diseases are not easily available.

At the time of independence in 1947, there were only three big mental hospitals in the country, each at Hyderabad, Lahore and Peshawar, with a total of 2000 beds. Patients were often brought in chains and the general state of health care at these hospitals was not very promising either. There have however been significant developments in the planning and establishment of mental health services during the last few decades. Opening of psychiatric departments in almost all medical colleges, start of mental health facilities at district headquarter hospitals and increased awareness about the needs for mental health programs at local level among the policy makers are being considered a step forward. The private sector is also becoming visible in the service delivery systems in big cities but is primarily involved in the provision of individual treatment care services. These services are delivered in parallel with public sector but there is no formal integration, or referrals system between the two sectors.

Unfortunately the number of trained mental health professionals in the country is still much less as compared to the demands and the traditional healers continue providing main mental health services in the community. The alternate practitioners are a major source of service provision in most of the rural areas in Pakistan. Commonly known as "baba" or "pir" or as "suufi", they are working in a big number in the community and a large number of people have faith in their healing powers, hence shrines and other holy places are flocked by the masses, irrespective of educational or ethnic background, seeking cure especially for mental illness (Gadit 2007).

The national mental health policy was first formulated in 1997 which addressed issues of advocacy, promotion, prevention, treatment, rehabilitation and inter-sectoral collaboration in mental health (Karim et al. 2004). It envisaged to train primary care providers, to establish resource centres at teaching hospitals and psychiatric and detoxification centres. There was provision for crisis intervention and counselling services, special facilities for mentally handicapped and up gradation of large mental hospitals. The national health policy of Pakistan concentrates on poverty alleviation, priority attention to primary and secondary sectors of health to replace the earlier concentration on tertiary care and seeing good governance as the basis of health sector reform to achieve quality care (Government of Pakistan Planning Commission 2005). It is a pity that the allocated mental health budget is still around 0.4 % of total health care expenditures but despite a number of financial and organisational limitations, many innovative programs to develop indigenous models of care like the 'Community Mental Health Program' and 'Schools Mental Health Program' have also been developed in some parts of the country. These programs have been found effective in reducing stigma and increasing awareness of mental illness amongst the adults and children living in rural areas but a lot is still needed to make a real difference (The World Bank 2005).

The Government of Pakistan has repealed the Mental Health Act of 1912 and promulgated a new mental health law in the year 2001. The ordinance looks as an ideal document emphasizing on protection of human rights of the patients and asking for a provision for the prevention of mental illnesses, promotion of mental health information, establishment of community based services but its implementation and use in day to day practice is still not visible. Current focus on Non Communicable Diseases (NCDs) is another step forward as these disorders impose the largest health burden in the developing countries (Nishtar et al. 2005). National Health Policy emphasized the importance of non communicable diseases. In 2003, the country was among the first few developing countries to develop an integrated national plan of action, which addressed the four major NCDs with common risk factors along with injuries and mental health. Both the policy and plan could not be implemented due to the change in government. In 2009, the Ministry of Health (MoH) proposed the establishment of a National Commission for Prevention of non communicable diseases, with public and private partnerships and volunteerism as its driving force. The process of creating the Commission has however come under legal question and has been halted.

The primary health care level is not well equipped and staffed to deliver mental health preventive or treatment services although the Ministry of Health (MoH) has developed a public sector health system with four major levels: primary care facilities for outpatients (basic health units and dispensaries), district hospitals for basic inpatient and outpatient care, tertiary hospitals in urban areas, and vertical programs. Public institutions lack core elements and capacity to manage integrated health programs and the population-based prevention is generally not addressed especially in mental health. Human resources are also not well-distributed and not trained to manage mental health programs in the country. The number of general physicians is not sufficient for population service delivery and the number of mental health professionals is even much less what is required for an effective mental health program. For mental health, the total number of psychiatrists is about 300, insufficient for such a large population. A National Essential Drugs List exists that contains 452 drugs (one of the largest in South Asia), however, only a quarter of primary health centres are stocked with basic medicines and availability of the drugs (especially for use in mental health) are problematic (Gadit 2007).

Contributions of NGOs and voluntary organisations in initiating mental health services are very promising in the country (Gadit 2007). Fountain House, Lahore, an institution being run by civil society is an example of such a service that has been functioning as a model rehabilitation centre for more than 25 years. This has emerged as a training facility not only for the country but a number of training programs have been offered to professionals in the region as well. Similar services have also been started in areas like mental retardation, drug abuse treatment and general counselling services for the victims of trauma and violence in many parts of the country. However the growing number of psychosocial problems, increasing stresses related to violence and terrorism in the country certainly require expansion of mental health services at all levels.

Bangladesh with a population of more than 150 million is an emerging developing country in the SAARC region. Most of the population live in the countryside in more than 64,000 small villages and in terms of having a better health care system, the population continue struggling for having an ideal health care delivery system. Mental health problems similarly do not get a priority and it is extremely difficult to determine prevalence figures for psychiatric disorders in Bangladesh since the majority of patients are contained within their families and will never seek any outside help (Mullick and Goodman 2005; Hosain et al. 2007). Certainly, chronic patients are not seen sleeping rough on the streets as in the west as many Sufi shrines in the country appear to attract populations of many chronic mentally ill who may find some conformity with their beliefs in the philosophy and culture that exists in the shrines. Certainly the members of the Sufi shrine culture are extremely tolerant of the behaviour and experiences of managing the sufferers of mental illnesses (Alam 1978; Islam 1993).

The number of mental health professionals is again not enough to meet the demands of such a big population. Most of the services are limited to big cities and the rural places are still struggling to get cover for mental health in many parts of the country. Bangladesh has a long tradition of specialty hospitals and foundations in both public and private/NGO sectors that provide individual-based clinical care for health but generally focusing less on preventive care. Most, including the poor, use private practitioners for first-line clinical treatment from the private sector and through pharmacies, both licensed and unlicensed.

With the introduction of mental health policy and different directions in achieving a better mental health system, Bangladesh aims to have a trained psychiatrist in each of the 64 districts to act as a referral centre for physicians and primary care workers and to be responsible for the development of mental health services in each region. Bangladesh is also investing a great deal of its resources in the recruitment and training of family welfare workers who will operate at the village level and will be involved in the identification and diagnosis of psychiatric cases, their referral to district specialist services and follow-up. Bangladesh also recognizes non communicable diseases as a major health threat and much progress has been made in policy development in this regard. A comprehensive national NCD plan, the Strategic Plan for Surveillance and Prevention of Non-communicable Diseases in Bangladesh, 2007–2010 (that also includes mental illnesses), has been adopted. However, implementation has been slowed down by issues including limited finances, absence of clear lines of responsibility and other priorities competing in health and social sectors. The current Strategic Investment Plan is notable for including prevention and control of major illnesses including mental health. The plan proposes publicly financed insurance and health vouchers to protect the poor against the costs of emergency care and catastrophic illnesses. However, efforts towards prevention and treatment have received lower priority.

Mental health needs in Sri Lanka are as high if not higher than in other parts of world. Sri Lanka has some of the best Primary Care Services in the world and the government is committed to achieving equally high standards in mental health care services as well. Sri Lanka has 19 million populations with around 50 psychiatrists providing the specialist services in mental health. The limited manpower is also seen with very few psychologists, occupational therapists, psychosocial workers (PSWs) and psychiatric nurses (Mendis 2003). The Sri Lankan health system is a government funded decentralised public health system, accompanied by a robust private sector. Sri Lanka has long emphasised prevention and public health, and this is reflected in the administrative divisions with different cadres, some of whom focus on public health and others focus on treatment of disorder, resulting in two parallel sets of health tiers, one for preventive work and the other for curative work. This emphasis on public health has probably contributed to Sri Lanka's high immunisation rate and good life expectancy relative to some other low income countries (Ministry of Health Sri Lanka 2010). The government does acknowledges that the country has one of the highest suicide rates of any country in the world and increasing substance misuse and post disasters psychosocial problems are putting extra pressure on the delivery of health services. There has not been a robust epidemiological study of mental disorder in Sri Lanka. After the Tsunami, studies have found an increasing rate of mental disorders whereas the long standing conflict in the country has also increased the rates of mental illnesses significantly (Ministry of Health Sri Lanka 2010).

Sri Lanka is on the way to becoming a model for mental health care for middle and low income countries (Mental Health Directorate 2006). With mental health policy of 2005, the Sri Lankan government seems to be heading in the right direction and the introduction of this mental health policy has been viewed as a step in the right direction. Already known for its good primary care, the country's maternal and infant health indicators are comparable to those of developed countries, in spite of a tense socio-political situation. In order to run the new health system efficiently, health professionals will be needed. Sri Lankan health policy makers have already taken appropriate steps towards training an increased number of mental health professionals. For instance, the Sri Lankan ministry of health, in association with the Sri Lankan National College of Psychiatry, has developed a one-year diploma course in psychiatry. An extensive psychiatric training module in medical school undergraduate programs has also been developed. With proper implementation of its new mental health policy, it is hoped that Sri Lanka will also develop some model initiatives for the other regional countries.

Nepal, situated between its two big neighbours China and India, is home to several ethnic groups. The majority of the 23 million population reside in the countryside. Nepal still remains one of the poorest countries in the world and despite this; it provides shelter to thousands of Bhutanese refugees in its land. Frequent natural disasters and recent violent conflicts in Nepal have further added hardship to life. Less than 3 % of the national budget is allocated to the health sector and mental health unfortunately receives insignificant attention as the Government spends not more than 1 % of the health budget on mental health (Regmi et al. 2004).

Nepal's National Mental Health Policy was formulated in 1997 but is yet to be fully operational. The traditional/religious healing methods still remain actively practiced, specifically in the field of mental health. The mental health service, comprising little more than two-dozen psychiatrists along with a few psychiatric nurses and clinical psychologists (mainly practicing in modern health care facilities) has started showing its impact—however this is limited to specific urban areas. Prevailing financial issues, low number of mental health facilities and poor infrastructure are the ongoing major constraints in the uplift of mental health in Nepal along with issues with mental health legislation that certainly require urgent attention (Acharya et al. 2006). The main challenges for mental health in Nepal are the provision of adequate manpower, implementing the mental health policy, developing the services across the country and increasing public awareness. There is however a gradual increase in acknowledgement of mental health in the general population and the number of people seeking treatment for mental health problems is increasing.

Bhutan has adopted the Primary Health Care (PHC) approach to the delivery of health care services and over the decades, important social and demographic changes have taken place in the health care system in Bhutan. Survey carried out in the in Bhutan has estimated that there are about 21,000 persons with disabilities in the country, amounting to 3.5 % of the total population of Bhutan. These figures are only suggestive and not definitive. A more detailed survey of the different regions may be necessary to correctly assess the extent and degree of disabilities and their causes in Bhutan. The Community Mental Health Program was prepared in 1997 coinciding with the beginning of the 8th Five-Year Plan. The program is integrated into the general health service and is being further strengthened through developing the manpower and other related resources (WHO 2011).

The Myanmar government formed a National Health Committee (NHC), the highest level policy making body for health matters. The committee worked out a new National Health Policy in 1993 adopting "Health for All" and recommended primary health care as the main approach and provision for general health care in the country. The role of NGOs and private sectors was also upgraded under the new policy. The National Health Plan (1996–2001) has also been prepared with Health being assigned as a priority in the national agenda. The Ministry of Health laid down the National Health Plan (2001–2006) under the guidance from the National Health Committee suggesting for implementing the national objective of uplifting of health, fitness and educational standards of the entire nation with an emphasis on developing a new health system in keeping with the political, economic and social conditions (Htay 2006). The indigenous medicine plays an essential role in health care delivery system by offering the communities access to alternative choices. In Myanmar, as in other countries, the role of indigenous medicine is now put on the forefront.

In summary while reviewing the status of mental health services in most of the South Asian countries, we do find some notable developments in planning and formulation of mental health care policies. However the implementation of these polices is not in place in its true sense in many of these countries, thus requiring urgent and continuous attention with reference to mental health delivery systems and programs in the region. Looking at some of the specific issues, the current status is still inconsistent. When we talk about stigma and discrimination, unfortunately the scene is not very promising in many of these countries. Social isolation of the mentally ill is worse than the real illness in many cases. It is sad and hard to accept that public identity of a person having a mental disorder ruin their social contacts and affect their sense of belonging to the society as an equal human being (Byrne 2000). There is still a "big" stigma around mental health that restricts the formation of any meaningful social pressure to change individual or governmental actions. Similarly ethical aspects of practices in South Asian psychiatry are faced with many dilemmas. Looking at ethical issues in general, the situation is not very promising either. How many countries follow ethical guidelines and directions in true sense? Just to add to the problem list, burden of care for the families, poverty, poor access to health care facilities, lack of availability of care for acute and long term psychiatric illnesses, misuse and abuse of mentally ill by quakes and traditional healers are equally important while talking about the mental health problems in these countries. Involvement of patients, carers and families is emerging as a global priority in the current practice of psychiatry (Wallcraft et al. 2011). Efforts in several countries to change community attitudes and general public perception and bring improvement in mental health care in partnership have produced resolutions and guidelines, but their wide use and the structural changes they call for are yet to be achieved especially in South Asian region. It is worth noting that many countries in South Asia still lack proper and appropriate mental health legislation (Kala and Kala 2008). In those countries where such laws are present, the implementation is still far from the real practice (WHO 2001). Teaching and training of psychiatry, another important issue in Asian countries still lacks many of the current directions in terms of needs and requirements in many countries in the region (Trivedi and Dhyani 2007). The

Postgraduate teaching programs in many countries again show similar trends. Such training is still relied on opportunities offered in the western and other developed countries although many countries have started their own higher postgraduate degrees and qualifications. This adds to the issue of brain drain and settlement of trained professionals outside the region adding to the already existing shortage of professionals and manpower strength (Gureje et al. 2009; Jenkins et al. 2010).

The recent studies also suggest that in South Asia, the number of people who commit suicide is higher than the number who dies because of road accidents, terrorism and HIV/Aids. It is among the top three causes of death in the young population and over 80–90 % of suicide cases relate to mental disorder and that more than two-thirds of all suicides are preventable. Nevertheless, mental health support barely exists in many regional countries to address the growing needs of the population (Vijayakumar et al. 2005; Understanding Suicide Terrorism in Bangladesh and Sri Lanka 2010). Management of psychological traumas associated with disasters does not get any priority although South Asian's takes the maximum misery and distress from earthquakes, tsunami, floods and natural or unnatural disasters (Carballo et al. 2005; Hussain et al. 2011). Similarly the impact of terrorism, internal displacement, rising number of refugees due to conflicts and wars do increase the psychiatric morbidity in the South Asian region and require extra resources for keeping the mental health services intact (Alexander and Klein 2005; Whalley and Brewin 2007; Murthy 2007; Panter-Brick et al. 2009).

However despite these limitations and inequalities, mental health scene in many of the regional countries is not that bleak. It is true that mental health and mental health care in the region are facing a number of problems but there are several strengths that cannot be ignored and indeed are worth praising. Intact family ties, religious and spiritual teachings towards care of the mentally ill and sense of belonging to the society are still sources of inspiration that can be admired and praised. Religion and spirituality continue to play a vital role in an individual's personal and social life and in terms of mental health they are part of a very powerful medium, and help in the healing process (Chaudhary 2008). Most psychiatric patients are visited by a large number of family members when admitted in hospitals on a daily basis and are taken back to their own homes when discharged. Rejection of the mentally ill is still uncommon in many places despite poverty and economic recession. Growing numbers of mental health non-governmental organizations (NGOs) in many countries present as models of care and are playing an important role in alleviating the sufferings of the mentally ill.

A number of innovations are also being tried in different areas of mental health care that show a changing trend in the practice of mental health in many South Asian regional countries. It is heartening to note that despite limited financial resources, many countries are taking a lead in many aspects of mental health. Community psychiatry movement is getting into the system in most of the countries in the continent and orientation about common mental health problems is emerging as an important agenda for the training of primary care physicians. Mental health legislations are being refined and use of family support, religious institutions and supportive socio-cultural traditions are being promoted in the treatment, management and prevention of mental health problems. This certainly adds to the list of achievements as useful resources in many settings. The issues about Patient's involvement in the care system is however limited but it is hoped that professional organisations will continue to work in making these changes an integral part of the mental health policies in these countries. There is a need of acknowledgement of their role by the policy makers who may require constant reminders. It is hoped that these issues will continue having a priority in the future professional deliberations.

Since the publication of World Health Reports in 2001 and 2005, WHO has produced a number of more reports highlighting the miserable social status of those who suffer from mental illnesses, but WHO member countries clearly lacks the strategy to translate these reports into action. The recommendations made in WHO reports are still a dream for most of the developing countries. There is a talk about the human rights of mentally ill people, about social inclusion and the need for resources, but the international agencies, foundations and governments are simply ignoring mental health in most of the developing world. There is still strong resistance against engaging with civil society to improve the system. These are the real issues to be debated for a better mental health, but what a pity that they are always forgotten. In 2008 WHO launched its most ambitious scheme—the Mental Health Gap Action Program—but it is doubtful that it will achieve its goals and change the will of the member countries and bring broader civil society in to action (WHO 2008). Much aid money has been used on communicable diseases but what about non communicable diseases (including mental health that is largely overlooked in this category) (Ebrahim and Smeteth 2001). Despite repeated observations that mental health is very much needed in international development and while almost surveys and studies estimate that mental and neurological disorders are among the leading cause of ill health and disability globally, there is still an appalling lack of interest from governments and the policy makers as well as the civil society.

We must remember that mental health is not just a psychiatric concern—but part of larger related problems (Prince et al. 2007). Mental illness adversely affects people's ability to work, creates potential burden on their families and carers and generally leads to further frustration, misery, agony and greater poverty and deprivation. It therefore has a significant economic impact upon every one of us. Despite this, half of all countries in the world have no more than one psychiatrist per 100,000 people and a majority of these countries face limitations in the practice of their mental health programs (South Asian countries being no exception). For this change to happen it is essential that the profile of mental health in international development is raised among the public, NGOs and governments. Without changing public perceptions of mental health and mentally ill, mental health will continue to be sadly neglected in the field of international development (Bruckner et al. 2011).

It can be concluded that despite vast cultural, religious, geographical and political diversities, many factors influencing mental health generally remain the same throughout the world and South Asian countries are not an exception. However countries in this region may present some differences in terms of their needs and requirements about mental health as compared to many other countries of the world. While sharing a number of similarities like scarcity of resources, financial limitations and increasing adversities there is a growing consensus among the regional countries in South Asia that the delivery of mental health care can be improved with better cooperation and collaboration. Partnership is of course needed in areas like research, sharing of epidemiological data, exchanges of training programs and planning & working jointly for formulation of policies for integration of mental health with their national general health care programs. Organizing health education and public awareness about community care and encouraging cooperation of private sector/non-governmental organizations (NGOs) can of course be other areas that can ensure enhanced participation of professionals and general public in dealing with the growing number of mental health problems (Patel 2009).

References

Acharya, L. et al. 2006. Mental health and psychosocial support aspects in disaster preparedness: Nepal. *International Review of Psychiatry* 18 (6): 587–592.

Alam, M. N. 1978. Psychiatric morbidity in general practice. *Bangladesh Medical Research Council bulletin* 4:38–42.

Alexander, D. A., and S. Klein. 2005. The psychological aspects of terrorism: From denial to hyperbole. *Journal of the Royal Society of Medicine* 98:557–562.

Bruckner, T. A. et al. 2011. The Mental health workforce gap in low- and middle-income counties: A needs-based approach. *Bulletin of the World Health Organisation* 89:184–194.

Byrne, P. 2000. Stigma of mental illness and ways of diminishing it. *Advances in Psychiatric Treatment* 6 (1): 65–72.

Carballo, M., B. Heal, and M. Hernandez. 2005. Psychosocial aspects of the Tsunami. *Journal of the Royal Society of Medicine* 98:396–399.

Chatterjee, P. 2009. Economic crisis highlights mental health issues in India. *The Lancet* 373 (9670): 1160–1161.

Chaudhry, H. R. 2008. Psychiatric care in Asia: Spirituality and religious connotations. *International Review of Psychiatry* 20:477–483.

Crisp, A. H. et al. 2000. Stigmatization of people with mental illnesses. *British Journal of Psychiatry* 177:4–7.

Desjarlais, R. et al. 1995. *World mental health: Problems and priorities in low-income countries.* New York: Oxford University Press.

Director General of Health Services (DGHS). 1982. *National mental health program for India.* New Delhi: Ministry of Health and Family Welfare.

Ebrahim, S., and L. Smeteth. 2001. Non communicable diseases in low income and middle income countries: A debate? *International Journal of Epidemiology* 35 (2): 494–495.

Gadit, A. A. 2007. Psychiatry in Pakistan: 1947–2006: A new balance sheet. *Journal of Medical & Biological Sciences* 1 (2): 1–19.

Gadit, A. M. 2006a. Mental health in pakistan: Where do we stand? *Journal of Pakistan Medical Association* 56 (5): 198–199.

Gadit, A. M. 2006b. Mental health research in Pakistan and the developing world. *Journal of the College of Physicians and Surgeons Pakistan* 16 (3): 169–170.

Gadit, A. M. 2007. Is there a visible mental health policy in Pakistan? *Journal of Pakistan Medical Association* 57 (4): 212–213.

Government of Pakistan. Planning Commission. 2005. The report of the subcommittee on mental health, substance abuse and the health of the elderly. 10th Five-Year Plan, 1–27.

Gureje, O. et al. 2009. Report on the WPA task force on brain drain. *World Psychiatry* 8:115–118.

Haq, M. Z. et al. 2008. National mental health program in india: An update. *International Journal of Social Psychiatry* 24:22–29.

Hosain, G. M. M. et al. 2007. Prevalence, pattern and determinants of mental disorders in rural Bangladesh. *Public Health* 121:18–24.
Htay, H. 2006. Mental health and psychosocial aspects of disaster preparedness in Myanmar. *International Review of Psychiatry* 18 (6): 579–585.
Hussain, F., et al. 2011. Prevalence of war-related mental health conditions and association with displacement status in post war Jaffna District, Sri Lanka. *JAMA* 306 (5): 522–531.
Isaac, M. 2011. National mental health program: Time for reappraisal. In *Themes and issues in contemporary Indian psychiatry,* eds P. Kulhara, A. Avasthi, and M. Thirunavukarasu. New Delhi: Indian Psychiatric Society.
Islam, R. 1993. Psychiatry in Bangladesh. *Psychiatric Bulletin* 17:492–494.
Jacob, K. S. et al. 2007. Mental health systems in countries: Where are we now? *The Lancet* 370:1061–1077.
Javed, A. 2007. *Mental health in developing countries*: *Some emerging priorities in Asia. WAPR Bulletin* 21:3–5.
Jenkins, R. et al. 2010 International migration of doctors, and its impact on availability of psychiatrists in low and middle income countries. *PLoS ONE* 5 (2): e9049. doi:10.1371/journal.pone.0009049.
Kala, A. K., and K. Kala. 2008. Mental health legislation in developing countries with special reference to south asia: Problems and Solutions. *Global Social Policy* 8:308–311.
Karim, S. et al. 2004. Pakistan mental health country profile. *International Review of Psychiatry* 16 (1–2): 83–92.
Khandelwal, S. K. et al. 2004. India mental health country profile. *International Review of Psychiatry* 16 (1–2): 126–141.
Lancet Global Mental Health Group. 2007. *Scaling up services for mental disorders: A call for action. The Lancet* 9594:1241–1252.
Lolas, F. 2006. Ethics in psychiatry: A framework. *World Psychiatry* 5 (3): 185–187.
Mendis, N. 2003. Importance of general hospital in the development of mental health care. *World Psychiatry* 2:10–17.
mhGAP. 2008. *Mental Health Action Program: Scaling up care for mental, neurological, and substance use disorders*. Geneva: World Health Organisation.
Ministry of Health Sri Lanka. 2010. National health policy—Sri Lanka. www.health.gov.lk/HealthPolicy.htm.
Mental Health Directorate. 2006. The Mental health Policy of Sri Lanka 2005–2015. Colombo. Ministry of Healthcare and Nutrition.
Mullick, M. S., and R. Goodman. 2005. Then prevalence of psychiatric disorders among 5–10 years old in rural, urban and slum areas in Bangladesh: An explonatory study. *Social Psychiatry and Psychiatric Epidemiology* 40:663–671.
Murthy, R. S. 2007 Mass violence and mental health–recent epidemiological findings. *International Review of Psychiatry* 19 (3): 183–192.
Niaz, U. 2003. Violence against women in south asian countries. *Arch Women's Mental Health* 6:173–184.
Nishtar, S. et al. 2005. Integrated population-based surveillance of non-communicable diseases: Pakistan model. *American Journal of Preventive Medicine* 29 (5 Suppl 1): 102–106.
Panter-Brick, C. et al. 2009. Violence, suffering, and mental health in Afghanistan: A school-based survey. *Lancet* 374:807–816.
Patel, V. 2009. The future of psychiatry in low and middle income countries. *Psychological Medicine* 39:1759–1762.
Patel, V. 2010. Research priorities for Indian psychiatry. *Indian Journal of Psychiatry* 52 (Suppl): 26–29.
Patel, V. et al. 2007. Treatment and prevention of mental disorders in low-income and middle income countries. *Lancet* 370:991–1005.
Prince, M. et al. 2007. No health without mental health. *Lancet* 370:859–877.
Regmi, S. K. et al. 2004 Nepal mental health country profile. *International Review of Psychiatry* 16 (1–2): 142–149.

Rothman, D. J. et al. 2009 Professional medical associations and their relationships with industry: A proposal for controlling conflict of interest. *JAMA* 301 (13): 1367–1372.

Saraceno, B. et al. 2007 *Barriers to improvement of mental health services in low-income and middle-income countries*. *The Lancet* 370:1164–1174.

Saxena, S. et al. 2003. Budget and financing of mental health services: Baseline information on 89 countries from WHO's project atlas. *The Journal of Mental Health Policy and Economics* 6:135–143.

Shafique, M. 1995 Psychiatry in pakistan (1947–1994): The balance sheet. *Psychiatric Bulletin* 19:567–570.

Sharma, V. K. 2006. Psychological support within disaster: Management in asian countries In: *Advances in psychological and social upport after disasters,* eds. O. Joseph, Diaz Prewitt, R. Srinivasa Murthy, and Rashmi Lakshminarayana, 29–35. India: VHAI.

Srinivasa, Murthy R. 2007. Mental health program in the 11th five year plan. *Indian Journal of Medical Research* 125:707–711.

The World Bank. 2005. *Public health surveillance system: A call for action.* Islamabad: Ministry of health, world bank, centres for disease control, World Health Organization.

Trivedi, J. K., and M. Dhyani. 2007. Undergraduate psychiatric education in south asian countries. *Indian Journal of Psychiatry* 49 (3): 161–165.

Trivedi, J. K. et al. 2007. Regional cooperation in South Asia in the field of mental health. *World Psychiatry* 6 (1): 57–59.

Understanding Suicide Terrorism in Bangladesh and Sri Lanka. 2010. (ed) S. Y. Surendra Kumar and Mostafa Mahmud Naser. Regional Centre for Strategic Studies, Colombo.

Vijayakumar, L., J. Pirkis, and H. Whiteford. 2005. Suicide in developing countries: Prevention efforts. *Crisis* 26 (3): 120–124.

Wallcraft, J. et al. 2011 Partnerships for better mental health worldwide: WPA recommendations on best practices in working with service users and family carers. *World Psychiatry* 10:229–236.

Whalley, M. G., and C. R. Brewin. 2007. Mental health following terrorist attacks. *British Journal of Psychiatry* 190:94–96.

World Health Organization. 1973. The development of comprehensive mental health services in the community. Copenhagen: WHO Regional Office for Europe.

World Health Organization. 2001. Meeting the needs of persons with mental disorders through legislation; Proceedings of a Regional Workshop on Mental Health Legislation Galle, Sri Lanka, 24–27 May 200: World Health Organization Regional Office for South-East Asia. New Delhi.

World Health organization. 2008. World Health Statistics. Geneva: WHO.

World Health Organization. 2008. WHO Mental Health gap Action. Programme (mhGAP). Geneva: WHO.

World Health Organization. 2011. *Mental health services atlas*. Geneva: WHO.

Chapter 7
Mental Health Programs and Policies in South Asia: Initiatives and Obstacles

Rahul Shidhaye, Siham Sikander, Mark Jordans, Prasantha De Silva and Sudipto Chatterjee

7.1 Introduction

South Asia is the most populous and, in economic terms one of the poorest regions in the world with enormous social, economic and health challenges, including pervasive poverty, inequality, violence, political instability and a high burden of communicable diseases (Patel et al. 2006). South Asia is experiencing rapid demographic and epidemiological transition; with ageing populations and non-communicable diseases contributing to a growing share of the burden of disease (Patel et al. 2006). Many parts of the region have weak public health systems and a growing, but inadequately regulated private sector with consequent health care costs rising at rates far exceeding inflation. Sri Lanka has been an exception in the South Asian region and enjoys health and education indicators superior to its relatively richer South Asian neighbours (CSDH 2008).

In this region 11 % of Disability Adjusted Life Years and 27 % of Years Lived with Disability are attributed to neuropsychiatric disorders (Lopez et al. 2006). The prevalence of mental disorders such as depression and anxiety reported in the region are amongst the highest in the world (Patel et al. 2006), and depression is projected

R. Shidhaye (✉)
Centre for Chronic Conditions and Injuries, Public Health Foundation of India, New Delhi, India
e-mail: rahul.shidhaye@phfi.org

S. Sikander
Research Division, Human Development Research Foundation, Islamabad, Pakistan

M. Jordans
Research and Development Department, HealthNet TPO and Institute of Psychiatry, Amsterdam, the Netherlands

P. De Silva
Family Wellness Unit, HEMAS Hospital, Wattala, Sri Lanka

S. Chatterjee
School of Population Health, University of Melbourne and Sangath and Parivartan, Goa, India

J. K. Trivedi, A. Tripathi (eds.), *Mental Health in South Asia: Ethics, Resources, Programs and Legislation,* International Library of Ethics, Law, and the New Medicine 58,
DOI 10.1007/978-94-017-9017-8_7

to be the single greatest contributor to the burden of disease by 2030 (Patel et al. 2011). Women, in particular those who live in socially disadvantaged circumstances, bear a disproportionate burden of these common mental disorders (Scholte et al. 2004; Shidhaye and Patel 2010). The South Asian region is one of biggest providers of specialist mental health human resources to rich countries (Mullan 2005), but paradoxically mental health systems in this region are highly inadequately resourced which has resulted in a huge treatment gap of around 90 % (Patel 2007).

In last two to three decades though, mental health has received increasing attention in national health policy and programming. In this chapter, we review the current status of mental health programs and policies in South Asia, highlight the progress made and identify the continuing challenges and potential solutions that need to be addressed urgently to close the treatment gap in the near future.

7.2 The Current Status of Mental Health Systems in South Asian Region

A **mental health policy** is the official statement of a government conveying an organized set of values, principles, objectives and areas for action to improve the mental health of a population (WHO 2011). A dedicated mental health policy is present in approximately 60 % of countries globally (WHO 2011). India does not currently have a dedicated mental health policy, though the National Health Policy 2002 incorporates most of the suggestions made in the draft of the National Mental Health Policy which was never finalized (WHO 2005a). Pakistan formulated its mental health policy in 1997 (WHO 2005a). The policy covers areas of advocacy, promotion, prevention; treatment and rehabilitation through inter-sectoral collaboration (1998). In the same year, a substance abuse policy was also formulated which included interventions for both reduction of supply and demand (1998; WHO 2005a). In Nepal a national mental health policy was developed in 1997 in which mental health is proposed as an element of primary health care, revised five years ago (Kortmann 2004) and a primary care implementation guideline submitted two years ago (Upadhyaya and Mahat 2007), yet formal adoption of the policy remains pending. Sri Lanka in recent years has witnessed strengthening in mental health systems and the boost has come from the cabinet-approved Mental Health Policy in 2005. National level mechanisms that contributed the success include establishment of Mental Health Directorate, National Mental Health Advisory Council, and National Institute of Mental Health (2005).

India was one of the first countries to start National Mental Health Program in 1982 with the aim of extending community-based mental healthcare through the existing primary healthcare system (Goel 2011). The District Mental Health Program (DMHP) was conceptualized during 1984–1990 to deliver mental health services in 4 districts and then to 25 districts in 20 states during 1995–2002 and over 125 districts in the next 7 years (Ministry of Health and Family Welfare and Government of India). The key approaches used were the training of primary health care personnel, the provision of neuropsychiatric drugs in peripheral institutions, establishing psychiatric units at the district level with streamlined referrals and encouraging community participation.

In Pakistan, mental health policy formulation was preceded by the National Mental Health Program of Pakistan which was established way back in 1986. Being part of the general health policy of the country, it aimed at incorporating mental health in primary care, removing stigma, caring for mental health and substance abuse across the country and maintaining principles of equity and justice in the provision of mental health and substance abuse services (1987).

Little progress has been made on implementing the policy framework in Nepal and this is partly due to the absence of a dedicated person responsible for mental health in the Ministry of Health and Population (MoHP). Especially given the impact of a decade of armed conflict that has scourged the country and a multitude of pre-conflict risk factors, there is an urgent need to address the high psychiatric morbidity in Nepal, especially for marginalized populations.

7.3 What has Been the Impact of National Mental Health Programs in South Asia?

Published literature and independent evaluation of the DMHP in **India** indicate that the DMHP is, to a large extent, ineffective in practice (Murthy 2011). Some of the reasons for this unsatisfactory state of affairs are: the top–down, 'one size fit all' approach to service delivery that cannot accommodate diverse ground realities, poor governance, managerial incompetence and unrealistic expectations from low-paid/poorly motivated primary healthcare personnel (Goel 2011). In India health is a State subject and implementation of public health programs is the responsibility of State Department of Health Services. The nodal agencies identified for implementation of mental health program are mostly departments of psychiatry in medical colleges which are managed by the Department of Medical Education. Lack of coordination between Department of Health Services and Department of Medical Education has been cited as one of the major barriers for the integration of mental health with primary healthcare (Murthy 2011).

In **Pakistan,** the failure of the National Mental Health Program to deliver expected results has resulted in Pakistan having one of the highest burdens of mental illness in the region (Mirza and Jenkins 2004; Husain et al. 2006; 2007). As is usually the case, the reasons for this are multi-factorial in nature. Firstly Pakistan allocates only 0.4 % of its total health budget to mental health (WHO 2005a). This translates into high out of pocket expenditure for patients and their families, which adds to their already considerable economic burden and often leads to discontinuation of treatment (Gadit 2004; WHO 2005a). Secondly the country has a shortage of trained mental health professionals, which cannot cater to the country's mental health needs (Khan 2004; Gadit 2007). Thirdly, there seems to be gaps in mental health integration into the primary health care system. Reflective of this, a recent survey showed that patients who were prescribed psychopharmacological medications, 50 % did not know their diagnosis; 86.3 % were not cautioned about the potential side effects of the drugs and overall knew less as compared to those patients that were prescribed other medications (Ganatra et al. 2009; Irfan 2010; Nishtar and Mehboob 2011). Lastly, Pakistan's parliament has signed off a constitutional amendment that has led to the abolition of the Ministry of Health in June 2011. This means that Pakistan does not have a ministry or a state department at the federal level that is in charge of national responsibilities for health. Unfortunately this amendment came at a time when there was a need to increase the capacity of the health system in view of many serious health challenges faced by Pakistan (Ministry of Health and Family Welfare and Government of India; Nishtar and Mehboob 2011). In effects, this also means that the National Mental Health Program is now abolished as provinces will now formulate their own mental health programs.

The situation in **Nepal** is not very different. Promisingly, a basic strategy for the integration of mental health in primary health care was formulated as part of the National Health Sector Support Program for the period 2010–2015. While time is remaining, few noticeable efforts have yet been made in the implementation of this plan. A loosely established national network for mental health has been lobbying for the following components: (a) integrate mental health within existing and future health and social programs; (b) set up a low-cost and sustainable system for providing mental health promotion, prevention and treatment within each district of Nepal; (c) accurate national mental health data from the Health Monitoring Information System and census data; (d) appoint a focal person for Mental Health within the MoHP.

7.4 New Initiatives in Mental Health Policy and Program Implementation in South Asia

In India, despite relatively bleak record for mental health care, there has been a considerable increase in political will and public interest in mental health care in recent years. Government of India has plans to roll out the District Mental Health Program to all districts under the 12th Five Year Plan (2012–2017), and evidence based models for mental health care delivery are being sought to inform the design of this roll-out. The Ministry constituted a Mental Health Policy Group in April 2011 to prepare a National Mental Health Policy and Plan. Based on the document reviews, consultations with key stakeholders and field visits, the Mental Health Policy Group proposed substantial changes to the design of the District Mental Health Programme (DMHP) for the Twelfth Five Year Plan period (2012–2017).

The South Asian region has been awarded a National Institute of Health (NIH) grant for the setting up of a collaborative hub for international research in mental health. The broad goal of this **S**outh Asian **H**ub for **A**dvocacy, **R**esearch and **E**ducation on mental health (**SHARE**) is to establish a collaborative network of institutions in South Asia to carry out and to utilize research that answers policy relevant questions related to reducing the treatment gap for mental disorders in the region.

The **Program** for **I**mproving **M**ental Health Car**e** (PRIME) is another research consortium in which Ministries of Health in India (Government of Madhya Pradesh) and Nepal are key collaborators. PRIME is a consortium of research institutions and Ministries of Health in five countries in South Asia and Africa (Ethiopia, India, Nepal, South Africa & Uganda), with partners in the UK and the World Health Organization (WHO). The goal of PRIME is to generate world-class research evidence on the implementation and scaling up of treatment programs for priority mental disorders in primary and maternal health care contexts in low resource settings.

Sri Lanka has banked upon a new initiative to introduce two intermediate cadres of mental health professionals, namely Diploma Holders, and Medical Officers in Mental Health to expand mental health services in rural areas (WHO 2007). With Tsunami, an unprecedented catastrophe in the country, a new category of Community Support Officers on contract basis with support of INGOs, and WHO, was introduced in the affected areas. Upon findings of an evaluation of the initiative, the Eastern province adapted the strategy by employing already existing cadre of lower grade hospital health workers as psychiatric social assistants for field work. Similar strategy is being considered in the Northern Province too.

Another initiative is introduction of Community Support Centers at field level attached to Medical Officer of Health areas. Since the Medical Officer of Health is in charge of overall preventive services in a designated area through public health workers, and inter-sectoral coordination with other health related officials such as social service, police etc, they provide relevant services to people with social as well as personal problems through Community Support Centers (2005).

Over the years, research groups with donor agencies launched series of coordinated research projects in rural Sri Lanka to reduce access to pesticides for suicide prevention. These projects generated substantial amount of knowledge which is critically useful not only for Sri Lanka but to other countries in the region (Manuweera et al. 2008).

Caregiver societies, as a strategy of empowering consumers and affected families, saw a renewed national push by way of setting a target of establishing at least one such society at district level at the 2010 National Mental Health Day. Almost all these initiatives have been actively supported by the WHO. As a way of improving evidence-based practice, Sri Lanka conducted first-ever National Mental Health Survey in 2007 with support of the World Bank.

7.5 Mental Health Program Initiatives Undertaken by the NGO Sector with Potential to Scale-up

Non-Governmental Organizations working in mental health sector in India have made tremendous strides in mental health promotion and care, against massive odds, ranging from low awareness about mental illness to the lack of motivation by donors (Patel and Varghese 2004). The strength of these organizations does not lie in their ability to reach out to the millions of people with mental disorders, but in evolving and perfecting quality programs and models which have the character of replicability. Through innovation and accountability, they can provide models for the public healthcare system to emulate and partner (Patel and Varghese 2004).

The most formidable challenge in developing comprehensive services for persons with schizophrenia and other psychotic disorders is the acute shortage of mental health specialists, particularly in rural and socio- economically disadvantaged areas. Though the WHO has long advocated that the solution to mitigate the problem is to integrate basic medical services for persons with schizophrenia in primary care facilities, there is little evidence that this strategy is even modestly effective and sustainable.

Community Based Rehabilitation (CBR) for Schizophrenia

CBR is a strategy within general community development for the rehabilitation, poverty reduction, equalization of opportunities and social inclusion of all people with disabilities. It is recommended that the implementation of CBR programs be through the combined efforts of people with disabilities themselves, their families, organizations and communities, and the relevant governmental and non-governmental health, education, vocational, social and other services.

Since the initial report of CBR services for persons with schizophrenia being more effective than OP based care in a rural and resource poor setting in India (Chatterjee et al. 2003), CBR services have been implemented in parts of many low income countries, largely through the efforts of community based organizations. By now, there is consistent observational evidence that this method is feasible to implement and acceptable to persons using services in meeting the medical and social needs of persons with schizophrenia (Chatterjee et al. 2009). In addition, an ongoing randomized controlled trial in three sites in India will provide high quality evidence of the clinical and cost effectiveness of community based treatments for persons with schizophrenia in a collaborative manner with specialists, lay community workers and persons with schizophrenia and their caregivers (Chatterjee et al. 2009). In recognition of the widespread use and success of this strategy in providing comprehensive services in non-specialist settings, the recent mhGAP guidelines issued by the WHO defining the treatments for psychotic disorders like schizophrenia endorses the adoption of the CBR strategy to address rehabilitation, developmental and social needs.

Mental health services for persons with schizophrenia in resource rich settings are delivered almost completely by qualified professionals working within a well-defined and prescriptive health system and are closely regulated in accordance with a strong policy and legislative framework. The complexity and gaps in integrating the various elements effectively with such highly articulated systems has often led to modest user satisfaction and outcomes for persons with schizophrenia, in spite of considerable human and financial resources. The CBR methodology, on the other hand (out of necessity!), uses socially inclusive framework in understanding the holistic needs and plugs the gaps in services with the active participation of the person with schizophrenia, family members and local community structures. This in turn allows for the creation of a more enabling social environment for the person to meet normative aspirations, as far as possible.

Pakistan's NGO sector, in collaboration with public sector institutions, has been working to address mental health needs of the vulnerable populations. One of such NGO's being Health Net TPO. In 2010, HealthNet TPO provided psychosocial support to communities in the Buner district, affected by the armed conflict between militants and government forces. Currently, HealthNet TPO works closely with knowledge centers and academia to develop and test new models for interventions. It is collaborating with Health Services Academy in Islamabad and the Institute of Psychiatry, Rawalpindi (http://www.healthnettpo.org). Similarly Human Development Research Foundation (HDRF), being a not-for-profit organization, in collaboration with international and national institutions, has been developing and testing community based psycho-social interventions, delivered at scale using Lady Health Workers Program of Pakistan; to address maternal mental health. HDRF has developed close linkages with public sector institutions like the Health Services Academy which is a degree awarding institution in public health (www.hdrfoundation.org.pk).

In the absence of tangible Government action in furthering mental health care in Nepal, several other initiatives have taken place. Few organizations with a specific focus on mental health and psychosocial support stand out, the Center for Victims of Torture, Nepal (CVICT), the Center for Counseling and Mental Health (CMC), the Transcultural Psychosocial Organization, Nepal (TPO) (affiliated with HealthNet TPO, the Netherlands), each with its own focus. CVICT has been specialized over the years in the provision of care to victims of torture and organized violence and has, together with TPO pioneered in long-term skills-based training courses for community counselors (Jordans et al. 2003). In addition, TPO has been implementing psychosocial support programs for especially vulnerable populations (refugees, former child soldiers, survivors of trafficking etc), combining development of packages of care with scientific research. CMC has a long track record in the implementation of mental health care in primary health care settings. Also, the Nepal Mental Health Foundation is a leading mental health rights institution in Nepal. It is dedicated to combat stigma and discrimination attached to mental health problems in all sectors of society. Several other smaller organizations focus on mental health or psychosocial wellbeing.

In Sri Lanka, especially after the tsunami, NGOs contributed significantly to the development of the Mental Health services. World Vision, Australia assisted the government through the WHO in revitalizing the services in Tsunami-affected areas particularly to strengthen community-based programs through Community Support Officers. International VSO program through its Sri Lankan counterpart has brought to the country many an experienced overseas volunteers such as psychiatric nurses, social workers, media communication experts. As an attempt of promoting evidence-based practice, the VSO in collaboration with the Ministry of Health conducted first-ever National Research Forum in January, 2011. Further, VSO formulated guidelines for medium-stay units and for media reporting mental health issues through extensive consultative processes. BASIC NEEDS pioneered empowering communities establishing self-sustainable care giver associations in Southern part of the country. An initiative from these organization, the "Theater of the Oppressed", which was originated in Brazil, is having a ripple effect in other parts of the country (Raja et al. 2010).

7.6 Scaling-Up of Mental Health Programs: Challenges and Potential Solutions

With this background, we would now describe some of the domains that are necessary to influence mental health policy, planning and service delivery; additionally, we highlight the barriers and potential solutions in this regard.

7.6.1 Political Commitment and Mental Health Policy

Political commitment and buy-in is essential to influence mental health policy making and further development of the national mental health plan. The commitment at the highest level of policy-makers ensures prioritization of mental health. In order to scale-up the mental health services it is almost a pre-requisite that political leaders and decision makers understand the importance of services and prioritise actions to address mental health needs (Thornicroft et al. 2010). In a recent publication, respondents in close to half of the countries in which the study was done, identified continuing poor awareness and low priority or poor commitment by political leaders as major barriers to development of mental health services (Eaton et al. 2011). Key barriers to the implementation of services are absence of a national government mental health policy, strategy, or program (Eaton et al. 2011).

How to get the political buy-in is a million dollar question. Success in implementation of any public health program rests foremost on political commitment at highest level.

In Sri Lanka, political commitment was demonstrated by the cabinet approval of the Mental Health Policy in 2005 and naming the only mental hospital as the National Institute of Mental Health. The process was expedited by an extra-ordinary catastrophe, the Tsunami. The time is now opportune for mid-term review of the Policy since the Policy was introduced for the period, 2005–2015.

As mentioned above, in India the commitment at the level of Ministry of Health, Government of India has resulted in formulation of mental health policy group and revision of the Mental Health Act of 1987.

7.6.2 Planning Process

For successful implementation of any mental health program, it is essential that the principles of planning cycle are followed while the program is designed. In absence of rational planning and budgeting the mental health service delivery would be inefficient and fragmented.

The WHO Mental Health Policy and Service Guidance Package (WHO 2005b) mentions four planning steps; (a) situation analysis of current mental health services and service funding, (b) assessment of needs for mental health services, (c) target-setting for mental health services and (d) Implementation of service targets through budget management, monitoring and evaluation.

It is crucial to have a good situational analysis for mental health systems as this will serve as an essential baseline for needs assessment and further planning of mental health service delivery. Situational analysis should provide information on mental health policy, governance and financing issues, organization of services and existing human resource in mental health and extent to which inter-sectoral collaboration has been achieved between mental health and education, police, judiciary and women and social empowerment sector. Situational analysis is not just about getting prevalence of priority mental health disorders and the exercise should go well beyond simple head-counting. We have a fairly good data on global estimates of priority mental health disorders and what we really need to know is the extent of the coverage of the services, treatment gap, pathways to care, referral pathways in public and private health system.

7.6.3 Integration with Existing Public Health Programs

In India, the National Rural Health Mission (NRHM) is a flagship program of Central Ministry of Health to improve health service delivery. Unfortunately, the NRHM pays little attention to mental health services; except for some passing statements like 'mental health should be prioritized' there isn't much in terms of mental health service delivery and program implementation. In India this is certainly a window of opportunity for integrating mental health service delivery with general health programs. A significant interest has been generated for promotion of mental health in other sectors, but mechanisms are necessary for formulating coordinated action plans and consensus is necessary between preventive and curative sectors for formulating objectives both at national and regional level.

7.6.4 Demand Generation

In order to initiate appropriate interventions, it is essential that the gap between community understanding and stigma related to mental illness and the 'scientific' view is addressed through close engagement and innovative programs. Mental health literacy refers to people's knowledge and beliefs about mental disorders which aid their recognition, management and help seeking choices. In India, the NMHP is funding advertisements about treatment of mental disorders and reduction of stigma associated with mentally ill in prime time television slots. While doing this, it is imperative to assess to what extent these activities impact population level mental health literacy and help-seeking behavior and whether they are cost-effective. Sometimes, mental health services are revamped in the wake of crises that demonstrate the acute need for such services. In Sri Lanka, the tsunami was the main catalyst for significant surge in demand for mental health. It accelerated introduction of the Mental Health Policy in 2005. Similar emphasis is required for strengthening the system in North-East Sri Lanka, the areas which had been repeatedly affected by three decades of protracted armed conflict.

7.6.5 *Organization of Services: Equity Issues*

In India there are around three and half thousand psychiatrists, over 70% of who work in urban areas where less than 30% of the population live, and over half of these psychiatrists work exclusively in the for-profit private sector. Vast populations in rural areas have no access to any specialist mental health services (WHO 2005a). Over 80% of the 23,000 psychiatric beds are located in 40 mental hospitals (WHO 2005a), most of which were built in colonial times, and where standards of care are characterized by over-crowding, lack of psychosocial interventions and, in some instances, serious human rights violations (Murthy and Sekar 2008).

In Nepal also mental health resources are scarce, with no formal mental health care in rural areas (Regmi et al. 2004) where more than 85% of the total population resides. In Sri Lanka, though significant improvements are evident in services for psychotic disorders, there are persistent inequities in resource distribution across the country; for example, 62% of psychiatric beds are located in Colombo.

7.6.6 *Task-Shifting/Sharing in Mental Health Service Delivery*

One of the most important reasons for the persistent treatment gap for mental disorders in South Asia is the lack of skilled human resources. Currently, there are less than 2 psychiatrists per 1,000,000 population in India (WHO 2005a), and the number of other specialist mental health workers—clinical psychologists, psychiatric nurses and psychiatric social workers—is even lower than the number of psychiatrists (WHO 2005a).

Country	Population covered/ psychiatrist	Population covered/MH professionals	Population covered by a psychiatric bed	% Psych beds in mental hospitals	Mental health budget (% total health budget)
India	500,000	322,581	40,000	80	2.05
Pakistan	500,000	113,637	41,667	25	0.4
Nepal	833,333	312,500	312,500	25	0.08
Sri Lanka	500,000	47,847	5,556	78	1.6

Source: WHO. Mental Health Atlas. Geneva: WHO. 2005

Thus, reducing the treatment gap needs innovative solutions emphasizing the delivery of care by non specialist health workers (NSHW), especially in rural areas. Task-shifting/sharing is the strategy of redefining the role of various members of the health workforce; by 'shifting' generic tasks to lesser trained workers, shortages of specialised staff are mitigated and there is greater efficiency in the use of available resources (Chisholm et al. 2007). NSHW, and the non-formal workforce, such as community volunteers and people with mental disorders and their family members, can be an essential human resource in South Asia; the key question is to identify their training and supervision needs to deliver quality assured treatments.

There is now growing evidence base on the effectiveness of NSHW in providing interventions for mental disorders, such as cognitive behaviour therapy by community-based primary health workers for perinatal depression in women in Pakistan (Rahman et al. 2003) and community based rehabilitation for schizophrenia (Chatterjee et al. 2003, 2009).

Manas

The recently reported (Patel et al. 2010a) results of the Manas study—a cluster randomized trial of integrating a collaborative, stepped care method of service delivery for common mental disorders through trained lay Health Counsellors in Goa, India provides some of the most compelling evidence that the task sharing approach is feasible, acceptable and effective in public primary care health facilities. In this study, lay health counselors were recruited and rigorously trained for 8 weeks to deliver a variety of psychosocial interventions for persons with depression and anxiety problems presenting to primary care health centres. This intervention (delivered by the lay health counselors working in primary care facilities with doctors and other staff and supervised by a psychiatrist) was more effective than a comparison 'enhanced usual care' intervention on measures of recovery from episodes, disability and work productivity over 12 months.

Thinking Healthy Program (THP)

The Thinking Healthy Program is a fully manualised intervention drawing on the principles of Cognitive Behaviour Therapy (CBT) techniques to address perinatal depression. This was developed in Pakistan with a view of getting it delivered by the Lady Health Workers (who are non-mental health professionals with an average eight to ten years of education) (Rahman 2007). THP was tested in a large cluster randomized controlled trial where these lay Lady Health Workers (LHWs) were trained to deliver it to mothers with major depression. The THP was fully integrated into the routine work of the LHWs. At 6 months post-partum, 77 % of mothers in the intervention group recovered from their depressive disorder compared to 47 % in the control group, effects which were sustained at 12 months. It also showed significant increase in infant immunization coverage, reduced diarrheal episodes among infants and increased uptake of family planning by mothers receiving THP (Rahman et al. 2008). The feedback from the 40 trained LHWs showed that almost all of them thought it was relevant to their day-to-day work and none of them considered it an extra burden. The training was short (2 days followed by a 1 day refresher after 4 months) and therefore feasible on a large scale.

Child Thematic Program in Nepal

This community-based psychosocial care system for children includes different overlapping levels of interventions to address varying needs for support. These levels provide assessment and management of problems that range from the social-pedagogic domain to the psychosocial, the psychological and the psychiatric domains. One component of the care package, the Classroom Based Intervention, is a 15-session classroom or community-based intervention, involving a series of highly structured expressive behavioral activities, which aims at increasing children's capacity to deal with the psychosocial problems, that having been/ being exposed to extreme stressors can cause. A cluster randomized controlled trial studied the efficacy of CBI as part of this program. Results show that CBI is moderately effective in reducing psychological difficulties and aggression among boys and increasing pro-social behavior among girls in Nepal (Jordans et al. 2010).

THP clearly showed it could be integrated into an existing public health program of Pakistan, however scaling up would have to tackle a few issues. Firstly THP had trained psychiatrists carrying out LHW trainings and supervisions. Secondly, the research team diagnosed maternal depression for LHWs. Therefore, to scale up, mechanisms for LHWs to identify maternal depression and using existing cadre of LHW trainers and supervisors needs to be developed.

In Sri Lanka, employment of intermediate cadres, Diploma Holders, and Medical Officers in Mental Health, contributed immensely in task-shifting process, particularly in underserved remote parts of the country. While the former receives one year of extensive training after a competitive screening examination, the latter receive 3 months of training soon after the internship. At present, approximately 65 Diploma Holders and 140 Medical Officers in Mental Health serve particularly in underserved areas. The National Institute of Mental Health trained nurses who are already in the service, as Psychiatric Community nurses.

In Nepal, there has been a growing focus on using a task-shifting approach to psychosocial and mental health care. In practice this has been operationalized through the training of community counselors emphasizing, (a) skills-based learning, (b) practical placements to apply learned skills supervised by a senior counselor/trainer, (c) post-training follow-up by supervisor, (d) long term training trajectories rather than brief courses (Murray et al. 2011).

7.6.7 Capacity-Building and Supervision

Capacity-building of primary health care workforce is an essential pre-requisite for de-centralized mental health care service delivery. In the Lancet publication, poor knowledge of mental illnesses among primary health-care staff and scarcity of mental health specialists for liaison and supervision have been identified as key concerns (Eaton et al. 2011). If the staff does not receive ongoing training and supervision, motivation to undertake mental health work is lost.

Task sharing always necessitates substantial training, but where there is high staff turnover, this investment might be wasted (Petersen et al. 2009). Some reports called for task sharing with families, carers, and volunteers, empowering them to play a more informed part in caring for people with mental illnesses in the community—a training investment less likely to risk so-called brain drain (Patel et al. 2010b). This peer support is also favoured by organisations of people using mental health services, families, and carers, but this strategy should avoid reducing choice by replacing proper provision of professional services on which people also rely.

Task-sharing or task-shifting should not be looked as task-replacement. Non-specialist health workers should be engaged in task-based work and not the entire management of the case. Continuous supervision and strong referral pathways of care should essentially accompany the task-shifting approach.

7.6.8 Availability of the Treatment Guidelines Specifically for Use in Low Income Countries

The WHO has initiated the mhGAP program (mental health Gap Action Program: Scaling up care for mental, neurological and substance use disorders) in 2009. This program is grounded on the best available scientific and epidemiological evidence about mental, neurological and substance use (MNS) conditions that have been identified as priorities. These priority conditions are depression, schizophrenia and other psychotic disorders, suicide, epilepsy, dementia, disorders due to use of alcohol, disorders due to use of illicit drugs, and mental disorders in children (WHO 2010). In 2010, WHO developed mhGAP intervention guide. The intervention guide is brief to facilitate interventions by busy non-specialists in low and middle income countries mhGAP intervention guide needs to be adapted for local use (WHO 2010).

7.6.9 Monitoring and Evaluation of Programs

In a systematic review of community mental health services in Africa, only a fifth of relevant programs included any evaluation (Hanlon et al. 2010) and the Lancet publication mentions that, of the 56 respondents who described new mental health programs in their countries, only 22 (39 %) reported completed evaluations (Eaton et al. 2011). Most research into scaling up of services emphasizes gaps in metrics and evaluation along with inadequate and incomparable primary data sources and analyses (Eaton et al. 2011).

Over the past years there has been a strong movement towards rigorous evaluation of the psychosocial and mental health programs being implemented, especially within the NGO sector, in Nepal. One study has demonstrated that a brief multidisciplinary treatment was moderately effective in reducing non-specific mental health consequences among torture survivors (Tol et al. 2009). A cluster randomized controlled trial has similarly showed that a school-based psychosocial intervention resulted in moderate beneficial gains (i.e. reduced psychological difficulties and aggression among boys, increased pro-social behavior among girls, and increased hope for older children) (Jordans et al. 2010).

For successful scaling-up of mental health program it is essential that assessment of performance at the ground level through continuous monitoring, ongoing audit by an independent agency and periodic review at the national level to identify areas of non-performance/reasons for the same at an early stage is undertaken and necessary corrective measures are introduced and relevant feedback is given for future planning. Determining objectives—which are specific, measurable, achievable, realistic, and time-bound (SMART), is the cornerstone for the success of any program. This is intricately related to quality of the information system.

7.6.10 Closer Collaboration Between the Private and Public Sectors

Most South Asian countries have health care systems characterized by very low public spending on health and unregulated provision of health services, mostly in the private sector with some of the highest out-of-pocket expenditures in the world.

Similar to other health conditions, the private sector in India is the largest service provider for mental disorders. However, at present, the public and private systems work as separate silos with little mutual communication. Given that the private sector is the largest specialist resource, a pragmatic collaboration is the way forward. There is scope for the involvement of specialists working in the private sector for working in medical colleges and district hospitals as honorary consultants; helping in the training of primary healthcare personnel; contributing to the district level mental health programs; supporting NGOs in their mental health initiatives and to advocate on behalf of persons with mental disorders (Murthy 2011). As an initial step in this direction, In India private psychiatrists are now involved as consultants in DMHP.

7.7 Conclusion

There is robust evidence for effective interventions for priority mental disorders like depression, psychosis and alcohol use disorders in South Asia. In order to address the huge, and largely unmet, burden of mental health disorders in the region, it is essential to scale-up these interventions by progressively strengthening existing mental health systems. Political commitment to improve public mental health resource allocation, a strong mental health policy framework with strategies for efficient implementation, contextualization of mhGAP guidelines and it's delivery through an emphasis on task sharing and active involvement of the NGO and private sector are some of the approaches that can address the barriers in scaling-up of mental health services.

References

A Report of the sub-working group on mental health care in Pakistan. 1987. *National program of mental health for Pakistan*. Islamabad: Planning Commission, Government of Pakistan.

Bridging the Gap, Integrating people with mental illness into the development process. 2007. Basic Needs, Sri Lanka.

Chatterjee, S., V. Patel, et al. 2003. Evaluation of a community-based rehabilitation model for chronic schizophrenia in rural India. *British Journal of Psychiatry* 182:57–62.

Chatterjee, S., A. Pillai, et al. 2009. Outcomes of people with psychotic disorders in a community-based rehabilitation program in rural India. *British Journal of Psychiatry* 195 (5): 433–439.

Chatterjee, S., M. Leese, et al. 2011. Collaborative community based care for people and their families living with schizophrenia in India: Protocol for a randomised controlled trial. *Trials* 12:12.

Chisholm, D., A. J. Flisher, et al. 2007. Scale up services for mental disorders: A call for action. *Lancet* 370 (9594): 1241–1252.

CSDH. 2008. *Closing the gap in a generation: Health equity through action on the social determinants of health. Final Report of the Commission on Social Determinants of Health*. Geneva: World Health Organization.

Eaton, J., L. McCay, et al. 2011. Scale up of services for mental health in low-income and middle-income countries. *Lancet* 378 (9802): 1592–603.

Gadit, A. A. 2004. Out-of-pocket expenditure for depression among patients attending private community psychiatric clinics in Pakistan. *The Journal of Mental Health Policy and Economics* 7 (1): 23–28.

Gadit, A. A. 2007. Psychiatry in Pakistan: 1947–2006: A new balance sheet. *Journal of the Pakistan Medical Association* 57 (9): 453–463.
Ganatra, H. A., H. Bhurgri, et al. 2009. Educating and informing patients receiving psychopharmacological medications: Are family physicians in Pakistan up to the task? *PLoS One* 4 (2): e4620.
Goel, D. S. 2011. Why mental health services in low- and middle-income countries are under-resourced, under-performing: An Indian perspective. *The National Medical Journal of India* 24 (2) 94–97.
Hanlon, C., D. Wondimagegn, et al. 2010. Lessons learned in developing community mental health care in Africa. *World Psychiatry* 9 (3): 185–189.
Husain, N., I. Bevc, et al. 2006. Prevalence and social correlates of postnatal depression in a low income country. *Archives of Women's Mental Health* 9 (4): 197–202.
Husain, N., I. B. Chaudhry, et al. 2007. Life stress and depression in a tribal area of Pakistan. *British Journal of Psychiatry* 190:36–41.
Irfan, M. 2010. Implemenation of mental health policy in Pakistan: A dream in search of reality? *Journal of Postgraduate Medical Institute* 24 (2):82–83.
Jordans, M. J., I. H. Komproe, et al. 2010. Evaluation of a classroom-based psychosocial intervention in conflict-affected Nepal: A cluster randomized controlled trial. *Journal of Child Psychology and Psychiatry* 51 (7): 818–826.
Jordans, M. J. D., W. Tol, et al. 2003. Training psychosocial counselling in Nepal: Content review of a specialised training program. *Intervention* 1:18–35.
Khan, M. M. 2004. The NHS international fellowship scheme in psychiatry: Robbing the poor to pay the rich? *Psychiatr Bull* 28:435–437.
Kortmann, F. 2004. *Proposal for an update of the Nepal mental health policy, strategy and plan of action*. Kathmandu: WHO.
Lopez, A., C. Mathers, et al. 2006. *Global burden of disease and risk factors*. Washington, DC: Oxford University Press and World Bank.
Manuweera, G., M. Eddleston, et al. 2008. Do targeted bans of insecticides to prevent deaths from self-poisoning result in reduced agricultural output? *Environ Health Perspect* 116 (4): 492–495.
Mental Health Policy Group constituted by Ministry of Health and Family Welfare, Government of India. http://mhpolicy.org. Accessed 8 July 2012.
Ministry of Health and Family Welfare and Government of India Mental Health Program. http://india.gov.in/sectors/health_family/mental_health.php. Accessed 6 Feb 2012.
Mirza, I., and R. Jenkins. 2004. Risk factors, prevalence, and treatment of anxiety and depressive disorders in Pakistan: Systematic review. *British Medical Journal* 328 (7443): 794.
Mullan, F. 2005. The metrics of the physician brain drain. *The New England Journal of Medicine* 353 (17): 1810–1818.
Murray, L. K., S. Dorsey, et al. 2011. Building capacity in mental health interventions in low resource countries: An apprenticeship model for training local providers. *International Journal of Mental Health Systems* 5 (1): 30.
Murthy, P., and K. Sekar. 2008. A decade after the NHRC quality assurance initiative: Current status of government psychiatric hospitals in India. In *Mental health and human rights,* eds. D. Nagaraja and P. Murthy, 101–142. Bangalore: National Human Rights Commission and NIMHANS.
Murthy, R. S. 2011. Mental health initiatives in India (1947–2010). *International journal of mental health systems* 24 (2): 98–107.
National Human Rights Commission. 1999. *Quality assurance in mental health.* New Delhi: National Human Rights Commission.
Nishtar, S., and A. B. Mehboob. 2011. Pakistan prepares to abolish ministry of health. *Lancet* 378 (9792): 648–649.
Patel, V. 2007. Redefining the role of psychiatry in South Asia. *Journal of Pakistan Psychiatric Society* 4 (1).
Patel, V., and M. Varghese. 2004. NGOs and mental health: Search for synergy. In *Mental health: An Indian perspective,1946–2003,* ed. S. P. Agarwal, 145–151, New Delhi: Elsevier.

Patel, V., A. Sumathipala, et al. 2006. South Asian region. Chapter for textbook of cultural psychiatry (eds Bhui and Bhugra).
Patel, V., H. A. Weiss, et al. 2010a. Effectiveness of an intervention led by lay health counsellors for depressive and anxiety disorders in primary care in Goa, India (MANAS): A cluster randomised controlled trial. *Lancet* 376 (9758): 2086–2095.
Patel, V., M. Maj, et al. 2010b. Reducing the treatment gap for mental disorders: A WPA survey. *World Psychiatry* 9 (3): 169–176.
Patel, V., S. Chatterji, et al. 2011. Chronic diseases and injuries in India. *Lancet* 377 (9763): 413–428.
Petersen, I., A. Bhana, et al. 2009. Planning for district mental health services in South Africa: A situational analysis of a rural district site. *Health Policy Plan* 24 (2): 140–150.
Rahman, A. 2007. Challenges and opportunities in developing a psychological intervention for perinatal depression in rural Pakistan–a multi-method study. *Archives of Women's Mental Health* 10 (5): 211–219.
Rahman, A., A. Malik, et al. 2008. Cognitive behaviour therapy-based intervention by community health workers for mothers with depression and their infants in rural Pakistan: A cluster-randomised controlled trial. *Lancet* 372 (9642): 902–909.
Raja, S., S. K. Wood, et al. 2010. Mapping mental health finances in Ghana, Uganda, Sri Lanka, India and Lao PDR. *International journal of mental health systems* 4:11.
Regmi, S. K., A. Pokharel, et al. 2004. Nepal mental health country profile. *International Review of Psychiatry* 16 (1–2): 142–149.
Report of the Subcommittee on Mental Health and Substance Abuse. Ninth Five- year Plan. 1998–2003. Prospective plan 2003–2013 Planning Commission, Government of Pakistan.
Scholte, W. F., M. Olff, et al. 2004. Mental health symptoms following war and repression in eastern Afghanistan. *JAMA* 292 (5): 585–593.
Shidhaye, R., and V. Patel. 2010. Association of socio-economic, gender and health factors with common mental disorders in women: A population-based study of 5703 married rural women in India. *International Journal of Epidemiology* 39 (6): 1510–1521.
South Asia: Ending poverty in a region of growing contrasts. http://siteresources.worldbank.org/SOUTHASIAEXT/Resources/223546-1226350812655/introduction.pdf. Accessed 6 Feb 2012.
The Mental Health Policy of Sri Lanka. 2005–2015. Mental Health Directorate, Ministry of Healthcare and Nutrition, Government of Sri Lanka.
The 18th Amendment to the Constitution of the Islamic Republic of Pakistan. 2010. http://www.pakistani.org/pakistan/constitution/amendments/18amendment.html. Accessed 3 Dec 2011.
Thornicroft, G., A. Alem, et al. 2010. WPA guidance on steps, obstacles and mistakes to avoid in the implementation of community mental health care. *World Psychiatry* 9 (2): 67–77.
Tol, W. A., I. H. Komproe, et al. 2009. Brief multi-disciplinary treatment for torture survivors in Nepal: A naturalistic comparative study. *International Journal of Social Psychiatry* 55(1): 39–56.
Upadhyaya, K. D., and P. Mahat. 2007. *Mental health guidelines for integration of mental health services in primary health care system*. Kathmandu: Management Divisions of Department of Health Services and Mental Hospital.
WHO. 2005a. *Mental health atlas*. Geneva: WHO.
WHO. 2005b. *Mental health policy, plans and programs*. Geneva: WHO.
WHO. 2007. Developing Community Mental Health Services. Report of the Regional Workshop, Bangkok, Thailand, 11–14 December 2006. New Delhi: Regional Office for South-East Asia.
WHO. 2010. *mhGAP Intervention guide for mental, neurological and substance use disorders in non-specialized health settings: Mental health Gap Action Program (mhGAP)*. Geneva: WHO.
WHO. 2011. *Mental health atlas*. Geneva: WHO.

Chapter 8
Mental Health Services in USA: Policies and Programs—What can India Learn from Western Models?

Anand K. Pandurangi, Antony Fernandez and Jagannathan Srinivasaraghavan

8.1 Introduction

The statement "there is no health without mental health" is by now well established in the West. Although there is a long history to the evolution of mental health care delivery systems in the USA, the comprehensive mental health care act of 1963 enacted under the Kennedy administration was a critical milestone in moving public mental health care delivery to the community, and establishing a minimum expectation of services for all individuals with mental disorders. A community service board was to be established in each county in every state and entrusted to be the primary public agency for care delivery (a typical county in USA is bigger than a Taluk and smaller than a district in India). In 2007, the last year for which complete and verified data for USA is available, there were 1067 community mental health centers (CMHC) with 745 sites that provided direct care to 1.4 million persons. They each serve an average of 1800 individuals. Overall, the public mental health system delivered services to 6.8 million people at a cost of US$ 37 billion, in 2007 (www.nasmhpd.org). The CMHC is part of a larger and diverse care delivery system consisting of federal, state, private, and not-for-profit mental health agencies including state hospitals, Veterans hospitals, private free standing psychiatric hospitals, general hospitals, academic medical centers, individual and group private practitioners, student counseling centers, and non-profit and charitable clinics. Nationally in the

A. K. Pandurangi (✉)
Virginia Commonwealth University, Richmond, VA, USA
e-mail: apandurangi@mcvh-vcu.edu

A. Fernandez
Department of Psychiatry, West Hospital, 1200 East Broad Street, 8th Floor, Room 303, P.O. Box 980710, Richmond, VA 23298, USA

J. Srinivasaraghavan
V. A. Medical Center, 2401 W. Main Street, Marion, IL 60559, USA
e-mail: Jag.Van@va.gov; jagvan@gmail.com

J. K. Trivedi, A. Tripathi (eds.), *Mental Health in South Asia: Ethics, Resources, Programs and Legislation,* International Library of Ethics, Law, and the New Medicine 58,
DOI 10.1007/978-94-017-9017-8_8

USA, it is estimated that annually 30 million individuals receive mental health services such as hospitalization, emergency and crisis services, outpatient care and (minimal) preventive care through these systems including children, women, elderly, and minority groups. Despite this array of services and agencies, it is estimated that only 50 % of the estimated 60 million persons with mental disorders in USA seek and receive mental health care. A major barrier was removed in 1996 when that a law was passed assuring parity for mental treatment services with medical/physical health services. In 2008, this law was updated with the Domenici-Wellstone Act to close previous loopholes. Even in a developed country such as the USA, there is a long way to go to ensure care for all who need it.

8.2 Brief History of Mental Health Care in the USA (http://www.mnpsychsoc.org/history.pdf Accessed June 10, 2012)

1773: First hospital for the mentally ill was opened in Williamsburg, Virginia.

1840: There were eight asylums for the insane. The first attempt to measure the extent of mental illness, and mental retardation was made through the Census of 1840

1900: Mental hygiene movement began championed by Clifford Beers, himself a patient, and author of "The Mind that Found Itself".

1930: Narcotics Division established within the US Public Health Service, leading to the Division of Mental Hygiene for research and treatment of nervous and mental disease, and drug addictions.

1946: President Truman signed the National Mental Health Act, largely drafted by William Menninger

1949: National Institute of Mental Health created to conduct and support basic and clinical research.

1946: Fountain House, a support group program was started by a group of former patients of Rockland State Hospital, NY, serving as a model for the later development of club houses and day care programs.

1955: Study by the Joint Commission on Mental Illness and report titled "Action for Mental Health" to establish and fund halfway houses, daycare, and aftercare.

1963: President Kennedy signed legislation to fund CMHC to substitute for custodial institutional care.

1965: Medicare and Medicaid were established. Medicare is a federal health insurance program for the elderly and the disabled. Medicaid is a joint state and federal health insurance program for the poor.

1972: Two new federal programs started: Supplemental Security Income and Social Security Disability Insurance providing support to the mentally disabled person to live more independently.

1974: California Supreme Court issues "duty to warn" decision in the Tarasoff case.

1975: Private health insurance plans start covering outpatient mental health services.

1979: Schizophrenia Foundation, a family advocacy group for persons with serious mental illness, was formed. It is now known as the National Alliance for the Mentally Ill (NAMI).

1980: Mental Health Systems Act signed by President Carter strengthening linkages between the federal, state, and local governments, and providing grant and contracts to provide services.

1986: Mental Health Planning Act passed to provide monies for case management through Medicaid.

1987: Medicare adds outpatient mental health benefit.

1988: Managed behavioral health care evolves with mental health carved out of general health services.

1990: Americans with Disabilities Act passed. This act establishes the rights of disabled individuals and prohibits discrimination on the basis of disability, including from mental disorders.

1991: CMHC authorized to bill for partial hospital services under Medicare.

1992: Alcohol, Drug Abuse & Mental Health Administration reorganized into Substance Abuse & Mental Health Services Administration, to provide MH/SA information, and support services research.

1996: Health Insurance Portability and Accountability Act (HIPAA) enacted to protect health insurance for workers during job changes, and safeguard privacy of health information.

1996: Social Security Administration begins terminating disability benefits if used for substance abuse.

1996: WHO reports Depression is a leading cause of worldwide disability.

1996: Federal Mental Health Parity Act compels companies that offer mental health benefits to employees to insure that coverage of mental and physical illness was reasonably equitable.

1999: Supreme Court rules on Olmstead v. L.C, holding it is a violation of the disabled person's rights to keep him/her in restrictive inpatient settings when more appropriate community services are available.

1999: Surgeon General Report emphasizes eradication of stigma, and research, innovation in treatment.

2004: President Bush's New Freedom Commission on Mental Health studies mental health service delivery systems, both private and public.

2009: President Obama signs Health Care Reform Act with support to the working poor to obtain health insurance.

8.3 Policies

Mental Health, Intellectual Disabilities and Substance Abuse Policies Policies derive from the laws, statutes and regulations discussed in the accompanying chapter titled "Mental Health Services in USA: Ethical and Legal Aspects, and

Human Rights—What can India Learn". They understandably vary from state to state in USA, although much uniformity is also noted. Further at the policy level, uniformity results from the fact that the accrediting agencies and their guidelines are the same nationally.

Policies may be thought of as the 2nd layer of the foundation of a successful enterprise, with laws and regulations being the first. Absence of policies, illogical and impractical policies, outdated policies, policies with inherent conflicts all lead to a dysfunctional organization and eventually its failure. Excessively detailed and cumbersome policies give a false sense of quality and safety. We provide below some important areas for policy development as adopted in the USA but applicable to health care systems everywhere. It is beyond the scope of this chapter to provide detailed templates of all the major policies relevant to a mental health agency, organization, hospital, clinic or practice. Resources provided in the bibliography especially the National Association of State Hospital Medical and Program Directors (NASHMPD, www.nasmhpd.org), the National Mental Health Association (aka Mental Health America, www.nmha.org) and the National Alliance for the Mentally Ill (www.nami.org) may be accessed for detailed policy areas and templates. We have also provided two appendices at the end of this chapter. The first appendix provides a subject-wise list of reports from NASHMPD covering a broad array of mental health and related topics sub-listed in a chronological order under each topic. The reader with specialized interest in administrative psychiatry including policy development and care delivery, and system performance may refer to these. The second appendix lists all mental health related agencies and resource centers. In 2005, the State of California passed a comprehensive act referred to as proposition 63 that aims to transform its mental health system into a comprehensive and progressive one. The reader may refer to this act for a summary of what is needed for a modern and community-based mental health system (http://www.dmh.ca.gov/Prop_63/mhsa/docs/MHSAafterAB100). Accessing these resources through the web or otherwise will yield a wealth of information for the reader in understanding the complexities involved in developing, organizing, and sustaining a healthy mental health agency, institution, hospital, clinic or practice, of any size and volume.

Below we briefly discuss the key areas within mental health policy development.

8.3.1 Mission

Typically a brief mission statement that captures the essence of the organization, its purpose for existence and its accountability to stakeholders is the first policy to be developed. It is traditionally displayed prominently at the entrance of the administration hall or building.

8.3.2 Physical Plant and Infrastructure

These cover the physical plant including building standards, (ex: window needed in each patient room and day room), minimum space requirements (patient rooms, day room, meeting rooms etc.), clean and soiled utility rooms, special rooms for interviews, seclusion and supplies, power supply and back-up power, roads and walkways on the property, transportation etc.

8.3.3 Environment of Care

Policies in this section pay attention to gender separation, age separation, disorder separation, metal detectors, secure units, alarms and panic buttons, one-way mirrors, privacy concerns, cleanliness and hygiene, infection control, hand hygiene, patient and staff comfort, healthy environment for work and clinical care, smoking policy, courtyards, etc.

8.3.4 Organization and Governance

Organizational structure and reporting relations, departments, programs and services, committees, reports, timelines, action-oriented follow-up, etc.

8.3.5 Human Resource

Management including workforce development and training, recruitment and retention, supervision, continuing education, leadership development, career ladders, vacation and sickness time off, benefits, retirement plans, grievance procedures and resolution, opportunities for minorities, avoiding discrimination, competitive recruitment, compensation, balancing family and work life, etc.

8.3.6 Clinical Care Services

Including: (i) access, (ii) scope of services including bio-psycho-social oriented multidisciplinary, multimodal assessments and treatments, (iii) professional, ethical, quality and safe services, (iv) sensitivity and customization to the uniqueness of the individual including physical, psychosocial, cultural, spiritual and linguistic needs, (v) integration of patient/client empowerment, person-centered and recovery oriented approaches into traditional models of care, and involvement of and liaison with family, (vi) continuity of care including seamless transition, and wrap-around services including medical and biopsychosocial services, integrated medical and psychiatric care, and pre-vocational, educational, occupational, recreational, transportation and housing services.

8.3.7 System Performance, Quality Improvement, and Safety

Services, programs, departments have to be integrated and coordinated to achieve optimal function and productivity. Services should not work at cross-purpose. Multidisciplinary meetings including between administration and clinical services are needed to carry out the mission successfully. For example the model of a triangular leadership that includes representatives of medical services, patient care services and administrative services works well to maintain quality, improve system performance, identify risks, plan for development etc.

Performance improvement projects help focus on specific issues such as fall risks, non compliance and recidivism, no-show rates, staff turnover rates etc.

Strong *safety monitoring systems* are needed to minimize adverse events and harm. For example, suicide assessment and precautions, medical clearance prior to psychiatric admission, etc. Systematic adverse event reporting, monitoring and follow-up can ensure that adverse events are captured early before they become a systemic problem. Staff should be trained to escalate communication whenever a potential adverse event is anticipated. Review of safety events on a periodic basis and comparison of nature and volume of events at different points in time enhances the safety of a system. Examples include elopement rates, self injurious behaviors, drug interactions, etc.

Regulatory Compliance Compliance with professional standards, institutional, agency, local, state and federal regulations as licensing agencies, accreditation agencies and third party payers is a major function of health care systems in the USA. Some of the well known accreditation agencies in the USA are the Joint Commission for Accreditation of Hospitals (JCAHO; www.jointcommission.org/accreditation/behavioral_healthcare.aspx), the National Committee for Quality Assessment (NCQA; www.ncqa.org), the Council on Accreditation (COA, previously known as Accrediting Body for Family and Children's Agencies; www.coanet.org) and The Commission on Accreditation of Rehabilitation Facilities (CARF; www.carf.org). Each of these has developed extensive manuals of recommended policies and procedures to gain their approval for accreditation. Such accreditation is often required to be eligible for third party payments, such as Medicare and commercial insurers, State funding, grant eligibility, etc. JCAHO provides extensive guidance and training on how to meet their requirements. JCAHO accredits hospitals by making periodic on-site visits and extensive review of a hospital's structure and functions. Hospitals are accredited for 2–3 year periods at a time. NCQA provides an elaborate list of variables on which to gather data and criteria that would meet their accreditation standards. This is referred to as the Health Care Effectiveness Data and Information Set (HEDIS); COA (www.coanet.org) is an international, independent not-for profit accreditation agency since 1977 founded by the child welfare league of America. It now provides accreditation for substance abuse services, adult day care, services for the homeless, foster care, inter-country adoption services. CARF (www.carf.org) was founded in 1966, and is an independent nonprofit accreditor of health and human services in Aging services, behavioral health, Opioid Treatment, Business and Services Management networks, Child and Youth Services, Employment and Community Services, Vision Rehabilitation, Medical Rehabilitation, Durable Medical equipment.

Whereas it may appear burdensome and even counter-productive to be so micro-managed by regulating agencies, such compliance indeed serves to maintain quality and safety, eligibility for funding, and more importantly to keep the public's trust in health care systems.

8.4 Human Rights

Including the right to treatment, right to refuse treatment, treatment in the least restrictive setting and by least restrictive means, without seclusion or restraint (unless needed for imminent safety of the individual or others). All major accreditation organizations have specific expectations on protecting the rights of our patients. The WHO guide and check list (http://www.who.int/mental_health/policy/mental_health_and_human_rights_october_2006) for organizations on human rights and policy development is a major resource in this regard. (Also see the HUMAN RIGHTS section in the accompanying chapter titled "Mental Health Services in USA: Ethical and Legal Aspects, and Human Rights—What can India Learn").

8.5 Fiscal Management

Policies and procedures in this area cover budgets, revenues including third party reimbursements, public funds (federal, state and local), private fee for service, grants and contracts, etc., and expenses such as capital and operational, fixed and variable costs, direct and indirect costs, etc. While health insurance companies used a model of "fee for service" until about 1980, since then a strong movement of "Managed Care" has emerged in the USA. While the stated intent of such a model is to rein in unnecessary and expensive costs of medical care, it has unfortunately had the effect of intruding on the traditional doctor–patient relationship and restricting care. Medical decisions are often made by care managers who have no clinical relation with the patient. Some corrections to this overreach by insurance companies have occurred in recent years including the most recent Affordable Care Act of 2010. For example, a company may not deny health insurance to an individual based on knowledge of pre-existing conditions. Reimbursement for psychiatry and cognitive services is generally lower than procedural and surgical services. Mental health payments are often "carved out" from general medical and surgical payment schedules and are handled by intermediate companies referred to as Behavioral Health companies. Between 1980 and 2000, such companies often developed elaborate and often idiosyncratic criteria of medical necessity, that the net result was one of care denial. In recent years efforts have been made at both the federal and the state levels to pass "parity" laws that mandate equal treatment of mental illness on par with physical illnesses.

Due to strong patient rights and high levels of litigation, physicians including psychiatrists often practice defensive medicine in the USA. Such practice often leads to unnecessary and expensive testing leading to inflation in health care costs. Sub specialization, direct access to specialty care (in contrast to a general physician serving as a gatekeeper), availability of technology, increased longevity and high morbidity in old age all contribute to high costs. Medical and management policies and procedures to provide balanced care thus assume enormous importance in achieving the fiscal health of an agency.

8.6 Information Technology

As in other fields, health care has been transformed by technological innovation. The electronic health or medical record (EHR, EMR) is now the norm in health care in USA. Remote access is quickly becoming the standard work mode with physicians and other health care providers being able to access patient information from home or on the road. Tele-psychiatry services are increasing and access to psychiatrists in underserved and rural areas and locations such as jails is increasing. While these have the enormous advantage of convenient and quick access to an individual's medical history, and providing service to vulnerable individuals, building the necessary technical infrastructure is costly. There is also the concern of loss of privacy. Policies are evolving in this arena, from templates for history and examination, to care sets, automated medication order sets, prohibitions against "cut and paste" and rules for electronic signatures, etc. For behavioral health, guarding the privacy of sensitive mental health information is of paramount importance. Many health systems have developed "segregated folders" for sensitive behavioral information.

8.7 Communication and Media Services

Timely and relevant communication within and outside the system is of paramount importance for the smooth functioning of the system. Current day communication systems use state-of-the-art technology such as virtual pagers for on-call physicians, smart phones, daily electronic bulletins etc. For external communications, many hospitals in the USA offer health channels on private television networks as well as utilize more traditional print media. Continuing professional education and public and community education is offered through these media. Mental illness awareness week (MIAW), depression screening as part of health fares, mental health walks, legislative day, etc. are observed by many hospitals, professional societies and advocacy groups. Such wide spread communication has made psychiatric disorders more acceptable to the public and less stigmatized. More people are willing to seek treatment for disorders such as depression and anxiety.

8.8 Advocacy

Despite the slogan "no health without mental health", unfortunately it sometimes feels like mental health is step-child within the health care field. Advocacy and education of the public and leadership are a necessary component of policy development. This is best illustrated by the emergence of the National Alliance for the Mentally Ill (NAMI) as a grass roots organization with significant political clout. NAMI focuses on many areas of education and advocacy and the key areas of their efforts are listed here: NAMI Basics on Mental Illness, Family-to-Family, In Our Own Voice, Peer-to-Peer, NAMI Connection, Hearts & Minds, Parents and Teachers as Allies, and Provider Education. NAMI has developed a grades report (http://www.nami.org/gtsTemplate09.cfm?Section=Grading_the_States_2009) for each state in the USA on how well they are doing in key areas, such as Health Promotion and Measurement, Financing & Core Treatment/Recovery Services, Consumer & Family Empowerment, and Community Integration and Social Inclusion. NAMI has identified the following as the pillars of a High-Quality State Mental Health System: (1) Comprehensive services; (2) Integrated services; (3) Adequately funded; (4) Focused on wellness and recovery; (5) Safe and respectful (of patients/consumers and families); (6) Accessible; (7) Culturally competent; (8) Consumer-centered and consumer—and family driven; (9) Well-staffed and trained; (10) Transparent and accountable.

NAMI recommends that states and the federal government undertake these five actions: (1) Increase public funding for mental health care services, (2) Improve data collection, outcomes measurement, and accountability, (3) Integrate mental and physical health care, (4) Promote recovery and respect, (5) Increase services for people with serious mental illnesses who are at most risk.

Mental Health Association of America is another advocacy organization with focus on common mental disorders and psychosocial therapies. They provide information on the following areas (in alphabetical order): Access to Medications, Children's Mental Health Services, Confidentiality, Criminal Justice, Cultural Competence, Death Penalty, Employment, Evidence Based Medicine, Financing, Healthcare Disparities, Health Information Technology, Healthcare Reform, Housing (PDF), Insurance Parity, Integrated Care, Involuntary Outpatient Commitment, Medicaid, Medicare, Medicare Prescription Drug Benefit, Mental Health Caucuses, Prevention, Psychiatric Advance Directives, Seclusion and Restraints, State Budget

Advocacy, State Level Rankings Resources, Universal Health Insurance, Veteran Issues.

The National Association of Program and Mental Health Directors (NASPMHD; www.nasmhpd.org) although not a consumer or family driven organization, has been an influential player in policy development within public mental health. Consensus conferences, White Papers and action papers from this organization are very informative. Areas of recent focus include: Campaign for Mental Health Reform; Community Mental Health Services Performance Partnership Block Grant; Co-Occurring Mental Illness and Substance Abuse Disorders; Criminal Justice/Mental Health; Health Insurance Portability and Accountability Act of 1996 (HIPAA); Medicaid; Medicaid Managed Care; Medicare; Mental Health Parity; President's New Freedom Commission on Mental Health; President's New Freedom Initiative; Reauthorization of the Substance Abuse and Mental Health Services Administration (SAMHSA); Seclusion and Restraint in Psychiatric Treatment Settings; Temporary Assistance for Needy Families (TANF); Veterans' Mental Health Issues.

8.9 Research, Development and Innovation

The continued success of an agency especially the development of new modalities of assessment and treatment requires strong research and development policies. Depending on the mission of an agency or institution, policies to encourage research and innovation are needed to ensure the agency offers state-of-the-art care. Innovation is necessary for new approaches to problems as well as cost-effective care. More and more clinics and practices now offer cognitive behavior therapy, dialectical behavior therapy, long acting injection clinics, clozapine clinics, transcranial magnetic stimulation, tele-psychiatry, web-based continuing education, etc. Some academic centers offer clinics for early detection of psychosis, genetic counseling, and training programs for psycho-education. Every state mental health department has a research wing that is responsible for encouraging translational research.

8.10 Conclusion

The aim of this review has been to provide a clinically oriented informational document regarding the history of mental health care in the USA, and enumerate the major policy areas for a successful mental health organization or system of care, as practiced in the USA. These policies serve as a framework for public, private, charitable and non-profit entities in developing and maintaining healthy systems of care, be it an individual, small group, clinic, institution or agency. The policies are described as practiced in the major system of mental health care in the USA. Whereas the systems in the USA are well developed and provide a high quality and wide range of services to many millions of people, they are far from being perfect.

Many limitations and deficiencies and unfortunately, even abuses exist. Much work is yet to be done.

The chapter has strived to be relevant to the broad variety of activities that psychiatrists and mental health professionals and administrators undertake, the diverse populations they serve, and the array of settings in which such service is provided. It is hoped that the reader has obtained a conceptual framework for the creation, development, and operation of one or more types of care delivery systems. Once such a framework has been created, the topical and institutional resources that follow in the appendices provide the more detailed guidelines, templates and metrics for a successful implementation of a system. The ultimate goal is reduce the suffering of persons afflicted with mental disorders, enable recovery, and strive towards mental health for all.

8.11 Appendix I

National Association of State Medical and Program Director (NASMPD) Reports

www.nasmhpd.org

Advocacy

- Consumer Involvement with State Mental Health Authorities: Part I Final Report Part II Appendices to Final Report, 2010

Ageing

- Planting the Seeds of Change: Developing Mental Health and Aging Coalitions To Improve Services for Older Persons with Mental Illness (description only online), 1997
- Promoting Emotional Health and Preventing Suicide: A Toolkit for Senior Living Communities, 2010

Children and Adolescent Services

- From Adolescence to Adulthood: Transition Services to Help Youth, 1999
- State Efforts to Expand Transition Supports for Adolescents Receiving Public Mental Health Services, 2001
- Mental Health, Schools and Families Working Together for All Children and Youth: Toward a Shared Agenda, 2002
- Service Systems Supports During the Transition from Adolescence to Adulthood: Parent Perspectives, 2002

Collaboration

- In the Public Interest: The Developing Alliance Between State and County Mental Health Authorities (description), 1996

- State-County Alliances, 1997
- Best Practices Symposium: State/Academic Collaboration, 2000
- Promoting Collaboration: An Exploration of Successful Partnerships Between Mental Health Planning and Advisory Councils and State Mental Health Agencies, 2001

Consent

- Involuntary Outpatient Commitment
- Consent Procedures and Form for Focus Groups, 2002

Continuum of Care

- Best Practice Symposium: Continuity of Care from the Hospital to the Community and Practice Guidelines Update, 2001

Criminal Justice

- Mental Health and Criminal Justice, 1998
- Issues Pertaining to the Development and Implementation of Programs for Persons Civilly Committed for Treatment Under Sexually Violent Predator Statutes, 1999
- Best Practices Symposium: Criminal Justice/Mental Health Interface, 1999
- Best Practices Symposium: Mental Health Courts, 2000
- Summer 2001—Mental Health and Juvenile Justice, Seeking Common Ground
- Criminal Justice Primer for State Mental Health Agencies, 2002
- Council of State Governments' Criminal Justice/Mental Health Consensus Report, 2003

Culture & Minorities

- Cultural Competence, 1997
- Exploring the Intersection Between Cultural Competency and Managed Behavioral Health Care: Implications for State and County Mental Health Agencies, 1997
- Meeting the Mental Health Needs of Gay, Lesbian, Bisexual and Transgender Persons: Executive Summary, 1999
- Meeting the Mental Health Needs of African Americans: Executive Summary, 1999
- Examples from the Field: Programmatic Efforts To Improve Cultural Competence in Mental Health Services: Executive Summary, 2000
- Creating Culturally Competent Mental Health Systems for Latinos: Perspectives from an Expert Panel, 2001
- Meeting the Mental Health Needs of Persons Who Are Deaf, 2002
- Meeting the Mental Health Needs of Asian Americans, 2003
- Meeting the Mental Health Needs of American Indians and Alaska Natives, 2004

Disaster Planning

- Responding to the Mental Health Impact of Major Disasters, 1998

- Mental Health All-Hazards Disaster Planning Guidance, 2003

Dual Diagnosis (with MI & ID)

- Developing Best Practice Guidelines for Treating People with Co-Occurring People with Mental Illness/Mental Retardation, 2002

Dual Diagnosis (with MI & Substance Abuse)

- National Dialogue on Co-Occurring Mental Health and Substance Abuse Disorders, 1998
- Best Practices Symposium: Dual Diagnosis, 1999
- Financing and Marketing the New Conceptual Framework for Co-Occurring Mental Health and Substance Abuse Disorders, 1999
- Promoting integration across service systems to meet the needs of persons with co-occurring mental illness and substance abuse disorders, and consumers with multiple needs: Making the Case—No. 1; Building Coalitions—No. 2; Involving the Private Sector—No. 3; Getting Started—No. 4; Finding the Money—No. 5; Making Systems Change Happen—No. 6; Evaluating Progress—No. 7; Core Qualities of the Change Agent—No. 8, 2000
- Exemplary Methods of Financing Integrated Service Programs for Persons with Co-Occurring Mental Health and Substance Disorders, 2002
- The Evolving Conceptual Framework for Co-Occurring Mental Health and Substance Use Disorders: Developing Strategies for Systems Change, 2005
- The Fifth National Dialogue of the Joint Task Force on Co-Occurring Substance Use and Mental Disorders: Final Report and Strategic Action Plan, 2006

Employment

- Employment, 1997

Health

- Suggested Model for Integration of Behavioral Health into Primary Care, 2003
- Summer 2003—Building Bridges: A Status Report on the Integration of Public Health and Public Mental Health
- Measurement of Health Status for People with Serious Mental Illness, 2009
- Obesity Reduction & Prevention Strategies for Individuals with Serious Mental Illness, 2009
- Morbidity and Mortality in People with Serious Mental Illness, 2005
- Smoking Policy and Treatment in State Operated Psychiatric Facilities, 2005
- Tobacco–Free Living in Psychiatric Settings: A Best–Practices Toolkit Promoting Wellness and Recovery, 2007
- Integrating Behavioral Health and Primary Care Services: Opportunities and Challenges for State Mental Health Authorities, 2009

Homelessness, Housing

- Housing, 1996

- Housing for Persons with Psychiatric Disabilities: Best Practices for a Changing Environment (description), 1996
- State of the States: Services for Persons Who Are Homeless and Mentally Ill, 2001
- Affordable Housing: The Role of the Public Behavioral Health System, 2011

Leadership

- Managed Care, 1996
- A Snapshot of Community Mental Health Services: Report of a State Mental Health Agency Survey: description, 1997
- Summer 1998—Planning Public Mental Health Services, 1998
- Law, the Role of Government and the Future of Public Mental Health: Executive Summary, 2000
- Conflict Resolution, 2000
- Offices of Consumer Affairs: A Pathway to Effective Public Mental Health Services: Executive Summary, 2000
- Bridging the Gap Between Research and Services with Evidence–Based Practices, 2001
- Seizing the Moment: Redefining State Mental Health Agency Role in Time of Budget Uncertainty, 2002
- Best Practice Symposium: Lessons for State Clinical Directors from Business Management, 2002
- Best Practices Symposium—Trends in the Public Mental Health System: An Environmental Scan, 2002
- State Mental Health Commissions: Recommendations for Change and Future Directions, 2002
- Answering the Challenge: Responses to the President's New Freedom Commission Final Report, 2004
- Brag and Steal Resource Guide: Promising Mental Health Programs From Around the World, 2008
- The Role of the Medical Director in a State Mental Health Authority—A Guide for Policy Makers, 2009
- Preparing the Adult Mental Health Workforce to Succeed in a Transformed System of Care, 2009
- Principled Leadership for the State Hospital CEO/Administrator Toolkit, 2009
- Responding to a High-Profile Tragic Incident Involving a Person with a Serious Mental Illness: A Toolkit for State Mental Health Commissioners, 2010
- The Oklahoma Enhanced Tier Payment System: Leveraging Medicaid to Improve Mental Health Provider Performance and Outcomes, 2011
- Cornerstones for Behavioral Healthcare Today and Tomorrow, 2012

Medications

- Algorithms and the Treatment of People with Serious Mental Illness, 1998
- Best Practices Symposium: Use of Antipsychotics, Medicaid Formularies, Seclusion/Restraint, State/Academic Collaboration, and Mental Health Courts, 2000

- Psychiatric Polypharmacy, 2001
- Psychiatric Medications, 2001
- Best Practice Symposium: Children's Psychopharmacy, 2001
- Focus on Utilization of NRI's Performance Measurement System Data, Reducing Medical Morbidity, Guiding Physician Practice to Ensure Open Formularies, and Effective Use of LOCUS and Treatment Planning, 2003
- Focus on: Pharmaceuticals…Pricing & Acquisition, 2004
- NASMHPD Issue Paper: Clinical Antipsychotic Trials of Intervention Effectiveness (CATIE) What Does It Mean for Practice and Policy?, 2005
- NASMHPD's Medical Directors' Statement on Comparative Effectiveness of Antipsychotic Medications and Individualized Treatment, 2008

Performance Improvement

- Performance Measures, 1996
- Best Practices Symposium—Focus on Computerized Performance Measurement Systems, 2001
- Use of Mortality Data, 2004

Person Centered Care

- Developing Trauma-Informed Behavioral Health Systems: Report from NTAC's National Experts Meeting on Trauma and Violence, 2002
- The Damaging Consequences of Violence and Trauma: Facts, Discussion Points, and Recommendations for the Behavioral Health System, 2004
- Responding to Childhood Trauma: The Premise and Practice of Trauma Informed Care, 2006
- Organizational Stress as a Barrier to Trauma-Sensitive Change and System Transformation, 2007

Prevention

- Parents with Psychiatric Disabilities, 2000
- Prevention Approaches for State Mental Health Authorities, 2004
- Prevention Strategies for SMHAs, 2004
- Suicide Prevention Efforts for Individuals with Serious Mental Illness: Roles for the State Mental Health Authority, 2008
- Primary Prevention in Behavioral Health, 2011
- Becoming a Preventionist: Making Prevention Part of Your Mental Health Practice; A Continuing Education Course, 2012

Recovery

- Best Practices Symposium: Psychiatric Rehabilitation/Recovery, 1999
- Recovery, 1999–2000
- Review of Recovery Literature: A Synthesis of a Sample of Recovery Literature, 2000
- Managing Conflict Cooperatively: Making a Commitment to Nonviolence and Recovery in Mental Health Treatment Settings, 2002

- Mental Health Recovery: What Helps and What Hinders? A National Research Project for the Development of Recovery Facilitating System Performance Indicators, 2002
- Emerging New Practices in Organized Peer Support: Report from NTAC's National Experts Meeting on Emerging New Practices in Organized Peer Support, March 17–18, 2003, Alexandria, VA, 2003
- Creating Recovery Facilitating Systems of Care, 2004
- State Efforts to Expand Transition Supports for Young Adults Receiving Adult Public Mental Health Services: Report on a Survey of Members of the National Association of State Mental Health Program Directors, 2005
- Phase II Technical Report, Mental Health Recovery: What Helps and What Hinders? A National Research Project for the Development of Recovery Facilitating System Performance Indicators, 2006
- Paving New Ground—Peers Working in In-Patient Settings, 2008
- Pillars of Peer Support, 2009

Seclusion & Restraints

- Seclusion & Restraint, 1999
- Reducing the Use of Seclusion and Restraint, PART I, 1999
- Best Practices Symposium: Seclusion/Restraint, 2000
- Summer/Fall 2002—Violence and Coercion in Mental Health Settings: Eliminating the Use of Seclusion and Restraint, 2002
- Reducing the Use of Seclusion and Restraint: PART III, 2002
- RESTRAINT AND SECLUSION: A Risk Management Guide, 2006
- Seclusion and Restraint Briefings, 2008
- Six Core Strategies to Reduce Seclusion and Restraint Use, 2008

Technology

- The Use of Telemedicine in Providing Mental Health Services, 1999

Terrorism

- Mental Health's Response to Terrorism, 2002
- Spring 2002—Responding to Terrorism: Public Mental Health Systems Move to the Forefront
- Best Practice Symposium: State Mental Health Authorities' Response to Terrorism, 2004

8.12 Appendix II

Sites for Mental Health Information in USA

- **Federal Web Sites**:
  - US Department of Health and Human Services (HHS)

- Substance Abuse and Mental Health Services Administration
  - Center for Mental Health Services (CMHS)
  - SAMHSA's Disaster Technical Assistance Center (DTAC)
  - Older Adults with Behavioral Health Needs Technical Assistance Center
- National Strategy for Suicide Prevention
- Adminstration for Children and Families (ACF)
- Administration on Aging (AoA)
- Centers for Disease Control and Prevention (CDC)
- Centers for Medicare and Medicaid Services (CMS)
- Health Resources and Services Administration (HRSA)
- National Institutes of Health (NIH)
  - National Institute on Alcohol Abuse and Alcoholism (NIAAA)
  - National Institute on Disability and Rehabilitation Research (NIDRR)
  - National Institute on Drug Abuse (NIDA)
  - National Institute of Mental Health (NIMH)

- US Department of Justice (DoJ)
  - Bureau of Justice Assistance (BJA)
  - Office of Juvenile Justice and Delinquency Prevention (OJJDP)
- US Department of Education
  - Office of Special Education and Rehabilitative Services
- US Department of Housing and Urban Development (HUD)
  US Department of Veterans Affairs (VA)
  - National Center for Post-Traumatic Stress Disorder (NCPTSD)
- US Social Security Administration (SSA)
- Americans with Disabilities Act (ADA) Home Page
- Office of the Surgeon General
- President's New Freedom Commission on Mental Health

- **CMHS/Other Federal Agency Research, Training and Technical Assistance Centers**:
  - Center for Psychiatric Rehabilitation
  - Center for Support of Mental Health Services in Isolated Rural Areas
  - The Evaluation Center at HSRI
  - National Center for American Indian and Alaska Native Mental Health Research
  - National Consumer Supporter Technical Assistance Center
  - The National GAINS Center for People with Co-Occurring Disorders in the Justice System
  - National Empowerment Center (NEC)
  - National Indian Child Welfare Association (NICWA)

- National Mental Health Consumers' Self-Help Clearinghouse
- National Alliance on Mental Illness
- Homelessness Resource Center
- National Research and Training Center on Psychiatric Disability
- National Technical Assistance Center for Children's Mental Health at Georgetown University
- Advocates for Human Potential, Inc.
- Research and Training Center for Children's Mental Health
- Portland Research and Training Center on Family Support and Children's Mental Health
- SAMHSA's National Mental Health Information Center
- Technical Assistance Partnership for Child and Family Mental Health
- Center on the Social and Emotional Foundations for Early Learning (CSEFEL)
- Technical Assistance Center on Social Emotional Intervention for Young Children (TACSEI)

• **Other Mental Health Web Sites:**

  - Alliance fo the Safe, Therapeutic and Appropriate Use of Residential Treatment (A START)
  - American Association of Community Psychiatrists
  - American Foundation for Suicide Prevention (AFSP)
  - American Psychiatric Association (APA)
  - American Psychological Association (APA)
  - American Psychiatric Nurses Association (APNA)
  - Bazelon Center for Mental Health Law
  - Campaign for Mental Health Reform
  - CHADD—Children and Adults with Attention-Deficit/Hyperactivity Disorder
  - Criminal Justice/Mental Health Consensus Project
  - Depression and Bipolar Support Alliance (DBSA)
  - Federation of Families for Children's Mental Health
  - First Nations Behavioral Health Association
  - Gift from Within—PTSD Resources for Survivors and Caregivers
  - Mental Health America (formerly the National Mental Health Association)
  - Mental Health Liaison Group (MHLG)
  - Mental Health Part D
  - Mental Health Statistics Improvement Program Online
  - Multi-State Disaster Behavioral Health Consortium
  - Nathan S. Kline Institute for Psychiatric Research
  - NAMI (National Alliance for the Mentally Ill)
  - National Asian American Pacific Islander Mental Health Association
  - National Association of County Behavioral Health and Developmental Disability Directors (NACBHD)
  - National Association of Mental Health Planning and Advisory Councils (NAMHPAC)

- NASMHPD Research Institute, Inc. (NRI)
- National Assocation of Reimbursement Officers (NARO)
- National Coalition for Mental Health Consumer/Survivor Organizations
- National Center for Mental Health and Juvenile Justice
- National Council for Community Behavioral Healthcare (NCCBH)
- National Healthcare Cost and Quality Association
- National Latino Behavioral Health Association (NLBHA)
- National Leadership Council on African American Behavioral Health, Inc.
- National Suicide Prevention Lifeline (1-800-273-TALK)
- Network of Care for Mental Health
- SAMSHA's Resource Center to Promote Acceptance, Dignity, and Social Inclusion Associated with Mental Health (ADS Center)
- Suicide Prevention Resource Center (SPRC)
- US Psychiatric Rehabilitation Association (USPRA, formerly IAPSRS)

• **Other Relevant Web Sites:**

- American Public Human Services Association (APHSA)
- Association of State and Territorial Health Officials (ASTHO)
- Campaign for Children's Health Care
- Child Welfare League of America (CWLA)
- Consortium for Citizens with Disabilities (CCD)
- Council of State Administrators of Vocational Rehabilitation (CSAVR)
- Council of State Governments (CSG)
    - Reentry Policy Council (RPC)
- The Finance Project
- Joint Commission on Accreditation of Healthcare Organizations (JCAHO)
- Legal Action Center
- National Alliance to End Homelessness
- National Association of State Alcohol and Drug Abuse Directors (NASADAD)
- National Association of State Directors of Developmental Disabilities Services (NASDDDS)
- National Association of State Directors of Special Education (NASDSE)
- National Association of State Medicaid Directors (NASMD)
- National Conference of State Legislatures (NCSL)
- National Disability Rights Network
- National Governors Association (NGA)

References

1. www.nasmhpd.org.
2. http://www.mnpsychsoc.org/history.pdf. Accessed 10 June 2012.
3. www.nmha.org.
4. www.nami.org.

5. http://www.dmh.ca.gov/Prop_63/mhsa/docs/MHSAafterAB100.pdf. Accessed 10 June, 2012.
6. www.jointcommission.org/accreditation/behavioral_healthcare.aspx. Accessed 10 June 2012.
7. www.ncqa.org.
8. www.coanet.org.
9. www.carf.org.
10. http://www.mentalhealthamerica.net/go/action/policy-issues-a-z. Accessed 25 March 2012.
11. http://www.who.int/mental_health/policy/mental_health_and_human_rights_october_2006.pdf. Accessed 10 June 2012.
12. http://www.nami.org/gtsTemplate09.cfm?Section=Grading_the_States_2009. Accessed 10 June 2012.

Chapter 9
Mental Health Services in USA: Ethical and Legal Aspects and Human Rights—What India can Learn from Western Models

Jagannathan Srinivasaraghavan, Antony Fernandez and Anand K. Pandurangi

9.1 Introduction: Ethics in the Practice of Psychiatry

Psychiatrists hold the same ethical ideals and principles as their medical and surgical colleagues, in their diverse roles as clinicians, treating physicians, therapists, teachers, scientists, consultants, and colleagues. However, mental illness is widespread and the suffering associated with mental illness is greatly increased by stigma, socio economic disadvantages, and coexisting medical and psychiatric conditions, making it imperative that the ethical standards for the psychiatric professions go beyond those applied to the practice of General Medicine. Further, mental illnesses directly affect thoughts, feelings, actions, behaviors, and relationships. For example consider this scenario where psychiatrists because of their unique clinical expertise, are entrusted with the professional obligation to protect the public by preventing patients from causing harm to oneself or others. This obligation may require them to breach the usual expectations of confidentiality and treat the patient against his/her wishes and/or share information with law enforcement. Another example is where Psychiatrists are called upon to assume duties of importance to society, such as legal or organizational consultation, that are outside the scope of customary clinical duties. For all these reasons, it is incumbent upon psychiatrists to be especially attentive to the ethical aspects of their work.

J. Srinivasaraghavan (✉) · A. Fernandez
Southern Illinois University, School of Medicine & V.A. Medical Center,
Marion, IL 60559, USA
e-mail: jagvan@gmail.com; Jag.Van@va.gov

A. Fernandez
Mc Guire V.A. Medical Center, Richmond, VA 23249, USA
e-mail: drtony fernandez@hotmail.com; antony.fernandez@va.gov

A. K. Pandurangi
Department of Psychiatry, Virginia Commonwealth University, West Hospital, 8th Floor,
Room 303, P.O. Box 980710, Richmond, VA 23298, USA
e-mail: apandurangi@mcvh-vcu.edu

J. K. Trivedi, A. Tripathi (eds.), *Mental Health in South Asia: Ethics, Resources, Programs and Legislation,* International Library of Ethics, Law, and the New Medicine 58,
DOI 10.1007/978-94-017-9017-8_9

9.2 Principles of Medical Ethics of the American Medical Association (American Psychiatric Association 2001a, 2001b)

This section presents the nine principles which serve as the foundation for ethics and professionalism in the field of medicine, including the specialty of psychiatry. The American Psychiatric Association conforms to these AMA principles in its Constitution and Bylaws.

- Provide competent medical care, with compassion and respect for human dignity and rights.
- Uphold the standards of professionalism, be honest in all professional interactions, and strive to report physicians deficient in character or competence, or engaging in fraud or deception, to appropriate entities.
- Respect the law and also recognize a responsibility to seek changes in those requirements, which are contrary to the best interests of the patient.
- Respect the rights of patients, colleagues, and other health professionals, and shall safeguard patient confidences and privacy within the constraints of the law.
- Continue to study, apply, and advance scientific knowledge, maintain a commitment to medical education, make relevant information available to patients, colleagues, and the public, obtain consultation, and use the talents of other health professionals when indicated.
- Be free to choose whom to serve (except in emergencies), with whom to associate, and the environment in which to provide medical care.
- Recognize a responsibility to participate in activities contributing to the improvement of the community and the betterment of public health.
- Regard responsibility to the patient as paramount.
- Support access to medical care for all people.

9.3 Ethical Principles in the Practice of Psychiatry

9.3.1 The Physician-Patient Relationship

The physician-patient relationship is the foundation of medicine and many ethical principles have a bearing on this relationship, which includes among others, respect for persons, beneficence, autonomy, honesty, confidentiality, and fidelity. The physician–patient relationship generally begins when a physician has a face-to-face interaction with a patient in which the physician is entrusted with the responsibility of applying his or her knowledge and clinical skills to achieve the health and well-being of the patient. In contemporary American society, the relationship has been viewed as a consensual agreement between two autonomous individuals who are free to enter, sustain, or discontinue the relationship unconstrained by discrimination, coercion, or fear of physician abandonment. In cases of severe mental illness, the disparity in the relationship creates an ethical obligation for physicians to place the needs of the patient above their own professional interests. As noted in the section below on Legal aspects of Psychiatry (Sects. 2.05 and 2.06), physicians must be vigilant for situations that can cause physical, sexual, psychological, or financial harm to the patient. There is a special sensitivity to the trust and dependence created by the communication of highly personal information and the physician has an ethical obligation to respect that trust.

In America, third party obligations and the clinical context also influence the ethical expectations of the physician-patient relationship. The consulting physician while retaining the fundamental responsibility to serve the interests of the patient will share clinical information with appropriate clinicians. The physician's obligation to preserve a patient's confidentiality may be limited because of obligations to a third party such as employment, forensic or military settings.

9.3.2 Professional Competence

Professional competence is the cornerstone of ethical psychiatric practice and is the ability to apply the accepted standards of clinical practice to patient care. Practicing within the bounds of professional competence requires attention throughout one's career and life-long learning must be emphasized as necessary for professional competence.

From an ethical perspective, it is expected that psychiatrists will through continuing education and supervision maintain a sufficient level of professional competence and practice within the bounds of their competence. It is further expected that psychiatrists will make referrals only to persons who are, in the psychiatrist's best judgment, competent to deliver the necessary treatment.

9.3.3 Dual Agency Roles (also see the section on Legal Aspects of Psychiatry 2.06)

The term "dual agency" refers to the competing allegiances and obligations, a psychiatrist may have in an interaction with a patient. For instance, a psychiatrist may have competing duties to an institution and an individual patient. Treating psychiatrists should generally avoid serving as expert witnesses for their patients or performing evaluations of their own patients for legal purposes. This can damage the therapeutic relationship and confound the validity and purpose of the evaluation.

One role is absolutely prohibited in all fields of medicine. Physicians may not ethically participate in any manner that supports, facilitates, or enacts human torture or the development and monitoring of interrogation techniques that involve torture.

9.3.4 Confidentiality

Confidentiality is the obligation not to reveal patient related information without the patient's explicit permission. Recognizing the difference between the ethical duty to keep confidences from the legal duty that governs the handling of private medical information helps the psychiatrist focus on the patient's interest rather than on mere compliance with privacy regulations. Patients' willingness to reveal sensitive information depends on their trust in the physician-patient relationship and its expectation of confidentiality. Besides a therapeutic rationale, there are ethical duties that arise from principles such as beneficence, and non-maleficence.

Explicit permission is important for the ethical disclosure of patient information by psychiatrists. However, psychiatrists may accept or receive information under many circumstances and psychiatrists should be sensitive to the feelings this kind of information disclosure may raise for patients.

There are several important considerations that guide the confidentiality of medical information. Patients should be told of the limits to confidentiality at the beginning of the physician-patient relationship and as events arise that create potential revelations. Disclosure of confidential information should occur only if informed consent has been given by the patient in a manner consistent with relevant legal statutes, as discussed earlier in this chapter. Disclosure of patient information should always be limited to the requirements of the situation. Psychiatrists should only record pertinent information necessary for continued patient care in the progress notes. It is advisable to have psychotherapy process comments, and such details as patient's dreams and wishes, and intimate personal details of patients or related individuals in separate psychotherapy notes rather than in the medical record.

9.3.5 Informed Consent (also see under Legal Aspects of Psychiatry Section 2.01)

Informed consent is an ethically and legally important process that involves information-sharing and authentic decision-making about an individual's health. In the United States the legal standard for information disclosure continues to evolve and varies by jurisdiction. As note in Sect. 3.01, many states apply the "professional standard," or the "reasonable practitioner standard", which is the amount and content of disclosure determined by what most physicians traditionally disclose. Another standard, is the "reasonable person standard." This standard has an emphasis on patient autonomy and requires that physicians disclose what a reasonable person would want to know. These standards include a description of the proposed treatment, its potential risks and benefits, any relevant alternatives and their risks and benefits, and also the risks and benefits of no treatment at all.

The requirement of voluntariness in informed consent affirms the autonomous decision making of the individual and it prohibits coercive influences in the consent process. Ethical practice requires that the practitioner be fully and constantly be aware that at times the experience of dependence, societal marginalization, and insufficient access to clinical care may create a situation of desperation that may interfere with voluntary decision-making. While these vulnerabilities need not confer incapacity, they should be explored in order to optimize a patient's decision-making.

Important exceptions to informed consent exist:

Genuine emergencies do not require informed consent. Emergency care occurs in the framework of implied or presumed consent.

Care for children or incompetent patients, requires consent from parents or surrogates. Assent of incompetent individuals is obtained whenever possible.

Therapeutic privilege is the doctrine that allows a physician to withhold information if it is truly damaging to the patient. Such an exception should be rare and withholding information about side effects, for instance, in the hope of increasing compliance is not acceptable.

9.3.6 *Decision-Making Capacity*

Decision-making capacity is the ability of an individual to reach an informed, reasoned, and free choice, when making a specific decision. Capacity is a consideration in psychiatric and non-psychiatric conditions that affect cognition, or affective regulation.

Assessment standards frequently expressed in ethics and law include evidencing a choice, understanding relevant information, rationally manipulating information, and appreciating the situation and its consequences. Psychiatrists may be asked to perform capacity assessments when patients exhibit cognitive deficits, or present atypical behaviors and uncommon decisions. Psychiatrists recognize that deficits in decision-making capacity may be overcome by targeted educational and clinical interventions such as the use of a single trusted clinician to communicate information. Interventions to diminish psychotic symptoms, reduce anxiety, or reduce sedating side effects are sometimes valuable in overcoming incapacity.

For high risk decisions psychiatrists may apply assessment standards on a "sliding scale", with more stringent assessments and higher thresholds of capacity required for decisions that are more complex, or risky. When incapacity persists surrogate decision makers may be involved in the decision making process in accordance with local statutes.

9.3.7 *Involuntary Psychiatric Treatment*

Involuntary psychiatric treatment most commonly involves inpatient psychiatric treatment or court-ordered outpatient treatment. Mandated treatment is justified by the doctrines of police power and of parens patriae. The state's enforcement apparatus is used to place individuals into medical treatment.

Mandated treatment creates inherent ethical tensions in clinical practice. It requires great sensitivity because psychiatrists are contributing to decisions directly controlling patient freedom of choice. Careful balancing of the principles of personal freedoms and community needs are necessary.

The justification for involuntary hospitalization is usually patients' imminent dangerousness to themselves or others, or their inability to meet basic needs. To meet these criteria, dangerousness must not only be imminent but related to a major mental illness. Due to the seriousness of depriving a patient of his or her freedom, involuntary commitment usually requires judicial review. Most states make access to legal counsel mandatory. Also consideration is given to the least restrictive alternative to hospitalization.

Historically, in the United States, prior to the ascendancy of dangerousness-based statutes in the 1970s, treatability was the most common legal criterion for involuntary psychiatric hospitalization. The APA's model law for civil commitment embodies a standard that adds treatment rationales to commitments using dangerousness criteria.

The APA's model law requires that:

The person is suffering from a mental disorder; and

There is reasonable prospect that the disorder is treatable at or through the facility to which patient is to be committed and such commitment would be consistent with the least restrictive alternative principle; and

The person either refuses or is unable to consent to voluntary admission for treatment; and

The person lacks capacity to make an informed decision concerning treatment; and as a result of the disorder, the person is a) likely to cause harm to himself or to suffer substantial mental or physical deterioration, or b) likely to cause harm to others.

9.3.8 Therapeutic Boundary-keeping (also see below, under Legal Aspects of Psychiatry, Sect. 9.4.5 Treatment Boundary in Psychiatric Practice)

Physicians must never exploit or otherwise take advantage of patients. Boundaries connote a professional distance and respect that ensure an atmosphere of safety and predictability. Appropriate therapeutic boundaries are also necessary for therapeutic efficacy. Psychiatrists are trained to examine and appreciate the significant psychological and social overtones of the treatment relationship. Physicians must limit the relationship with patients to the therapeutic context and avoid patient interactions that are aimed at gratifying the physician's needs and impulses.

Professional boundaries limit both sexual and non-sexual behavior. Inappropriate physical contact, perhaps the most obvious form of boundary violation may cause harm by exploiting the inherent power differential between the patient and the psychiatrist. Any sexual activity, with current or former patients, is unethical.

Likewise, any occasion in which the physician interacts with a current or former patient in ways that may suggest an intimate, non-professional relationship should be avoided.

Non-therapeutic interactions within the treatment relationship such as financial and business dealings, or trustee and guardianship roles should be avoided. These boundary violations have the potential to impair the physician's judgment, and to diminish the effectiveness of treatment.

Psychiatrists are encouraged to address boundary issues with the patient in therapy and to seek consultation with colleagues. Some boundary violations may only be resolved by termination of the therapeutic relationship.

Some specific applications of boundaries in psychotherapy include:

Money—Psychiatric fees exemplify the business nature of the therapeutic relationship. Barter (i.e., allowing the patient to trade or work for the therapist in order to pay for treatment) is at best confusing and ill-advised. Barter in some jurisdictions is illegal.

Gifts—Small gifts from patients, especially small handmade gifts, are acceptable. Their meaning and symbolism are appropriate for discussion in therapy. Large personal gifts should be avoided. The appropriateness of accepting such gifts should be determined in consultation with colleagues or ethics committees. The restrictions on receiving gifts from industry (e.g. pharmaceutical companies), are well defined in the AMA Council on Ethical and Judicial Affairs (CEJA) code. In general, gifts from industry should benefit patients, relate to the physician's work, and be of minimal value.

Self-disclosure—Self-disclosure from the therapist is not, in general, conducive to the therapeutic relationship and should be avoided. Therapists should not burden patients with their own personal issues, and they should not be used to influence the patient in any way not directly relevant to the treatment goals. The disclosures required by general standards of truth and honesty are expected

Behavior with family and other patient intimates—Personal relationships between the therapist and the patient's family (or individuals intimately associated with the patient) should be avoided during the course of therapy and usually even after it ends.

9.3.9 *Financial Conflicts of Interest in Relations with the Pharmaceutical Industry*

The APA endorses the codes of conduct outlined in recent documents of the American Medical Association (AMA, Opinion of Council on Ethical and Judicial Affairs 2013; American College of Physicians, Ann Intern Med 2002). The practice of psychiatry may bring competing values into conflict. Failure to recognize and actively manage conflicts of interest does constitute a serious compromise of professional integrity.

The FDA sets minimal standards because it expects objective, independent physicians—motivated by the welfare of their patients and an allegiance to academic integrity—to exercise their own scrutiny. Therefore integrity and true professional self-regulation require that the standards be set at a higher level than mere regulatory compliance. Such integrity cannot be externally imposed; it should be the aspiration of individual practitioners as well as of professional societies.

It is important to note that disclosure does not eliminate a conflict. It only shifts the responsibility for managing negative consequences to the recipient of the disclosure. Routine disclosures may consequently lull the profession into failing to recognize real conflicts when they arise.

Financial conflicts of interest in interactions with the pharmaceutical industry pose a danger to the practitioner's independence. Rules, although important, serve only as minimal standards of conduct. Disclosure, a common institutional rule, merely shifts the responsibility for the conflict to others without addressing the potential dangers to integrity and objectivity. Therefore, avoiding persistent conflicts of interest remains an important ethical obligation for the psychiatric practitioner.

9.3.10 Ethical Issues in Developing Nations

Developing nations pose special ethical challenges to psychiatrists because many developing nations face great limitations of health care resources and heightened barriers to care. Psychiatrists are in short supply and it is not uncommon for psychiatrists to be responsible for large populations. These features of developing nations may therefore create situations where patient needs cannot be met completely.

Cultural factors sometimes lead to difficulties separating personal and professional relationships. Psychiatrists in developing nations may experience greater difficulty in protecting the health information of their patients. Also the consequences of confidentiality breaches may be serious and enduring, particularly given the stigma associated with mental illness.

Certain communities may also require sensitivity to cultural practices that are unique to the group. Practices, rituals, and conceptualizations of fundamental medical principles may require psychiatrists to obtain consultation or education on their role in these interactions. Respecting values can be useful in improving the relationship with the patient as well as the entire community.

Psychiatrists in developing nations may consequently provide care of broader scope than psychiatrists in the developed world. Psychiatrists in these countries should make special efforts and adopt specific practices to assure that their patients are provided appropriate care to the full extent possible.

9.4 Legal Aspects of Psychiatry

Psychiatric practice in the United States needs to be compliant with the legal regulations imposed by the state or the Federal Government depending on the jurisdiction. There are large common elements in these statutes. However, each state in the Union also imposes unique statutes that one has to be cognizant of. All psychiatric practices need to be familiar with certain basic legal aspects as noted below. There are other interfaces between psychiatry and the law that are more specialized that only a subspecialist, such as a Forensic Psychiatrist is qualified to deal with. The American Board of Psychiatry and Neurology recognized Forensic Psychiatry as a sub specialty with certification examination starting in 1994. There are currently about 2000 Psychiatrists who are Board certified in Forensic Psychiatry in the United States. The American Academy of Psychiatry and the Law, was started with about 15 members in 1969 and now has about 2000 members, and is the premier organization for forensic psychiatry in the U.S. It has offered model guidelines for the ethical practice of forensic psychiatry (American Academy of Psychiatry and the Law 1995).

9.4.1 Informed Consent (also see above, under Sect. 1.15 Ethics: Informed Consent)

Informed consent reflects the basic human right of self-determination and was articulated in modern times in response to some egregious human rights violations during the twentieth century. In the landmark case Schloendorf v. Society of New York Hospital, Justice Benjamin Cardozo concluded "every human being of adult years and sound mind has a right to determine what shall be done with his own body…". Another predecessor to current notions of informed consent is the Nuremberg Code of 1947 created in response to inhumane research practices committed by Axis scientists during World War II. The three tenets of the Code relating to informed consent are: (i) voluntary consent is essential for human participants in research, (ii) the human subject must be free to discontinue participation if desired, and (iii) the principal investigator must be prepared to end the research procedures if there is probable cause to believe that continuation might result in injury, disability, or death of a subject.

The term Informed Consent itself was first used in a 1957 California case Salgo v. Leland Stanford Jr. University Board of Trustees (Salgo v. Leland 1957). Since then it has evolved to a greater degree. The standard to which a practitioner was held was initially the "reasonable practitioner standard" based on a case Natanson v. Kline in 1960 (Natanson v. Kline 1960). In this case, the patient claimed he was inadequately informed about the risks of cobalt irradiation following mastectomy. Subsequently, in a 1972 case in Washington D.C., Canterbury v. Spence (Canterbury v. Spence 1972). a new standard referred to as the "reasonable person standard" was imposed. In essence, the court proposed the standards of disclosure be based on that which a reasonable person would find material to his or her clinical decision making. Currently some states continue to follow the "reasonable practitioner standard" or "professional standard", while others follow the "reasonable person standard". In order to be valid the informed consent has to be truly voluntary. For example, in a case Kaimowitz v. Department of Mental Health (Kaimowitz v. Michigan Department of Mental Health 1973), the court ruled that a prisoner is in an inherently coercive environment and that it would be impossible for him/

her to be giving a truly voluntary consent especially to a high risk procedure of experimental psycho surgery. Informed consent can only be given by a competent person. All adults are legally considered to be competent. The law however, defines children and adolescents as not fully competent, and the age of consent for different procedures as well as for certain medical treatment varies from state to state.

There are differences among authors whether the competency to consent should be uniform or variable and based on the complexity of treatment procedures including inherent risks. Exceptions to voluntary informed consent by the patient himself/herself are legal incompetency, medical emergency, patient waiver and therapeutic privilege. Who may provide consent when a person is deemed incompetent varies. Generally it is the legal next of kin, and sometimes it could be the hospital administrator or Chief of staff, and at other times a court order may be required. Again, the states differ in regards to the basis by which an incompetent person's right are considered as having been protected. Some states follow the "best interest model" in which it is to be determined as to which of the treatments would best serve the patient's interest, and others use a "substituted judgment model" in which a decision is made based on what the patient might have chosen if he or she were competent. In Psychiatry, for most common treatments such as common medications and therapy, a separate and written informed consent is not routinely obtained but medications such as clozpine, and procedures such as Electro Convulsive Therapy (ECT), deep brain stimulation, or other invasive treatments require a written informed consent. In geriatric patients, a Psychiatric Advance Directive such as living will and/or durable power of attorney is frequently utilized for treatment planning. Participation in research, especially with any potential for harm has extensive guidelines for obtaining consent. Beyond mental illness, other vulnerable conditions requiring greater scrutiny of "voluntary" participation in research apply to children, pregnant women and prisoners (National Commission for the Protection of Human Subjects of Biomedical and Behavioral Research, The Belmont report 1979; U.S. Department of Health and Human Services 1983).

9.4.2 Voluntary and Involuntary Hospitalization

Most state statutes allow for two types of admission on a voluntary basis. An informed admission would be one when the patient has the right to leave any time and comes into the hospital with a written or verbal agreement. This may be suitable for patients who get admitted for substance abuse treatment or rehabilitation programs, but may not be suitable for treatment of severe psychiatric disorders. For the latter, even when it is a voluntary admission, the statutes provide for a mandatory period of time, usually up to five days that the patient may be required to stay, and during this time, usually used for observation, assessment and initial treatment, the hospital could decide either to release the patient or to keep the patient longer. If patient needs continued stay and refuses consent, commitment procedures are implemented in order to keep the patient involuntarily. The involuntary stay or commitment is initiated with a petition to the magistrate. Such a petition may be made by any citizen of the state, family members, police or the hospital authorities. It varies from state to state as to who can certify such a petition. It may be a licensed physician, a clinical psychologist or a psychiatric nurse. In most states, to make a patient stay, a second professional has to concur. Typically, this second person is a licensed psychiatrist or a clinical psychologist. All states allow a probable cause hearing by the local court to make the decision. Although the procedures vary from state to state, the principle by which an involuntary hospitalization is authorized is based on 2 principles. The first principle is "Parens Patriae" or "father of the country" meaning that the state assumes charge for a person who is unable to take care of self, and the second principle is that of "police powers" which authorizes the state to prevent imminent harm to self or others. In the case Addington v. Texas (1979), the US Supreme court decided the standard of proof required to commit a person has to be "clear and convincing". This standard is higher than the "preponderance of evidence" standard required in civil cases but less than the standard of "beyond a reasonable doubt" required in criminal proceedings.

9.4.3 Involuntary Civil Commitment to Outpatient Treatment

Before the 1980s even though most states allowed for outpatient commitment there were no statutory procedures to practically enforce the order and thus very few people were committed. Since then, there has been a significant increase in the number of people who are committed to outpatient treatment, in several jurisdictions across the U.S. while enforcement challenges continue. These laws are referred to as "mandatory outpatient treatment" or "assisted outpatient treatment". Most jurisdictions use re-hospitalization as the major consequence of non-compliance with mandated outpatient treatment, while some others allow for bringing the patient to the outpatient clinic by a police officer. There is modest evidence from New York State that mandatory outpatient commitment works to reduce hospitalization and violence, especially when implemented as part of a package of outpatient services including housing and day treatment.

9.4.4 Right to Refuse Treatment

Historically civil commitment for hospitalization was for the purposes of treatment. However, with the changing criteria of civil commitment based on "dangerousness to self or others" and/or "unable to care for self", it was argued that a person who is confined based on dangerousness still can be competent to make treatment decisions and thus can decide about their own treatment even while committed to the confines of the hospital. Some states more concerned about the individual's autonomy and rights choose a judicial course to decide when the right to refuse treatment may be overturned. Again, the jurisdictions vary in that some courts apply the *best interest model* in competent patients and the others use the *substituted judgment model*. There are other states as well as federal jurisdictions who choose an *administrative review model* by which the decision to overturn the right of an individual to refuse treatment is decided by a designated administrative team including clinicians, laymen, community organizers and possibly legal scholars. In the jurisdictions using administrative review model, nearly 90–95% of treatment refusals were overturned. However, in jurisdictions where a judicial rather than administrative decision is made the results are quite variable. In a study done in the state of Illinois the overturn rate was rather large, from 0 to 60% (Srinivasaraghavan et al. 2002).

9.4.5 Confidentiality and Privilege

The Psychiatrist's responsibility to keep information confidential is even more paramount than that of the general physicians, and the privilege belongs to the patient. Without his/her consent the information cannot be released to any third party. Only military law is without privilege regardless of whether the psychiatrist works for the military or is a civilian. The hospitals are also entrusted with maintaining the confidentiality of its patients. Even the mere fact stating that a patient is in the hospital can be considered a disclosure of information. Upon death, usually a decedent's executor has the right to waive the deceased patient's privilege, but in some states even this requires a court order. Even though the psychiatrist—patient interaction is confidential, there are exceptions to this standard. For example, child abuse is reportable by statute. In 1974, the supreme court of California in its ruling in the case of Tarasoff (Tarasoff v. Regents of the University of California 1974) imposed a "duty to warn" and subsequently in 1976 this was replaced with "duty to protect" (Tarasoff v. Regents of the University of California 1976). In the Tarasoff case, Prosanjit Poddar after sharing with his therapist that he has thoughts of harming Tatiana an ex-girlfriend of his, went ahead and murdered her. A court case ensued as to whether the therapist should have violated his patient's confidentiality to inform the intended victim. The court ruled in 1974 that the therapist had such a "duty to warn" and in 1976 this was further revised to a "duty to protect" on the part of the therapist (Tarasoff v. Regents of the University of California 1974, 1976). Different jurisdictions have chosen different paths since then. While some states do follow the duty to protect, several other states have not imposed this duty. The majority of the jurisdictions only impose a duty to protect if it is an identifiable victim, although in other jurisdictions liability has been found even when the therapist did not have information beforehand on the later victim.

9.4.6 Treatment Boundary in Psychiatric Practice (also see above under Sect. 9.3.8 Ethics: Therapeutic Boundary-Keeping)

The Psychiatrist always has to be cognizant of keeping objectivity and also appropriate interpersonal distance from the individual who he/she is treating.

Boundary violations can happen in different ways (see Ethics section below). Of legal significance are the following: (i) Monetary exploitation: This is one form of boundary violation. This could happen if the therapist uses insider information derived or obtained from the patient for personal gain in terms of stock offers and mergers, and such, (ii) Double agent situations: These refer to psychiatrist's conflicting loyalties when they have to serve an institution as well as the patient. For example, a psychiatrist working in the military has to serve the needs of the military to the best extent while also trying to serve the patient's best interest. A psychiatrist working in the correctional facility has to be aware of the rules and regulations in terms of safety that are imposed by the correctional centers while trying to serve the best interest of his patients. The school psychiatrist in a similar way has to take into account the interest of the patients, the parents and also the school administration and other students. (iii) Another form of boundary violation was claimed in Roe v. Doe (Roe v. Doe, 93 Misc2d 201, 400 N.S.Y2d, 668, Sup Ct 1977). This case illustrated a psychiatrist getting sued by his patient for publishing a book using the information obtained from the patient regarding his thoughts and feelings.

Boundary violations in therapist-patient sexual relation often end up in court. Typically such violation starts well before the actual sexual act. Contrary to common notion that it is the male therapist violating the boundary with female patients, it has been shown in a survey that both male and female psychiatrists have admitted to sexual contacts with their patients. Different jurisdictions have defined under what circumstances it is not unethical for a therapist to have sexual relationship with a *former patient*. In some states it is allowed after a cooling period of 1 or 2 years after termination of the treatment relationship. However, many psychiatrists and the American Psychiatric Association have opined that one should never have sexual contact with one's patient, current or former. In this type of boundary violation, other than malpractice litigation, there is also professional disciplinary action taken by the state Board of Medicine for involving oneself sexually with the patients.

9.4.7 Psychiatric Malpractice

A basic concept in psychiatric malpractice is "the standard of care", which is now a well established legal concept. This refers to the prevailing standard of healthcare in the community in which the professional practices. Rule 702 of the *federal rules of evidence* allows for technical, scientific or other specialized knowledge to be offered as expert testimony in assisting during the judicial process to determine the occurrence of malpractice. In order for a malpractice suit to be successful, the plaintiff has to show that (i) there was a duty to provide care derived from a doctor- patient relationship, (ii) that a deviation from the standard of care occurred, and (iii) damages resulted, and (iv) that the damages are directly as a result of the deviation from the standard of care by the psychiatrist. Psychiatrists are not sued as often compared to other physicians, such as surgeons and obstetricians. It is estimated that 22 % of psychiatrists in the USA will be sued at least once in their life time, although the plaintiffs are successful only 20–30 % of the time.

Common causes for suing a psychiatrist are: negligence in suicide cases, improper diagnosis, improper treatment, and injuries resulting from restraints at seclusion. The law only requires that a practitioner provide average or reasonable prudent care. As long as it can be demonstrated that reasonable care (standard of care) was provided, even if the patient is harmed, no malpractice has occurred. One exception to this standard occurs, when the negligence is so egregious that it is "res ipse loquitor" meaning "the thing speaks for itself". Expert opinion is always sought in malpractice cases, and often there are experts who approach the subject from different angles, with one or more experts supporting the plaintiff and one or more supporting the defendant.

9.4.8 Electro Convulsive Therapy

Forty three states in USA have legislation that regulates electro convulsive therapy (ECT) in some way or other. Most of these regulations address the administration of ECT. In some the regulations only deal with treatment in general. Informed consent is specifically mentioned in many statutes. Many states also address how to provide ECT when the patient is considered incompetent to make a decision.

ECT is largely practiced in academic medical centers, general hospitals and private psychiatric hospitals. Although a few public psychiatry hospitals make ECT available to their patients, most prefer to refer out the patients.

Modified ECT with monitoring for cardio-respiratory safety and EEG seizure is the standard. There is no unmodified ECT in the USA.

9.4.9 Child Custody

Divorces are very common in the United States and 60 % of divorce cases involve children. Child custody becomes a major issue. Historically the child's custody was given to the mother. However, the trend has now changed. In the 1970's, the mother gained sole custody in 80 % of the cases, the father in 10 % and joint custody occurred in the remainder 10 % of cases. With less of a stigma and more acceptability, gay and lesbian couples assuming the role of parents has also been an issue. Psychiatric evaluations are sought when one or both the parents are mentally ill. Child custody battles sometimes bring issues such as false allegations of child sexual abuse, and psychiatric evaluations are sought to establish or rule out such abuse. The child abuse legal literature contains many cases where the authors found varying degrees of unsubstantiated allegations. Parental kidnapping of a child or children may occur just prior to the finalization of the divorce, and bring forth various psychiatric issues to be assessed by the professional for the benefit of the court. An interesting offshoot of psychiatric evaluations in child custody cases has been that the evaluation itself has been alleged to be harassment by the parent.

9.4.10 *Sexual Offenders*

The public has been more concerned about sexual offenders in the past two decades. As of 1994, there were nearly 234,000 offenders convicted of rape or sexual assault, among them 60% were under conditional supervision and the rest in the care or custody of correctional departments (Greenfeld LA: Sex Offenses and Offenders: Bureau of Justice Statistics 1997). Civil commitment under the Sexual predator laws was initially passed by the State of Washington by which using the 'police power' principle, individuals with non-psychotic disorders, personality disorders and paraphilias, though not requiring immediate mental health treatment could be admitted to mental hospitals and kept until deemed safe for release. The Supreme Court upheld the constitutionality of such admissions in Kansas v. Hendricks (Kansas v. Hendricks, 521 U.S. 346, 138 L. Ed. 2d501, 117 S. Ct. 2072; 1997) and at least 15 states have followed with similar statutes. Some states have required specific registration of sexual offenders modeled after the so-called Megan's Law and states are obliged to meet certain requirements in order to get federal law enforcement funds. The Jacob Wetterling Crimes Against Children and Sexually Violent Offender Registration Act of 1994 required registration of all the individuals convicted of criminal sexual offense against a minor or convicted of sexually violent offense.

The laws vary from state to state but all have generally four elements that are common: (i) A conviction, (ii) Relation to a sexual offense, (iii) Mental abnormality, and (iv) Predisposition to sexually dangerous acts in the future. The Supreme Court in Kansas v. Crane (2002) has required both a psychiatric diagnosis and serious difficulty in controlling behavior to justify indefinite commitment. States have differing and poorly defined terms in predicting future sexual dangerousness. Penile Plethysmography (PPG), Abel Assessment for Sexual Interest (AASI) and the Polygraph test are tools that may be used in evaluation of sexual offenders. In prediction of future sexual dangerousness, clinical and actuarial approaches are used (Zonana et al. 2004). Each approach has its pros and cons and it is beyond the scope of this chapter to discuss those issues.

9.4.11 *Post-Traumatic Stress Disorder*

Pitman & Sparr in a 1998 seminal article (Pitman and Sparr 1998) examined developments and state of the art in law, forensic evaluation, and expert testimony. They provided specific forensic guidance that continues to be relevant today. Enthusiasm for PTSD as the basis of a criminal defense seemed to wane until recently but the wars in Iraq and Afghanistan have renewed emphasis once again (Cullen 2005). A dramatic behavioral scenario cited in the literature is the experience of a flashback during which aggression or a criminal act could occur when a Veteran thinks he or she is in danger again. In contrast O'Brien (1998) noted that there is no evidence of an association between criminal behavior and PTSD caused by non-combat trauma. The association of criminal behavior with PTSD does not establish a causal relationship between PTSD and criminal behavior: Antisocial personality and substance abuse have been found to be important intervening variables.

Sparr and Pitman (2007) noted the infrequent use of PTSD as a defense in criminal cases. The recent case of Jesse Bratcher was a much-discussed exception to the decline in use of the insanity defense. Bratcher, who was being treated and compensated for service-connected PTSD when the crime took place, argued that his killing of an unarmed man occurred while he was having a flashback. A Grant County, Oregon jury found him guilty but insane due to PTSD; instead of a possible 25-year prison term, Bratcher was committed to the Oregon State Psychiatric Security Review Board for evaluation on October 7, 2009 (State of Oregon v Jessie Bratcher Grant County case number 08-08-219CR).

In addition to the use of PTSD as the basis of an insanity defense, recent cases have seen the stress associated with combat exposure considered as a mitigating factor at sentencing Porter v. McCollum (2009). Since 2008, a new treatment court model has addressed Veteran defendants' mental health and substance use issues; there are now 50 operational Veterans Treatment Courts (VTC) (http://www.nadcp.org/JusticeForVets), modeled after the first such court in Buffalo, New York. Although generally not limited to defendants with a particular diagnosis, the perceived prevalence of PTSD among justice-involved Veterans is often cited as the impetus for these courts' formation (Russell (2009).

State legislatures have been active in proposing legislation that directs their court systems to address mental illness of Veterans in their courts: Six of 10 states that have recently considered such legislation have specifically mentioned PTSD as one of the target mental illnesses. Holbrook (2010) has argued that jurisdictions embracing the VTC model have shifted from their traditional focus on victims' interests (retributive justice), toward defendants' interests (therapeutic justice).

Young and Yehuda (2006) cautioned against the use of a PTSD diagnosis as a shorthand for a fully defined, universal experience, since multiple mechanisms might contribute to outcomes of the disorder. They also emphasize that individual variability frustrates any effort to describe a standard course or prognosis, a fact about which courts require education.

Criminal behavior and violence and their relationship to mental disorders including PTSD is beyond the scope of this chapter. It is only noted that the relationship is complex and often contains intermediate variables, especially substance abuse and socio-economic factors. The Department of Veterans Affairs and the Department of Defense have developed specific practice guidelines for timely intervention to avoid development of aggressive behaviors (The Management of Post-Traumatic Stress Working Group 2004). Models of assessment and intervention have emerged as part of a multi-step sequential program in mentally ill persons vulnerable to criminal behaviors and vice-versa. Returning veterans are a target group with vulnerabilities to substance abuse and in a smaller sub-group to aggression (Munetz and Griffin 2006).

In the civil legal context, PTSD provides straightforward causal relationship that plaintiffs' lawyers welcome. The removal from the DSM criteria for PTSD of the requirement that the traumatic event be beyond the range of ordinary human experience led to an increase use of PTSD as a factor in civil litigation. Individuals involved in motor vehicle accidents were able to seek damages for PTSD. Even in workers compensation cases although the DSM-IV requirement of actual or threatened physical injury or assault or a threat to physical integrity is more difficult to establish, special features of PTSD (stressors, symptoms, functional impairments), have helped overcome legal barriers to worker's compensation cases, and mental harm claims in particular have increased dramatically. It is not surprising then that the various components of PTSD diagnosis have an increased potential for fraudulent claims.

9.5 Human Rights

The evolution and establishment of human rights stemming from the Nuremberg Convention, the Geneva Convention and the Helsinki Declaration have been a major force in shaping MH policies and care delivery methods in the USA. Landmark cases have included the Olmstead act which specifies an individual's right to treatment in the least restrictive setting. This concept of treatment in the least restrictive setting has led to a sea change in care delivery systems and their locations, from the traditional large asylum or state mental hospital located many miles away from population centers to a community model of care. Today, the state hospital system is considered a dying breed, and is increasingly being utilized for the forensic population rather than civilian patients with mental illness. In tandem with these developments, the emergence of managed care in 80's and 90's, largely to control spiraling health care costs has led to a closure of many free-standing private psychiatric hospital and shifted the locus of care to the doctors' offices, outpatient counseling and community clinics. Similar trends are evident in the care of individuals with intellectual disabilities (mental retardation) and developmental disorders as well as those with substance use disorders.

In recent years, patient centered and recovery models and approaches have significantly influenced methods of assessment and development of treatment plans in all settings especially in the community service boards. These models appropriately emphasize the centrality of the patient, empower the patient (client, consumer, person) and enhance autonomy.

A significant conflict namely the interests of the individual versus interests of society especially in the area of safety and violence has always guided the laws, policies and procedures in mental health care. While cases with dramatic violence often capture media attention and lead to defensive practices on the part of the clinician and stricter rules of discharge, individual rights, humane traditions and evidence based practices lead us away from stereotyping and stigmatizing all individuals with mental illness and "locking them away". The following basic rights exist for every person with a mental disorder in the USA.

1. Every individual has a right to mental health care
2. Consent of the individual is required prior to assessment and treatment

   1) An individual has a right to accept or refuse treatment, and to withdraw from treatment
   2) Health care information is private
   3) Treatment and services should be provided in the least restrictive setting.
   4) Treatment and services should be provided in the patient's own community.
   5) The State has a duty to protect its citizens. In carrying out this duty, the State may curtail the rights of an individual, with reason.
   6) Services should be comprehensive in nature (as permitted by resources)
   7) No individual may be denied treatment based on race, gender, national origin, religion, lifestyle including sexual orientation.
   8) Individuals should be assisted with transition of care
   9) Individual choice should be respected.
   10) Individuals should be actively engaged in treatment planning
   11) Individuals with one or more disabilities should be provided suitable assistance and accommodation in assessment and treatment including treatment setting.
   12) Individuals in institutional settings should be provided all basic amenities to include self care and hygiene and visitation and communication.
   13) No treatment may be rendered which may be considered punitive.
   14) Restraint and seclusion may be used only for the physical protection of the individual or others. Each episode of restraint or seclusion should be justified separately within an hour of such an occurrence.
   15) Involuntary commitments to an institution and/or treatment shall have specific time limits, and shall automatically expire unless properly renewed.
   16) Assessments and treatment plans should be reviewed and updated at specified intervals.
   17) Special permission/consent is needed for the use of antipsychotic medication and electroconvulsive therapy.
   18) No procedure such as sterilization, lobotomy etc may be performed on an individual with mental illness or intellectual disability with the consent of the individual. A court, guardian, family member, power of attorney, institutional representative or any other individual, agency or entity may permit such a procedure. The above procedure may never be performed against an individual's wishes.
   19) Medical care should include both physical and psychological aspects.
   20) Individuals have a right to safe housing.
   21) Preventive and early intervention should be provided (resources permitting)
   22) Treatment approaches should promote recovery for individuals with hope, empowerment, respect, social connections, self-responsibility, and self-determination.
   23) Treatment approaches and environment of care should be customized to the unique cultural, ethnic, and racial diversity of the individual (resources permitting).

24) Treatment models should include the patient/consumer and family members, and be provided by cultural competent staff.
25) A patient advocate shall be available for all patients.
26) A local human rights committee (LHRC) be available in all localities and institutions.
27) A state human rights committee (SHRC) be available in each state. The local human rights committee shall review and investigate as necessary any complaint by a patient/consumer as to his/her concerns over violation of his/her rights. The LHRC shall report its findings to the SHRC. The SHRC may ask the commissioner of mental health (or equivalent) and the state licensure agency to take necessary corrective steps if it finds any violation of a patient's rights.
28) States may additionally have a department for the rights of disabled persons that will advocate for the person with disability due to mental or physical disorder. Such an agency should be independent of the executive branch of the government.

References

Addington v. Texas. 1979. 441 U.S. 418.

AMA, Opinion of Council on Ethical and Judicial Affairs. Nov. 2013. E-8.061 Gifts to Physicians from Industry. http://www.ama-assn.org/ama/pub/article/4001-4236.html. Accessed 30 Sept 2012.

American Psychiatric Association. 2001a. *The principles of medical ethics with annotations applicable to psychiatry*. Washington, DC: APA.

American Psychiatric Association. 2001b. *Opinions of the ethics committee on the principles of medical ethics with annotations especially applicable to psychiatry 2001 edition, Ethics primer*. Virginia: American Psychiatric Association.

American College of Physicians, American Society of Internal Medicine Position Paper. 2002. Physician-industry relations. Part I: Individual physicians. *Annals of Internal Medicine* 136:396–402.

American Academy of Psychiatry and the Law. 1995. Ethics guidelines for the practice of forensic psychiatry. American Academy of Psychiatry and the Law, Bloomfield CT.

Canterbury v. Spence. 1972. 464 F2d 772, 787.

Cullen, F. T. 2005. The twelve people who saved rehabilitation: How the science of criminology made a difference. *Criminology* 43:1–42.

Greenfeld, L. A. 1997. Sex offenses and offenders: An analysis of data on rape and sexual assault (NCJ 163392). Washington, DC, U.S. Department of Justice, Bureau of Justice Statistics.

Holbrook, J. G. 2010. Veterans' court and criminal responsibility: A problem solving history and approach to the liminality of combat trauma. In *Young veterans: A resilient community of honor, duty and need,* eds. D. C. Kelly, D. Gitelson, and S. H. Barksdale. New York: Springer. (Forthcoming; Widener Law School Legal Studies Research Paper No. 10–43).

http://www.nadcp.org/JusticeForVets. Accessed 25 March 2012.

Kansas v. Hendricks. 521 U.S. 346, 138 L. Ed. 2d501, 117 S. Ct. 2072. 1997.

Kaimowitz v. Michigan. Department of Mental Health. Div no. 7319434-AW, Cir. Ct. of Wayne County, Michigan. Reprinted in Brooks, A.D. 1974: Law psychiatry and the mental health system. 1973. Boston: Little, Brown and Co.

Munetz, M. R., and P. A. Griffin. 2006. Use of the sequential intercept model as an approach to decriminalization of people with serious mental illness. *Psychiatric services* 57 (4): 544.

Natanson v. Kline. 1960. 300 P.2d 1093, 1104, 1106.

National Commission for the Protection of Human Subjects of Biomedical and Behavioral Research. 1979. *The Belmont report: Ethical principles and guidelines for the protection of human subjects of research.* Washington, DC: Government Printing Office.

O'Brien, L. S. 1998. *Traumatic events and mental health: Chapter 9, Medico-legal aspects of post-traumatic illness. Psychiatry & medicine series*. New York: Cambridge University Press.

Pitman, R. K., and L. F. Sparr. 1998. PTSD and the law. *PTSD Research Quarterly* 9 (2): 1–6.

Porter v. McCollum, 130 S. Ct. 447. 2009. United States v. John Brownfield, No. 08-cr-00452-JLK (D. Colo. 2009).

Roe v. Doe, 93 Misc2d 201, 400 N.S.Y2d, 668 (Sup Ct 1977).

Russell, R. T. 2009. Veterans treatment court: A proactive approach. *New England Journal on Criminal and Civil Confinement* 35:357–372.

Salgo, v. Leland. 1957. Stanford junior university board of trustees. 154 Cal. App 2d 560.

Sparr, L. F., and R. K. Pitman. 2007. PTSD and the law. In *Handbook of PTSD: Science and practice,* eds. M. J. Friedman, T. M. Keane, and P. A. Resick, 449–468. New York: Guilford.

Srinivasaraghavan, J., S. Andrew, N. Watkins, and A. Mahalik. 2002. Variations in court ordered medications in Illinois, USA. XII world congress of psychiatry abstracts Volume 2 PO27-1, 187 in Yokohama, Japan, August 24–29.

Tarasoff v. Regents of the University of California. 1974. 529 P.2d 553, 118 Cal. Rptr. 129.

Tarasoff v. Regents of the University of California. 1976. 17 Cal. 3d 425.

The Management of Post-Traumatic Stress Working Group. 2004. VA/DOD clinical practice guidelines for the management of post-traumatic stress. Version 1.0 Department of Veterans Affairs, Washington, DC. http://www.healthquality.va.gov/ptsd/ptsd_full.pdf/. Accessed 30 Sept 2012.

U.S. Department of Health and Human Services, Code of Federal Regulations, Title 45: Public Welfare. 1983. Part 46: Protection of Human Subjects Regulation Governing Protections Afforded Children in Research (Subpart D).

Young, G., and R. Yehuda. 2006. Understanding PTSD: Implications for court. In *Psychological knowledge in court: PTSD, pain, and TBI,* eds. G. Young, A. W. Kane, and K. Nicholson, 55–69. New York: Springer. doi:10.1007/0-387-25610-5_3.

Zonana, H., J. A. Roth, and V. Coric. 2004. *Forensic assessment of sex offenders: The American psychiatric publishing textbook of forensic psychiatry*. Washington, DC: American Psychiatric Publishing.

Part III
Resources and Opportunities/Community Psychiatry

Chapter 10
Mental Health Programs at Community Level in South Asian Countries: Progress, Problems and Prospects

R. Srinivasa Murthy

As per the broad Regional Strategy of the South East Asian Regional office (SEARO) of the World Health Organisation, in SEAR Member Countries, mental health programs have generally concentrated on hospital-based psychiatry. However, there is an increasing awareness in these countries of the need to shift the emphasis to community-based mental health programs. The WHO Regional Office for South-East Asia is now concentrating on supporting the Member Countries for the development of community-based mental health programs and also for programs on prevention of harm from alcohol and substances of abuse. The programs will be culturally and gender appropriate and reach out to all segments of the population, including marginalized groups. There are many barriers to the implementation of community mental health projects and programs. While some countries have developed mental health policies, there has not been adequate implementation. Governments urgently need to be sensitized on the importance of mental health and to clearly define the goals and objectives for community-based mental health programs. Mental health services should be integrated into the overall primary health care system. At the same time, innovative community-based programs need to be developed and research into relevant issues and traditional practices be promoted. Communities have to be educated and informed about mental and neurological illnesses to remove the numerous myths and misconceptions about these conditions. But most importantly, the stigma and the discrimination associated with mental illnesses must be removed. The Regional Office is developing strategies for community-based programs based on five 'A's: Availability, Acceptability, Accessibility, Affordable medications and Assessment.

10.1 Introduction

Community psychiatry is an important approach to the organisation of mental health care in both economically rich and in low and middle income (LAMI) countries (Bhugra et al. 2005; Thornicroft et al. 2011; IOM 2001; Ghodse 2011). The development of this community psychiatry movement all over the world, is a part

R. S. Murthy (✉)
The Association for the Mentally Challenged, Hosur Road, Bangalore, India
e-mail: smurthy030@gmail.com

J. K. Trivedi, A. Tripathi (eds.), *Mental Health in South Asia: Ethics, Resources, Programs and Legislation,* International Library of Ethics, Law, and the New Medicine 58,
DOI 10.1007/978-94-017-9017-8_10

of series of phases of development of mental health care during the last two to three hundred years, starting from setting up special institutions for the care of the persons with mental disorders (asylums), the humane treatment of the ill persons, deinstitutionalisation, recognition of the rights of the ill persons with mental disorders (WHO 2001; UNCRPD 2006). For the countries of the Region, there is not only the lack of mental health resources, there are additional burden of mental health needs arising from manmade disasters (industrial disasters, conflict, war) and natural disasters (earthquakes, tsunami).

The chapter presents the background of the development of community mental health programs in the countries of the South Asia; the significant developments in each of the member countries; lessons learnt from the country experiences, the challenges and future directions.

10.2 Common Challenges for Mental Healthcare

There are a number of common challenges in developing mental health care programs in the SEA Countries, namely, (i) there is a large 'unmet need' for mental health care in the community (Kohn et al. 2004; World Mental Health Survey 2004; Wang et al. 2007); (ii) there is poor understanding of the psychological distress as requiring medical intervention in the general population (Wig 1997; Thornicroft et al. 2009); (iii) there is limited acceptance of the modern medical care for mental disorders in the general population (Sebastia 2009); (iv) there are severe limitations in the availability of mental health services (professionals and facilities) in the public health services (WHO Atlas 2011); (v) there is poor utilisation of the available services by the ill population and their families (Udomratn 2011); (vi) there are problems in recovery and reintegration of the person with mental illness (Mendis 2011) and (vii) institutionalised mechanisms (policy, legislation, funding etc) for organisation of mental health care are not adequate in most of the countries (NHRC 1999; Saraceno et al. 2007; Mercier 2007; Minhas 2007; WHO Atlas 2011).

10.3 International Development of Mental Health Services

The development of mental health care all over the world is best described as a developing process. In the developing story, at this point in history, there is focus on community based mental health programs in all the countries of the world (Thornicroft et al. 2011) and rights of the person with mental illnesses. The World Health Report, 2001 recognised the changes over the last two centuries as follows:

> Over the past half century, the model for mental health care has changed from the institutionalization of individuals suffering from mental disorders to a community care approach backed by the availability of beds in general hospitals for acute cases. This change is based

> both on respect for the human rights of individuals with mental disorders, and on the use of updated interventions and techniques. The care of people with mental and behavioural disorders has always reflected prevailing social values related to the social perception of mental illness.

The reasons for these changes were:

> During the second half of the Twentieth century, a shift in the mental health care paradigm took place, largely owing to three independent factors, namely (i) psychopharmacology made significant progress, with the discovery of new classes of drugs, particularly neuroleptics and antidepressants, as well as the development of new forms of psychosocial interventions; (ii) the human rights movement became a truly international phenomenon under the sponsorship of the newly created United Nations, and democracy advanced on a global basis, albeit at different speeds in different places and (iii) social and mental components were firmly incorporated in the definition of health of the newly established WHO in 1948. These technical and socio-political events contributed to a change in emphasis: from care in large custodial institutions to more open and flexible care in the community.

The WHO 2001, described community care as follows:

> Community care is about the empowerment of people with mental and behavioural disorders. In practice, community care implies the development of a wide range of services within local settings. This process, which has not yet begun in many regions and countries, aims to ensure that some of the protective functions of the asylum are fully provided in the community, and the negative aspects of the institutions are not perpetuated.

The changes in mental health services in economically rich countries, during the second half of the last century and especially during the last ten years, have been largely driven by the movement towards ensuring human rights of the person with mental disorder. This is reflected in deinstitutionalisation, care in the community and greater voice to users and the carers (UNCRPD 2006; IASC 2007).

Thornicroft and Smuzler, in their Textbook of Community Psychiatry (2001), summarised the situation as follows:

> The practice of psychiatry in the second half of the 20th century, and especially in its last decade, has changed fundamentally. Mentally ill people have been moved out of the relative 'simplicity' of the large institution, with its clear structures and hierarchies and into the community. This necessitated new types of relationships between 'health' and 'social' care. A range of new facilities has been required for the treatment, care and support for people with mental health problems in the community, replacing many of the functions previously provided in hospitals. More agencies and staff (professional and non-professional) have declared an interest and entered the scene, often bringing new and quite different perspectives on the needs of those with mental disorders. Among these new voices in the community have been those of service users themselves. Increasing cultural diversity and respect for social difference have added to the range of value systems to be taken into account. At the same time, governments have taken increasing direct interest in mental health issues, formulating ever more specific strategies, guidance, directives and legislation. (Thornicroft and Smuzkler 2001).

Another reflection of the shift in focus, is the large number of initiatives in describing the mental disorders, understanding the 'disability' of the mentally ill persons, 'impact' of the ill person on the family, 'coping' by family, 'user movement', the recently accepted UN Convention on the Rights of Persons with Disabilities (UNCRPD 2006) and the 'recovery model' for mental health care.

Szmukler and Thornicroft (2001) define community psychiatry as follows:

> Community psychiatry comprises the principles and practices needed to provide mental health services for a local population by (i) establishing population-based needs for treatment and care; (ii) providing a service system linking a wide range of resources of adequate capacity, operating in accessible locations and (iii) delivering evidence based treatments to people with mental disorders.

It is salient to note, in the above definition, the significant parts are **needs of the population**, **wide range of services** and **accessibility of services**.

Community psychiatry in the rich countries has come to represent a wide range of initiatives beyond that provided by mental health professionals. For example, a recent book **'Empowering people with severe mental illness'** (Linhorst 2006) covers patient empowerment includes treatment planning, housing, organisational decision making, policy making, employment, research and service provision.

Similarly, the **'Textbook of Community Psychiatry'** (Thornicroft and Smuzkler 2001) and the second edition titled, **'Oxford Textbook of Community Mental Health'** (Thornicroft et al. 2011) covers a wide variety of subjects. For example, under the service system, the following areas are included: integration of components into the systems of care (multidisciplinary teams, sectorisation, generic versus specialist teams, training for competence). Under the service components (case management and assertive community treatments, emergency psychiatric services, partial hospitalisation, day care and occupation, residential care, out-patient treatment, in-patient treatment). Under the interfaces between mental health services and the wider community (primary care, integrated health and welfare services, community alliances); and users and carers as partners.

10.4 Development of Mental Health Services in South East Asian (SEA) Countries

In contrast to the economically rich countries, the development of community psychiatry in SEA countries, occurred against the background of almost no organised mental health services (Swift 1972; de Jong 1996; Bhugra et al. 2005; Roderigo et al. 2005; Trivedi et al. 2007; Ghodse 2011). There were special challenges of lack of awareness in the community, existing systems of traditional care, stigma, poorly functioning institutions (considered in detail in the Chapter of Dr. K.S. Jacob) (NHRC 2008). Almost all of the persons with mental disorders are living in the community, most often without any organised services, with the family providing the care in whatever form they are able to do (ranging from isolation, restraint to committed care). In a way, community psychiatry has developed in these countries as ***'the service'*** and not an 'alternative' to institutionalised care. This distinction of the development of community psychiatry is important, to understand the developments in the countries under consideration in this chapter.

The following section presents the community mental health initiatives of each of the countries of the Region. During the last few years, WHO developed country mental health reports as part of **'Assessment of Mental Health Services'** (WHO-AIMS 2003) which provide comprehensive account of mental health services. The following country reports are drawn from these reports along with other publications.

10.4.1 Afghanistan (WHO 2006)

Afghanistan has been in a conflict situation for the last few decades. Afghanistan has a national mental health plan, policy and legislation since 1987, a National Mental Health Strategy (NMHS) since 2009, which addresses the main mental health issues (MOH 2009).

The NMHS outlines the mental health needs of the population.

> Recent data indicate that among the people of Afghanistan, 50 % of them are suffering from mental problems due to thirty years of conflict.

Further, the NMHS, recognises the impact of these development as follows:

> The effects of trauma, the rise of substance abuse, suicidal tendencies, and violence can affect psychological functioning on all levels. These effects can also prevent normal development in a significant number of children. Without appropriate and timely intervention, it is possible that society at large will become even less functional over time. The disintegration of basic values as a result of prolonged war and conflict has contributed to a loss of the social and cultural identity, changing values and gender roles. Issues of honor and dignity, as well as the feeling of having little control regarding these situations as well as over one's own emotions possibly due to traumatic war experiences, often strengthens the restrictive side of tradition. As a result, domestic violence rises and excessive control of women and children easily produces new trauma. This cycle continues to the extent that reconciliation and the prospect of a peaceful society seem to be out of reach. Somatization and a lack of awareness of the origin of the suffering can lead to the belief that only medication can help. This problem is reinforced by the limited knowledge of medical staff regarding mental health and psychosocial issues. This limited knowledge leads to misidentification and misdiagnosis and in turn poses a significant barrier for client access to appropriate services. It adds to the victimized state of the Afghan population and creates new dependencies. People drift into social isolation and the family as a self-helping system breaks down.

The main aims of the National Mental Health Strategy are:

- To promote mental health of the people of Afghanistan;
- To minimize the stigma and discrimination attached to mental disorders;
- To reduce the impact of mental disorders on individuals, families, and the community;
- To prevent the development of mental health problems and mental disorders, wherever possible;
- To provide quality, integrated, evidence and rights based care for individuals suffering from mental disorders at all levels of health system.

The vision of the Mental Health Strategy (2011–2015) is to access and utilize community based mental health care services for all Afghans in all levels of the health system till 2020.

The expected benefits of the integration of psychosocial care are fewer chronic patients and fewer suicides; increased cost effectiveness; reduced dependency on medication and reduced hospital admissions; prevention of ill health; reduced domestic violence; positive impact on addiction; and reduction of social stigma connected to mental health problems.

The majority of users are treated in outpatient facilities and in inpatient units. Schizophrenia and other disorders are most common in inpatient units, while in the mental hospital substance use disorders and schizophrenia are seen most frequently. Less than one percent of training for doctors is devoted to mental health in comparison of 2% of nurses. Four percent of primary care doctors and 1% of nurses have received at least 2 days of refresher training in mental health in 2004. Only doctors can prescribe psychotropic medications in primary care settings. There are only two qualified psychiatrists, 61 other doctors, 37 nurses and 40 other mental health workers in the mental health services in country. There is no financial or legislative support for people with psychiatric problems. Almost all mental health outpatient facilities are located in major cities. In rural areas, primary healthcare staffs have either limited training in mental health or mental health services are entirely unavailable. Limited resources are available for mental health and much of these resources are directed towards the mental hospital, leaving inpatient and outpatients facilities under funded. Most of the resources have been spent for training of primary care staff while no supervision and monitoring systems have been established.

The Ministry of Public Health supports a standard training program based on the "Basic Package for Health Services" or BPHS for primary care. There is a mental health module. However, the training for primary cares staff is inadequate. Although an essential psychiatric medicines list is available in the country, these medications have not been available in primary health care facilities due to postponement of implementation of the mental health component of BPHS. Family and consumers associations do not exist in the country while links between the mental health sector and other sectors are very limited. The interaction between mental health providers and primary health care staff is limited to trainings conducted only occasionally. There is only one-day treatment facility available in the country. There are 5 community-based inpatient units available in the country for a total of 0.28 beds per 100,000 populations. Twenty percent of the community-based psychiatric inpatient units had at least one psychotropic medicine of each therapeutic class available in the facility or a nearby pharmacy. There are no community residential facilities available in the country. At the present time, there is no legislative or financial support for persons with mental disorders for provisions for employment, provisions against discrimination at work, or provisions for housing. Extensive efforts have been made to develop the manuals, records and other mental health support materials during the recent years by the International Medical Corps (IMC) and the European Union support. There are postgraduate training centres for three year training of psychiatrists in Kabul and other Regions.

A striking aspect of Afghanistan is the availability of information about the prevalence of mental disorders (Scholte et al. 2004; Srinivasa Murthy 2007), and stresses of daily life in the conflict situation (Miller et al. 2008; Miller and Rasmussen 2010; Panter-Brick et al. 2008, 2009, 2011). There have been a number of psychosocial care initiatives to address the mental health needs of adults and children. One of the important initiatives taken in 1998 was to train non-specialist in mental health care (Mohit et al. 1999). Recently, a very important innovative study comparing routine care and counselling by trained counsellors has been completed (Ayoughi et al. 2012). At 3-month follow-up, psychosocial counselling patients showed high improvements with respect to the severity of symptoms of depression and anxiety. In addition, they reported a reduction of psychosocial stressors and showed an enhancement of coping strategies. At the same time, the severity of symptoms, the quantity of psychosocial stressors and coping mechanisms did not improve in patients receiving routine medical treatment.

10.4.2 Bangladesh (Karim et al. 2011)

Bangladesh has severe limitations in the availability of trained professionals and psychiatric facilities. There are only 50 outpatient mental health facilities and no facility provides follow-up care in the community. There are no day treatment mental health facilities in the country. There are 31 community-based psychiatric inpatient units for a total of 0.58 beds per 100,000 population. There are 11 community residential facilities in the country and 55 % of the beds in these facilities are for children and adolescents and 81 % of admitted patients are female and 73 % of them are children. There is one 500 bedded mental hospital in the country and on average patients spend 137 days in the hospital. There are 15 beds for mentally disordered people in forensic inpatient units and 3900 beds in other residential facilities such as homes for persons with mental retardation, detoxification inpatient facilities, homes for the destitute, etc. The density of psychiatric beds in or around Dhaka, the largest city, is 5 times greater than the density of beds in the entire country. Four percent of the training for medical doctors is devoted to mental health, in comparison to 2 % for nurses. Most of the primary health care clinics are physician based and they make few referrals to mental health professionals. No health staffs except doctors are allowed to prescribe psychotropic medications.

There is no consumer association in the country but there is 1 family association with 40 members. There are about 10 NGOs in the country involved in individual assistance activities in mental health. There are no legislative and financial provisions to protect and provide support for mental health service users in respect of employment and protection of rights. There are 11 community residential facilities available in the country for a total of 0.92 beds/places per 100,000 population. Both physician based primary health care (PHC) and non-physician based PHC clinics are present in the country. In terms of physician-based primary health care clinics, all or almost all clinics (81–100 %) have assessment and treatment protocols for key mental health conditions available, in comparison to only a few clinics (1–20 %) in non-physician based primary health care. A few (1–20 %) of physician-based primary health care doctors make on average at least one referral per month to a mental health professional. Some of non-physician based primary health care clinics (1–20 %) make a referral to a higher level of care. In terms of professional interaction between primary health care staff and other care providers, some (21–50 %) of primary care doctors have interacted with a mental health professional at least once in the last year. There is no consumer association. There is no interaction between mental health facilities and consumer or family association. In addition to consumer and family associations, there are 10 NGOs in the country involved in individual assistance activities such as counselling, housing, or support groups. There is a coordinating body (National Institute of Mental Health, Dhaka) to oversee public education and awareness campaigns on mental health and mental disorders.

10.4.3 Bhutan (WHO-AIMS 2006)

Organised mental health is less than two decades old (I had the opportunity to be part of the initiatives in 1987 and 1999). There is no mental hospital in Bhutan. Overall, resources for the mental health system are very scarce, and the only way to improve and sustain mental health services is to integrate mental health care with general health care services. The mental health care system is community oriented. However, due to a lack of trained health workers at the primary health care level, patients have been treated mainly in the district and referral hospitals during the last few years. With the completion of training of all non-physician health workers in the coming few years, it is expected that mental care services will reach the community in the whole country. There are 63 outpatient mental health facilities available in Bhutan. All 63 facilities are fully integrated with mental health in-patient units. Mental health outpatient facilities treat 5,266 people or 783 users per 100,000 population. The primary diagnoses of users treated through mental health outpatient facilities are mood (affective) disorders (22 %) and neurotic, stress-related and somatoform disorders (17 %). All outpatient facilities provide medication. In terms of available treatments, only some (21–50 %) of the patients in outpatient facilities have received one or more psychosocial interventions in the past year.

Diagnoses of patients admitted to community-based psychiatric inpatient units are primarily from the following three diagnostic groups: mood disorders (32 %), mental and behaviour disorders due to psychoactive substance use including alcohol (27 %), and schizophrenia (19 %). 23 % of all admissions to community-based psychiatric inpatient units are involuntary. The average number of days spent in the community-based psychiatric inpatient units per discharge is 27.71. Of all admitted patients, less than 1 % was physically restrained or secluded at least once in the past year in these facilities. A majority (51–89 %) of the community-based psychiatric inpatient units received one or more psychosocial interventions in the past year. All community-based psychiatric in-patient units (hospitals and a few

Basic Health Units) had at least one psychotropic medicine of each therapeutic class (anti-psychotic, antidepressant, mood stabilizer, anxiolytic and antiepileptic) and 30 Basic Health Units had all psychotropic drugs, with the exception of mood stabilizer drugs available in the facility. There are no community residential facilities in Bhutan. Additional residential facilities include: a 50-bed facility for juvenile delinquents, a 15-bed facility for drug detoxification and rehabilitation, and a five-bed residential facility for the elderly. Non-doctor/non-nurse primary health care workers are only allowed to prescribe the psychotropic medications supplied by the government to the health centres. In contrast, primary health care doctors are allowed to prescribe without restriction. There are no consumer and family associations in Bhutan. Nonetheless, families of mental health patients are involved with the treatment and management of individual cases. There are two NGOs in the country involved in individual assistance activities such as counselling, housing, or support groups. One NGO is working with abused women, while the other is working with substance abuse patients. At present no legislative and financial provisions exist to protect and provide support for person with mental disorder in Bhutan.

There has been a recent innovative experience in providing mental health care in the community. Acceptance of modern medicine, particularly psychiatric treatment is a huge challenge in a traditional society like Bhutan, where most people prefer traditional forms of treatment. However, traditional treatment is not effective in treating severe mental disorders such as schizophrenia. To address this problem, an innovative pilot project to introduce modern psychiatry to Bhutan was launched in 2002. As a first step, focus group discussions with community leaders, traditional healers, and health providers modeled on international best practices (including case identification according to ICD 10 diagnostic criteria) helped identify persons with mental illness in the community. This represented the first time when community leaders and traditional healers were involved in case identification. Local health workers followed up on these cases by providing a diagnosis and offering treatment under the supervision of trained psychiatrists. This pilot project was conducted in three distinct geographical regions in Bhutan. The experiences and knowledge gained during the pilot is presently being expanded to the remaining 17 districts in Bhutan. The team created sustained awareness and advocacy among community leaders and traditional healers about the prevalence of mental illness in the community and the unmet needs of patients. Community leaders and traditional healers in particular started referring severely ill mental patients to modern doctors for treatment. This resulted in traditional and modern health workers working together successfully and understanding each other better. The pilot project brought the practitioners of the two systems together, appreciating each others strengths and weaknesses and collaborating and consulting each other for a common cause—to provide the best treatment and avoid duplication and confusion among patients. The project identified, diagnosed, and initiated treatment in more than 300 severely mentally ill patients in the community who otherwise would not have received treatment. The primary health workers had the opportunity to learn from real patients and get hands-on training to diagnose and provide treatment to these patients in their own homes and communities.

Through this process of advocacy and by generating data on mental illness, the political leaders and government bureaucrats were sensitized about the mental health needs of the population and that mobilizing resources to address those needs can have a sustainable impact. The pilot project took about a month to complete in each district. The first few days were spent on training local health workers. The next stage involved extended focus-group discussions to identify potential cases in their community. The third phase was the visit to households to confirm diagnosis and initiate treatment. More than 300 severe mentally ill patients were identified and put on medication. More than 60 health workers were trained on mental health through this project. More than 500 community leaders, traditional healers and government civil servants were oriented to mental health concepts through this method.

As a result of these efforts, awareness of mental health, mental disorders and help seeking behaviour have increased among the Bhutanese population. In the three pilot districts, treatment and social support to patients has increased significantly. Health workers have become more open and receptive and government funding to mental health programs has increased. This initiative has received recognition of an international award as one of the innovations in mental health care (Dorji 2008, Personal communication)

10.4.4 India

The community mental health initiatives of India cover (i) family involvement in mental health care; (ii) setting up of general hospital psychiatry units; (iii) integration of mental health with general health care; (iv) utilisation of non-specialists for mental health care; (v) community level rehabilitation; (vi) public mental health education; (vii) use of traditional practices and (viii) mental health research (Agarwaal et al. 2004; Thara 2002, 2007; Thara et al. 2008; Kapur 2004; Jadhav and Jain 2009; Srinivasa Murthy 2011; Patel and Saxena 2011).

During the last 50 years, the place of mental health as part of the general health has significantly changed. From a situation of no organised mental health care at the time of Independence, over the last seven decades, mental health issues are actively seen as part of the public agenda in the various areas. Some of the examples are (i) the formulation of the National Mental Health Program (Government of India 1982), (ii) integration of mental health with primary health care at the district levels (Government of India 2000), (iii) the adoption of Mental Health Act, in 1987, and Persons with Disabilities Act in 1995, (iv) Supreme Court judgements about mental hospitals and attempted suicide, (v) voluntary agencies initiatives in the areas of self-help groups, rehabilitation, drug dependence and suicide prevention (Patel and Thara 2003) and (vi) wide media coverage of mental health issues. The overall effect has been the recognition of mental health as an important issue in the community and services to move beyond mental hospital care to care in the community. Mental health professionals in India view community mental health not as care in the community but as care utilising the community resources, or in other words, care by the community.

At the time of India's Independence, there were almost no mental health services in the country. The initial period of 1947–1966 focussed on doubling of the psychiatric beds, along with development of training centres in order to train psychiatrists, clinical psychologists, psychiatric social workers and psychiatric nurses. The period of 1960–70 saw the emergence of general hospital psychiatric units in a big way both in terms of service provision and training (Wig 1978). The community psychiatry initiatives were taken up initially in the 1970s and in a big way from the 1980s, following the adoption of the National Mental Health Program (NMHP) in August 1982 and the community level initiatives starting from the 1980s (Patel and Thara 2003).

A striking aspect of the development of mental health services in India, is as much the location of the care in the community (where most of the ill persons were already living) as the utilisation of a wide variety of community resources of the community. For example, in the initial phase, family members were the focus, followed by the utilisation of the existing general health care infrastructure through integration of mental health services with general health services. This was followed by involvement of school teachers, volunteers, counsellors, mentally ill persons, survivors of disasters, parents of children with mental disorders, the personnel of education system (Srinivasa Murthy 2006).

At the time of Independence in India, when most of the ill patients were living with their families or in the community, what challenged the psychiatric community was the need to provide care with almost no specialised personnel and facilities. Recognising the cultural feature of commitment of the families to their ill family members, psychiatrists looked to the family members of the ill persons as the answer. India is a pioneer in involving the family members in the care of their ill relatives from the early 1950s. This occurred first at the Amritsar Mental Hospital (Carstairs 1974; Srinivasa Murthy 2007) soon followed by the Mental Health Centre, at Vellore and mental hospital at Bangalore. Indian initiatives relating to families and mental health care have depended on the cultural practice of family support for the mentally ill persons. Since 1950s families have been formally included to supplement and support the psychiatric care by professionals. During the initial period, literally family members were admitted along with mentally ill to be part of the care of the patients. Following this initial effort, during the 1970s and 1980s, efforts were made to understand the functioning of families with an ill person in the family and their needs. Two centres, namely Postgraduate Institute of Medical Education and Research (PGIMER) Chandigarh, and National Institute of Mental Health and Neurosciences, Bangalore systematically studied the needs of the families, the role of non-medical professional to provide support to the families (Pai and Kapur 1982, 1983; Pai et al. 1983, 1985). Research into the special needs of mentally retarded and their families has also been studied. During the last twenty years, a more active role for families is emerging in the form of formation of self-help groups and professionals accepting to work with families in partnership (Srinivasan 2008; Srinivasa Murthy 2006). It is very interesting that during the last decade the value of family involvement in mental health care in developed countries has been recognised.

The setting up psychiatric services in general hospitals is an important initiative in India. The development of organised mental health care is essentially a post-independence phenomenon. Though, the first 15 years of the Independence saw doubling of the mental hospital beds to 20,000, the pharmacological advances in the treatment of mentally ill persons and the closing down of the mental hospitals in the western countries, gave a big push to the development of general hospital based psychiatric services. The initial GHP Units in Calcutta and Bombay came in the 1930s and 1940s. The big spurt in the GHP units occurred in the 1960 at academic centres at Chandigarh, Delhi, Madurai, Lucknow. These centres also became centres for training of psychiatrists and for mental health research (Wig 1978). It has been a slow and silent change but in many ways a major revolution in the whole approach to psychiatric treatment. The general hospitals psychiatric unit offered numerous advantages over traditional mental hospital. They are more accessible, easily approachable and above all less stigmatized. They are situated right in the community and they are more accessible and easily approachable. Families can frequently visit and relatives can even stay with disturbed patients. There is no stigma of mental hospital. There are limited legal restrictions on admission or treatment. Ambulatory treatment on out-patient basis is available with the use of drugs, ECT and psychotherapy. Proximity of other medical facilities ensures thorough physical investigations and early detection of physical problems. In the last one decade, psychiatric units in all major hospitals have become a reality. This shifting of the place of care to the general hospital setting has contributed significantly to the process of de-stigmatisation of psychiatric illnesses and psychiatric care and made psychiatric services more accessible and acceptable to the population.

The most important Indian initiative was to integrate mental health with general health services. The initial stimulus to this approach came from the recommendations of WHO in 1975, in the Expert Committee report ***'Organisation of Mental Health services in developing countries'*** (WHO 1975). The chief recommendations were to (i) develop country mental health plans; (ii) to choose priorities for mental health care; (iii) include mental health tasks in all health care personnel; (iv) provide essential psychiatric drugs in health care facilities; and (v) develop appropriate legislative support for these initiatives. India was the first developing country to formulate a national mental health program (NMHP) in 1982. Twenty five years later, WHO again reemphasised the approach through the recommendation in the World Health Report, 2001 to ***'provide treatment in primary care'.***

Training primary health care workers for mental health was started in 1975 at Chandigarh and Bangalore centres and integrating mental health with general health care (Srinivasa Murthy et al. 1978; Wig and Srinivasa Murthy 1980; Wig et al. 1981; Parthasarathy et al. 1981; Issac and Kapur 1980; Climent et al. 1980; Giel et al. 1981a, 1981b; Chandrashekar et al. 1981; Issac et al. 1982, 1988, 2009; Kapur 1975, 1997, 2004; Wig et al. 1980; Ladrido-Ignacio et al. 1983; Sartorius and Harding 1983; Harding et al. 1983a; Srinivasa Murthy and Wig 1983; Harding et al. 1983a, b, Ladrido-Ignacio et al. 1983). These initiatives not only brought to focus the unmet needs of the rural mentally ill persons and the feasibility of reaching services through general health services. These initial experiences formed the basis of the National Mental Health Program (NMHP) formulated in 1982. Currently the government supports over 125 district level programs in 22 states, covering a population of over 200 million. Following initial studies, other efforts to understand the integration of mental health with primary health care have occurred (Chisholm et al. 2000; James et al. 2002). Further even the funds allotted were not fully utilized. It was only in the 9th Five year plan that a substantial amount of Rs. 28 crores was made available and it was increased in the 10th five year plan to about Rs. 140 crores. The availability of funds in 1995 for the DMHP has shown that once funds are available States are willing to take up intervention programs and professionals are willing to be part of integrating mental health with primary health care and taking up a wide variety of initiatives for mental health care.

Involvement of 'non-specialists' in mental health care has been another important initiative of India. Limited human resources in terms of mental health specialists have been a perpetual barrier to providing mental health care to all the needy. Recognising the need to develop services to reach the total population, against the background of limited trained professionals, professionals have utilised a large variety of community resources for delivery of focussed mental health care (Srinivasa Murthy and Wig 1983; Srinivasa Murthy 2006). These have included health workers, school teachers, volunteers, lay workers with specific training to care for specific groups like persons with dementia (Ranganathan 1996, 2011). A large number of mental health resources have been developed for the training of non-specialists (Issac et al. 1984, 1994; Sharma 1986; Srinivasa Murthy et al. 1987). A recent document bringing together over three dozen experiences of 'Mental health by the people' (Srinivasa Murthy 2006) the initiatives have not only the health and education sectors but also included the family/carer initiatives for mentally ill and mentally retarded, the parent movement for learning difficulties, initiatives to reach the elderly population, suicide prevention by volunteers, disaster mental health care by non-professionals, efforts by voluntary agencies to fight stigma and discrimination.

One other important development is the increasing role of the voluntary organisations in developing small size locally relevant community based psychiatric care facilities like day care centres, vocational training centres, sheltered workshops, half-way homes and long stay homes (Nagaswami 2012; Patel and Thara 2003). These facilities have the advantage of limiting long term institutional care, incorporating the cultural sensitivities of the clientele, and utilisation of the local resources.

In the absence of modern mental health care, majority of the population has taken the help of traditional healers. The attitude of professional to traditional mental health care has been one of significant ambivalence (Sebastia 2009; Verghen et al. 2010). It is significant that no studies have been made in the recent years.

Developing programs to educate the general population about the modern understanding of mental disorders and their treatment has been an important activity of professionals. These efforts have been directed not only to fight stigma and discrimination but also to promote mental health, through mental health literacy efforts. There is a wide use of the mass media for these efforts in addition to folk measures. Notable are the program 'DATE' in the 1980s, the currently running national level TV based program, 'Mann Ki Baat'(over 30 episodes) and the state level program in Karnataka, titled 'Manochintana'(over 70 episodes) are taking mental health information to the general population. The efforts of individual psychiatrists and other professionals to write books for the general public have been very impressive (Wig 2006). The next frontier, in this area is for the wider use of the information technology and communication (mass media, mobiles, print media). This area has great potential to bring about changes in the general population along with a potential to stimulate 'self-care' and 'informal care'.

Research has formed an important part of the psychiatry movement in India. The efforts were linked to the goals of national mental health programs (ICMR 1982; Srinivasa Murthy 2004; ICMR 2005; ICMR-DST 1987). Two notable examples are the 'ICMR-DST Severe Mental Morbidity' study in the 1980s, and the setting up of the 'Advanced Centre for Research on Community Mental Health' by ICMR (1985–1991). The other efforts have been towards an understanding of mental disorders and the role of biological and psychosocial factors. The Indian Council of Medical Research provided valuable support with a large number of research projects directly and indirectly related to the emerging mental health program during the 1970s and 1980s. Research on course and outcome of schizophrenia, acute psychoses, old age psychiatric problems and community psychiatry added the local knowledge to influence the NMHP (Verghese et al. 1989; Chatterjee et al. 2003; Srinivasa Murthy et al. 2004; Thirthahalli et al. 2006, 2009, 2010).

10.4.5 Indonesia

The main community mental health approach adopted in Indonesia is the integration of mental health with primary health care. The other area of community mental health activity is the disaster mental health initiatives following the 2004 tsunami (WHO 2005).

10.4.6 Maldives (WHO 2006)

Health services are organized in Maldives in a four-tier system. At the central level there are two large general hospitals in Male. There are 6 regional hospitals, 10 atoll hospitals, 63 health centres, and 127 health posts. A large percentage of doctors are expatriates, with a high turnover.

There is one psychiatric day treatment facility available in the country. There is one community-based psychiatric inpatient unit available in the country for a total of 1.38 beds per 100,000 population. There are no community residential facilities available in the country. Government agencies and NGOs have promoted public education and awareness campaigns in the last five years. These campaigns have targeted the general population, adolescents, trauma survivors, and other vulnerable or minority groups. In addition, there have been public education and awareness campaigns targeting professional groups including healthcare providers, teachers, and social services staff. In terms of provisions for housing, priority is given to people with a physical disability, but not a mental disability.

The mental health facilities in the Maldives are generally community based and include a community-based psychiatric inpatient unit, a mental health outpatient facility, and a day treatment program. However, the geographical layout of the Maldives makes it difficult for users outside of Male to access these services. Strength of the mental health system in the Maldives is that psychotropic medicines are provided free of cost to the entire population. Essential psychotropic medicines of each therapeutic class are available in the community-based inpatient and in some (21–50 %) of the primary health care clinics, unit but are not available in the mental health outpatient unit. A primary weakness of the mental health system is that the integration of mental healthcare into the primary health care system is limited. There is little training on mental health issues for primary care staff and treatment protocols on key mental health conditions are not available in primary care clinics. The availability of human resources in Maldives is also limited. In particular, there are no psychosocial staffs working in the mental health facilities. The Maldives also lacks a mental health policy, plan, and legislation. A disaster preparedness plan for mental health and psychosocial care has been drafted (in 2005). A human rights review body exists, but has no authority to impose sanctions.

10.4.7 Myanmar (WHO 2006b)

There are 25 outpatient mental health facilities, 2 day treatment facilities, 17 community based psychiatric inpatients units and 2 mental hospitals. The majority of beds in the country are provided by mental hospitals, followed by residential units. The diagnosis of schizophrenia and neurotic disorders are the most frequent diagnoses in out-patient facilities, and schizophrenia and mood disorders are the most common diagnoses in mental hospitals. Essential psychotropic drugs from each therapeutic class are available in inpatient units, mental hospitals and out-patient facilities. In terms of refresher training on mental health, 1 % of primary health care doctors, 3 % of nurses, and 2 % of non-doctor/non-nurse primary health care workers have received at least two days of training. Non-doctor/ non-nurse primary health care workers are not allowed to prescribe psychotropic medications in any circumstance. Nurses are allowed to prescribe, but with restrictions; they are not allowed to initiate prescription but are allowed to continue prescription. Primary health care doctors are allowed to prescribe psychotropic medications without restrictions. There have been public education and awareness campaigns targeting professional groups including health care providers (traditional medicinal medicine, conventional, and modern).There was an awareness campaign in Nyaungdon Township for epilepsy with the support of WHO. There is one functioning day treatment facilities available which is reserved for children and adolescents only. The functioning day treatment facility is The School for Disabled Children, which is for children with autism, Down syndrome, mental retardation, and cerebral palsy. There are 17 community-based psychiatric inpatient units available in the country.

One percent of the training for medical doctors is devoted to mental health, in comparison to 13 % for nurses and 0.28 % for non-doctor/non-nurse primary health care workers. In terms of refresher training, 1 % of primary health care doctors have received at least two days of refresher training in mental health, while 3 % of nurses and 2 % of non-doctor/non-nurse primary health care workers have received such training. There are neither users/consumers associations, nor family members associations. The government does not provide economic support for either consumer or family associations. There are two also other NGOs, Myanmar Anti Narcotic Association (MANA) and Myanmar Maternal and Child Welfare Association (MMCWA) which are involved in individual assistance activities such as counselling and support groups. There is a coordinating body to oversee public education and awareness campaigns on mental health and mental disorders. The government agencies have promoted public education and awareness campaigns. The NGO's also involved in this work are MMCWA and MANA.

10.4.8 Nepal (Regmi et al. 2004)

Community mental health care initiatives started in the 1980s (Wright et al. 1989). Nepal's mental health policy was formulated in 1996. Key components of the policy include:(1) to ensure the availability and accessibility of minimum mental health services for all the population of Nepal; (2) to prepare human resources in the area of mental health; (3) to protect the fundamental human rights of the mentally ill; and (4) to improve awareness about mental health.

Mental health services are not organized in terms of catchment/service areas. There are 18outpatient mental health facilities available in the country. There are 3 day treatment facilities available in the country. There are 17community-based psychiatric inpatient units (i.e. general hospitals and teaching hospitals) available in the country. There is one mental hospital with a total of 0.20 beds per 100,000 population. The majority of users are treated in outpatient facilities. Both physician based primary health care centres (PHC) and non-physician based PHC clinics provide primary health care (negligible mental health services) services in the country. In terms of training for primary health care staff, two percent of the training for medical doctors is devoted to mental health, and the same percentage is provided for nurses. In terms of mental health financing, less than one percent of health care expenditures by the government are directed towards mental health. There is no human right review body with the authority to inspect mental health facilities and impose sanctions on those facilities that persistently violate patients' rights.

One NGO is running a community mental health service in 7 of the 75 districts of the country. In these 7 districts, primary health care workers have received mental health training and refresher trainings. In other districts, community mental health services are not available, as mental health service is not yet integrated in the general health service system. The distribution of human resources between urban and rural areas is disproportionate. The density of psychiatrists in or around the largest city is 8.52 times greater than the density of psychiatrists in the entire country, while the figure for nurses is 6.56. There are no active consumer associations fighting for mental health issues in the public arena. Two percent of the training for medical doctors is devoted to mental health, and the same percentage is provided for nurses. In terms of refresher training on mental health, none of the primary health care doctors have received such training. Both physician based primary health care (PHC) and non-physician based PHC clinics are present in the country. In terms of physician-based primary health care clinics, some of the clinics (between 21–50 %) have available assessment and treatment protocols for key mental health conditions. In comparison, only a few clinics (between 1–20 %) in non-physician-based primary health care have these protocols.

Primary health care nurses, non-doctor/non-nurse primary health care workers are allowed to prescribe but with restrictions. For example, they are not allowed to initiate prescription but are allowed to continue prescription. In addition to that, primary health care doctors are allowed to prescribe without restriction. As for availability of psychotropic medicines, a majority (51–80 %) of physician-based PHC clinics have at least one psychotropic medicine of each therapeutic category, in comparison to some (21–50 %) of the non-physician based clinics. There is no coordinating body to oversee public education and awareness campaigns on mental health and mental disorders. Government agencies, NGOs, and professional organizations have all promoted public education and awareness campaigns in the last five years. These campaigns have targeted the general population, as well as children and adolescents. In addition, there have been public education and awareness campaigns targeting professional groups, including teachers and healthcare providers. There is no mental health legislation protecting the right of patients with mental disorders as yet. However, a draft of the mental health legislation has been prepared and at present it is in the Ministry of Health and Population for revision and finalisation (Shyangwa and Jha 2011).

10.4.9 Pakistan (WHO 2009; Mubbashar 2001, 2011)

Pakistan has been a pioneer in developing community mental health programs since 1986. The chief approaches used are (i) integration of mental health with general health services; (ii) school mental health program; (iii) public mental health education and (iv) collaboration with traditional healers. Pakistan's mental health policy was last revised in 2003. The mental health plan was also last revised in 2003. The disaster/emergency preparedness plan for mental health was last revised in 2006. The revised mental health legislation was enacted in 2001 and it focused on the access to mental health care including access to the least restrictive care; rights of mental health service consumers; family members, and other care givers; competency, capacity, and guardianship issues for people with mental illness; voluntary and involuntary treatment; accreditation of professionals and facilities; law enforcement and other judicial system issues for people with mental illness; mechanisms to oversee involuntary admission and treatment practices; and mechanisms to implement the provisions of mental health legislation. A national mental health authority exists which provides advice to the government on mental health policies and legislation. The mental health authority also is involved in-service planning, service management and co-ordination and in monitoring and quality assessment of mental health services. Mental health services are not organized in terms of catchment/ service areas. Five mental hospitals are available in the country; these are organizationally integrated with mental health outpatient facilities.

The government provides economic support for both consumer and family associations. Family associations have been involved in the formulation or implementation of mental health policies, plans, or legislation in the past two years. Government agencies (e.g., Ministry of Health or Department of mental health services); NGOs; professional associations, private trusts; and foundations, International agencies have promoted public education and awareness campaigns in the last five years. 1 % of all admissions to community-based inpatient psychiatric units and 30 % of all admissions to mental hospitals are involuntary. Over 20 % of patients were restrained or secluded at least once within the last year in community-based psychiatric inpatient units, in comparison to 6–10 % of patients in mental hospitals. The density of psychiatric beds in or around the largest city is 2.15 times greater than the density of beds in the entire country. Such a distribution of beds facilitates access for rural users.

There is a coordinating body to oversee public education and awareness campaigns on mental health and mental disorders. Government agencies (e.g., Ministry of Health or Department of mental health services); NGOs; professional associations, private trusts; and foundations, International agencies have promoted public education and awareness campaigns in the last five years. These campaigns have targeted the following groups: The general population; children; adolescents, women, trauma survivors; ethnic groups; and other vulnerable or minority groups. In addition, there have been public education and awareness campaigns targeting professional groups including health care providers (conventional; modern; allopathic); complimentary/ alternative/ traditional sector; teachers; social services staff; and other professional groups linked to the health sector. The following legislative and financial provisions exist to protect and provide support for users: (1) provisions concerning a legal obligation for employers to hire a certain percentage of employees that are disabled, (2) provisions concerning protection from discrimination (dismissal, lower wages) solely on account of mental disorder, (3) provisions concerning priority in state housing and in subsidized housing schemes for people with severe mental disorders, and (4) provisions concerning protection from discrimination in allocation of housing for people with severe mental disorders. All of these provisions exist but are not enforced.

In terms of support for child and adolescent health, 3 % of primary and secondary schools have either a part-time or full-time mental health professional, and none of the primary and secondary schools have school-based activities to promote mental health and prevent mental disorders. The percentage of prisoners with psychosis is 2–5 %, while the corresponding percentage for mental retardation is 6–10 %. Regarding mental health activities in the criminal justice system, a few (1–20 %) of prisons have at least one prisoner per month in treatment contact with a mental health professional. As for training, a few (1–20 %) police officers and no judges and lawyers have participated in educational activities on mental health in the last five years. In terms of financial support for users, no mental health facilities have access to programs outside the mental health facility that provide outside employment for users with severe mental disorders.

The national mental health program of Pakistan was developed, in 1986 at a multidisciplinary workshop, and incorporated in the seventh, eighth and ninth five-year national development plans. It aims at universal provision of mental health and substance abuse services by their incorporation in primary health care. The strategies for realizing this aim were, (i) teaching and training of personnel at all tiers of primary health care and incorporation of mental health and behavioural sciences in the curricula of health, education, social sciences and law enforcement institutions; (ii) strengthening of existing centres and establishment of new psychiatric centres; (iii) streamlining adequate referral services and provision of essential drugs; (iv) a multidisciplinary approach, intersectional collaboration (with social services, nongovernmental organizations and the private sector) and linkage with community development and (v) rapid expansion and development of specialized human resources base.

Pakistan has been home to a large number of innovative community mental health programs (WHO-EMRO 2006). Development of a model of mental health care delivery integrated within primary health care was initially developed in two sub-districts of Rawalpindi, and is being replicated in parts of all provinces of the Pakistan (Mubbsher et al. 1986). Most of the policy-level and field-level administrators have been provided with orientation in the field of mental health, including those from the armed forces, which has resulted in the setting-up of mental health training programs as part of the ongoing in-service training program of district health development centres. These centres have been set up to build the capacity of primary care personnel to handle common health problems by organizing on-the-job training for them. Mental health has been included in the regular programs of training being run by these centres in Punjab and over the coming years it will be expanded to the other provinces. More than 2000 primary care physicians have so far been trained in mental health. Similarly training manuals have been prepared for local health volunteers, local health workers and multipurpose health workers, and so far more than 40.000 have received training all over the country, in a decentralized manner, under the District health development centres initiative. So far, more than 65 junior psychiatrists have been trained in community mental health in order to act as resource persons in development of community mental health programs in their areas and to provide the training, referral and evaluation support to integration of mental health in primary health care. A national essential drugs list has also been formulated which includes all the essential neuropsychiatric drugs included in the WHO list. Another major development has been the acceptance in principle to include indicators for mental illnesses as part of the national health management information system. This is a crucial development for integrating mental health into primary health care (Mubbasher et al. 1986, 1999; Yousaf 1997; Mumford 1997; Husain et al. 2000; Chisholm et al. 2000; Ahmad et al. 2001; Gater and Rehman 2001; Gadit and Khalid 2002; Karim et al. 2004).

Mental health professionals of Pakistan realized that schools can play an effective role in stimulating community efforts for mental health care provision. This realization led to the development of a school mental health program. The program is both child- and environment-centred and works through a series of four phases: familiarization, training, reinforcement and evaluation to achieve its objectives. During 2000, a mental health component was included in teacher-training programs at national level. More than 150 education administrators from all provinces have been provided with orientation training.

Training of master trainers from all provinces (batches of 40 for four months each) started from January 2001. Text book boards of all provinces have included of mental health issues in the school curricula being prepared by them (Mubbashar 1989; Saeed et al. 1999).

Faith healers and religious leaders are the first port of call for the majority of the mentally ill patients. Thus the potential benefits of involving the faith healers, rather than antagonizing them, in the provision of mental health services are manifold, foremost being the perception by the community that services are in line with their health belief system. After the initial reservations were overcome, a relationship beneficial to the mentally ill in the community was forged. One particular research project in this regard is worth mentioning which shows that about 25 % of the patients presenting to faith healers in a sub-district of 0.5 million were given "medical diagnoses" and referred to the nearest health facility, a significant departure from past practices (Saeed et al. 2000).

Nongovernmental organizations are playing an increasingly important role in developmental activities. The National Rural Support Program is an organization active in the field of income generation, education, agriculture, forestry, tourism and health, having access to about 20.000 village-level organizations. The program and its sister organizations have agreed to include mental health among all its activities and about 20.000 community activists will be trained each year through this initiative, highlighting the role of mental health in national development activities. Research has focussed on mental health policy research, epidemiological research, health systems research, economic evaluation of models of mental health care delivery, development and validation of research instruments, evaluation of inter-sectoral linkages, and clinical research (Mumford et al. 1996, 1997, 2000; Rahman et al. 2008; Khan and Reza 1998, 2000). The government of Pakistan has repealed the mental health act of 1912. A new mental health law embodying the modern concept of mental illnesses, treatment, rehabilitation, and civil and human rights was promulgated on 20 February 2001.

One of the most recent community mental health project relates to post-partum depression. The treatment of perinatal depression is a public-health priority because of its high prevalence and association with disability and poor infant development. The project integrated a cognitive behaviour therapy-based intervention into the routine work of community-based primary health workers in rural Pakistan and assessed the effect of this intervention on maternal depression and infant outcomes (Rahman et al. 2008).

10.4.10 Sri Lanka (Mendis 2011)

The Sri Lankan health system is a government funded decentralised public health system, accompanied by a robust private sector. Sri Lanka has long emphasised prevention and public health, and this is reflected in the division of labour for different cadres. 1.6 % of total health budget is spent on mental health. Sri Lanka has 19 million population and 48 psychiatrists. There are no psychiatric nurses but 800 nurses attached to psychiatric units having the same experience in psychiatry, 3 psychologists only attached to the university units and 57 occupational therapists and 20 psychosocial workers (PSWs) and 60 assistant PSWs. This means that there is approximately 1 psychiatrist per 500,000 with these psychiatrists placed at the national, provincial and district levels in the health system. There is a national mental hospital with 1013 inpatient beds (Angoda) staffed by 8 psychiatrists, 300 psychiatric nurses, and 8 occupational therapists but with no psychologists. There are 9 provincial psychiatric inpatient units with around 600 beds. There are 25 district hospitals. The district hospitals deploy staff for public health tasks, general curative services and for mental health. The district medical officer of health (generally each responsible for a catchment population of approximately 70,000) takes referrals of people with mental disorders. The psychiatric nurses are general nurses who have been given 1 month of training and then regular refresher courses, and are largely employed in the psychiatric wards at national, regional and district levels. At the sub district hospital level there are both public health staff and general treatment staff. The general treatment staffs are Registered Medical Practitioners (with 3 year general training) who treat general patients of all kinds. There are also 110 medical officers of mental health (MOMHs) in the country. This cadre is a relatively recent development in Sri Lanka. This was developed to assist in the decentralisation of mental health services. Medical officers of mental health (MoMHs) are qualified doctors who have been selected either for a one month course in psychiatry or a twelve month diploma in psychiatry, leading to a certificate from College of Psychiatrists, leading to the title of medical officer of mental health (MMOH). The intention is to have one MoMH in each of the 276 subdistricts in which case there will eventually be one MoMH per 70,000 population; at present there are 110 MoMHs, resulting in an actual ratio of one per 173,000 population. The rural hospitals are staffed by registered medical practitioners, 1–2 nurses and 1 midwife.

There are five important community mental health initiatives in Sri Lanka. First of these is the integration of mental health with general health services, described above. The training and deployment of over 100 medical officers and supporting them through decentralised psychiatrists is an important development. The second innovation is related to the long conflict situation. A number of innovative approaches have been developed to provide disaster mental health care (Somasundaram 1994, 1997, 1998, 2001, 2002, 2005, 2006). The third initiative relates to the 2004 tsunami and the aftermath (2006). The fourth initiative is in the area of suicide prevention (de Silva and Jayasinghe 2003). The fifth area is the rehabilitation of the chronic mentally ill persons (Mendis 2011).

10.4.11 Thailand (WHO-SEARO 2007; Udomratn 2011)

Thailand has mental health policy and plans. In Thailand there are three different social insurances, which provide free access to essential psychotropic medicines to 93 % of the population. A human rights review body exists, but it does not oversee regular inspections and has no authority to impose sanctions. The Mental Health Department (MHD) is the national mental health authority. There are122 outpatient facilities in the country, located in mental hospitals and in general hospitals. Eleven percent of these facilities are for children and adolescents only. There are no day treatment facilities in Thailand. The only residential facilities are for people with mental retardation and substance abuse. There are 25 community-based psychiatric units with 0.4 beds per 100,000 population and 17 mental hospitals with 13.8 beds per 100,000 population. The majority of patients admitted to mental hospitals have a diagnosis of schizophrenia. Mental health care has been integrated into primary health care by MHD for many years. Primary health care doctors have limited training and interaction with mental health services. There is a disproportionate amount of resources concentrated in the main cities, which limits access to mental health services for rural users. There are 5 user associations and 3 family associations present in the country interacting with a few mental health facilities. Public education and awareness campaigns are overseen by coordinating bodies. There are legislative provisions for employment, but not for housing.

10.5 Lessons from the Country Experiences

There are a number of common themes that emerge from the review. Lack of mental health **awareness** in the general population is one of them. **Stigma** about persons with mental disorders, the mental health services and service providers is another issue that faces the professionals and persons/families with mental disorders (Srinivasa Murthy 2005). **Limited infrastructure and professionals for care** has been addressed innovatively in the different countries. The most common approach has been the utilisation of the primary health care personnel. In some of the countries, families, community leaders have been part of the solution to address the need. **Legislation** is another issue to ensure human rights of the persons with mental disorders. There are few countries with no mental health legislation and in those countries where there are laws they are not in line with UNCRPD. There is also the challenge of harmonising the rights of the ill persons and that of the families. The countries of the Region, has contributed richly to understanding and responding to mental health needs of survivors of **disasters**, in developing self-care, involving non-professionals to provide essential care. S**uicide prevention** is an area in which volunteers have played an important role in providing essential services (Vijayakumar 2004). **Research** to support the above activities also has been an important response of the countries.

10.6 International Developments

The last four decades, from the time of the WHO Expert Committee Report of 1975, and more specifically the last two decades have been extraordinary period for mental health care all over the world (WHO 1975; Desjarlais et al. 1995; WHO 2001; Bhugra et al. 2005; Ghodse 2011). The most striking aspect of this movement for community mental health in the world is the continuing efforts to develop and support the movement like the Lancet initiative of 2007 (Horton 2007; Saraceno et al. 2007; Saxena et al. 2007; Jacob et al. 2007; Prince et al. 2007; Chisholm et al. 2007a, b) and other international initiatives (Thornicroft et al. 2008; Mari et al. 2009) and WHO efforts to develop policies to support the movement (WHO 2003; WHO-WONCA 2008). The development of mental and psychosocial interventions in emergency situations in 2007 is a major support for countries of the Region (IASC 2007). This movement is in an early phase and the full story will unfold in the coming years and decades. These initiatives can result in a better quality of life to the persons with mental disorders and their families and protection of their rights.

10.7 Personal reflections

I have been associated with the community psychiatry movement in a number of countries of the Region (Bhutan, India, Nepal, Pakistan, Sri Lanka) from the 1970s. I have been a participant of many of the country level initiatives and witness to other developments. An overview of the community psychiatry developments of the last few decades, presents a picture of a large number of initiatives. These initiatives have been largely the response to a specific need at a specific time period in a particular country. For example, in the 1950s, the lack of human resources in mental hospitals in India was addressed by bringing the families to become part of the care programs. In the 1960s, the availability of the psychopharmacological agents for the treatment of mental disorders and the growing general hospitals, resulted in the setting up of general hospital psychiatric units. During the 1970s, the growth of the public sector health services, the influence of the Alma Ata declaration guided the development of the community mental health programs and the formulation of the national programs of mental health. During the decades of 1980s and 1990s, the need for non-mental hospital facilities for rehabilitation resulted in setting up of a number of community care facilities, mainly by the voluntary organisations. The recognition of the human rights of the mentally ill persons is reflected not only in the improvement of the mental hospitals, but also with revision of the mental health legislation in a number of countries. Each of these initiatives have been started and guided by visionary professionals and have occurred at a particular time period and to address a specific need perception. One striking aspect is the innovativeness of the professionals and voluntary agencies to address the multiple needs using the available community resources. This has occurred in a number of countries. This I consider is the strength of mental health movement of the Region. The negative side of these developments is the lack of depth in most of the initiatives. Even when the initial results have been very positive (e.g. nurse involvement in community care) the innovations has not received the type of expansion and in-depth understanding that should have occurred. The lack of evaluation is seen uniformly in all the programs. The other aspect of significance is the largely person/centre specific nature of the initiatives. There have been limited efforts to join hands, carry forward to work beyond the initiators. A result of all of these factors is the lack of theory building and influencing the policies at the national levels. It is important that the next phase of development will address some of these needs.

I would like to illustrate the above need with the lacunae in three important initiatives of the countries of the Region, namely, integration of mental health with general health care, family involvement and disaster mental health care (Ustun and Sartorius 1995; Sartorius 1997; Srinivasa Murthy 1998; Cohen 2001; Patel et al. 2007; Srinivasan 2006).

Recently, Issac and Guruje (2009) have reviewed the integration of mental health with primary health care and pointed out the following:

> The large unmet need for mental health services in many LAMIC, despite the availability of effective and relatively affordable interventions, calls for an urgent effort to scale up primary care service in those countries. Efforts to scale up services must include a comprehensive review of the training provided for primary care providers in the recognition and treatment of mental health problems and a reorganization of the primary health care system. Assumptions made about the relative autonomy of the primary health care system have led to an unsupported and unmotivated health workforce. A reorganization of primary health care system in the LAMIC must recognize the need for an effective secondary care level., with a sufficient number of specialist mental health workers to provide training and support for primary care providers and back up for difficult cases requiring specialist interventions.

Reviewing the progress in the area of family involvement, Shankar and Rao (2005) conclude:

> It is important to recognise that there are large unfinished tasks to make families a part of the community mental health movement…'professional inputs have not kept pace' and conclude that the 'family movement in India is one of 'unfulfilled promises or great expectations for the future… the vision for the family movement in India would see families from passive carers to informed carers, from receiving services to proactive participation, from suffering stigma to fighting stigma. And it is the responsibility of the mental health system to facilitate this journey of care givers from burden to empowerment. (p. 285)

The countries of the Region have experienced a many situations of natural disasters (earthquakes, tsunami) and man-made disasters (war, conflict, industrial disasters) and there have been many innovative approaches to address the mental health needs of the population (WHO-SEARO 2005; Diaz et al. 2006). However, there is a need for evaluation of the interventions. Most of the assessment of impact is qualitative and does not allow for definite conclusions. For programs to be integrated to the larger activities of different stakeholders (NGOs, Govt, Professionals, people) there is need for greater effort at evaluation of the outcome. Processes of care alone are not adequate. The cultural dimension and the need for adaptation of experiences to suit the needs of the population is another lesson from the experiences. Some of the cultural factors are religion, place of women, expectations of survivors from help providers, the barriers of administration to coordinate the efforts, the rituals surrounding loss and recovery.

One other important development in community psychiatry in the countries of the Region, is the increasing role of the voluntary organisations in developing small size locally relevant community based psychiatric care facilities like day care centres, vocational training centres, sheltered workshops, half-way homes and long stay homes (Patel and Thara 2003). These facilities have the advantage of limiting long term institutional care, incorporating the cultural sensitivities of the clientele, and utilisation the local resources. However, there is need for evaluation of psychosocial care in community settings about the following aspects: (1) Characteristics of the clients—age, sex, literacy, occupation, income, social background, diagnosis, duration of illness, the past treatment; (2) Reasons for seeking 'institutional' PSR- the reasons could be a complex mixture of four factors, namely, (i) the nature of illness (e.g. Chronic schizophrenia, personality disorder etc); (ii) Specific therapy (eg. Supervised medication, therapeutic community, social skills training, vocational training etc); (iii) family factors (elderly parents, single parent, siblings living abroad etc); and (iv) community factors (stigma limiting the reintegration of the recovered back into the community). The reason for collecting this information and analysing this is to direct interventions (either therapies, the social changes or stigma) depending on the chief factors. It can also be that the different centres can organise services differently for the different reasons for 'institutional' care. Work in this area could also give information for the Government to take up appropriate actions, rather than depending only on the private/NGO sectors; (3) Duration of stay: How many are terminated/discharged prior to completing the admission goals; (4) Outcome of the stay in PSR facilities: This will be both in terms of the client and the family. To what extent are the goals realised and if they were not reached the reasons for the same- illness, therapeutic setting, staff problems, social factors etc; (5) Therapeutic processes during the stay of the client: This is vital to record both for human resources development, that is appropriate to the clients and to understand the benefits of different interventions. At present, most of the centre reports talk of counselling, group therapy etc without specifying what it really means; (6) Staff issues: This could include, the roles of the different categories of staff, the training needs, the turnover of staff etc. (7) Crisis handling: What are the types of crisis- suicide attempts, violence etc and their frequency and procedures for handling them; (8) Human rights: The study from BAPU TRUST, in India, has analysed the type of human rights abuses and lack of clarity of this area. It is important to record and develop norms for informed consent, 'restrictions', admissions against willingness of clients, guardianship, remedial measures etc; (9) Changes in the practices over the life time of the organisation (last 10–20–30 years) and how the Institution has built on its experiences; and (10) Lessons learnt and needs for PSR for future development.

There are many country specific challenges such as in Afghanistan, " to increase the capacity of the mental health sector will remain huge for the coming years" (Ventovogel et al. 2011); in Bangladesh, "psychiatrist services tend to be available only in big cities....psychotherapy is not widely available....Bangladesh lacks a mental health act" (Karim et al. 2011); in India, " to facilitate the training of general psychiatrists in specialist areas such as substance abuse, child psychiatry, and forensic psychiatry"(Patel and Saxena 2011); in Nepal, " the near-term future of Nepalese psychiatry does not look bright....people affected by the decade-long Maoist civil war, especially women and children, may present with trauma-related psychiatric problems requiring culturally sensitive interventions"; in Pakistan," lack of indigenous research has been a major hindrance to the rational planning and allocation of resources" (Mubbashar 2011); in Sri Lanka, " there is an urgent need to provide accessible basic services of good quality to meet the emerging needs of people living in the community"(Mendis 2011); in Thailand, " mental health problems are not well recognised by general practitioners. Patients poor understanding of psychiatric disorders causes a delay in seeking help and frequently early discontinuation of drug treatment" (Udomratn 2011).

10.8 Future of Community Psychiatry in the Region

In order to address the total mental health needs in the countries of the Region, there are a number of requirements, namely, professional level, community level and policy level (Srinivasa Murthy and Kumar 2006).

10.9 Professional Challenges

Professional leadership has been an important force for the many community psychiatry initiatives. There is a need to simplify mental health care skills and continually review and develop innovative approaches to deliver them, in order to meet the reality of the community needs and expectations. For mental health care to be undertaken by health workers, teachers, volunteers, family members, there is need for simple interventions. Professionals have to develop the appropriate information in a simple format and identify the 'level of care' and 'limits of care' to be provided by these personnel. These should include choosing priority mental disorders to be addressed in training, limiting the range of drugs to be used by the general practitioners, develop strong referral guidelines and the non-pharmacological interventions to be used by non-physician personnel. There should be both a willingness to share the mental health caring responsibilities with non-specialists, and overcoming the fear of some professionals of losing their work, identity and income. The method should be not to convert the non-specialist into a mini- psychiatrist, but to identify what is relevant, feasible and possible for the specific non-specialist to undertake. There is a need to decrease the amount of time devoted by specialists mental health professionals to individual clinical care and increase the time for training, support and supervision of other personnel. This is a big challenge for clinicians who value directly caring for ill people by themselves. This change in role becomes meaningful when it is recognised that training of other personnel has a multiplier effect in providing mental health services to the population. There is need to devote significant time to periodic support and supervision of the non-specialists. Fortunately, the easy and inexpensive availability of mobile phones, internet and satellite communication for tele-psychiatry, allows for distant support to the non-specialists on a continuous and interactive basis. There is a need for professionals to acquire the skills to work with the community, education sector personnel, welfare sector personnel, voluntary organizations, and policy makers. This includes understanding the planning process, fighting for priority for mental health in health programs, becoming familiar with legislations and budget procedures, and developing skills to negotiate with different stakeholders.

10.10 Community Level Challenges

In the countries of the Region, there is a paradoxical situation of limited services and poor utilisation of the available services, due to problems of stigma and lack of information in the general population. There is a need for bringing about a major shift in the thinking of the community in terms of understanding of the mental health and mental disorders, decreasing the stigma and discrimination of persons and families with mental disorders, and the creation of a wide range of community care facilities and services (WFMH 2009). There is also need to develop simple self-care information modules. For those requiring long-term care there is need to develop measures (for example using the mobile phones, internet, community radio) to help in monitoring the progress of mental condition at the home level. In addressing these needs to cover the total population and in a manner that requires limited travel, there is need to use fully the modern technologies like the world wide web, mobile phones, telemedicine, community radio to reach and continuously support the persons and families with mental disorders. The successful use of information technology in spreading the agricultural information should give hope for similar success in the mental health area.

10.11 Policy Level

At this level, the needs are: (i) greater amount of allocation of funds for mental health program, similar to Thailand; (ii) recognition of human rights of the persons and families of persons with mental disorders in all development programs, especially in the areas of education, welfare, housing, employment; (iii) strengthening of the programs to support the families; (iv) legislative support for non-specialists to provide mental health care and ensuring the human rights of the persons with mental disorders and (v) building of a range of community based care facilities to meet the total needs of the population.

10.12 Conclusion

Development of mental health services all over the world, countries rich and poor alike, have been the product of the larger social situations (political, social, economic and human rights), specifically the importance society gives to the rights of disadvantaged/marginalised groups (Wig 1989). Each generation has to find its own solutions as pointed out by Goldberg (1992).

> During the asylum era we knew nothing about the causes of mental disorder, and the proportion of the population who could expect to receive are from a mental health professional was very low. However, our predecessors had a very clear notion of the structures-both administrative and architectural-that are required, and these structures secured resources to maintain services. There is now an explosion of knowledge about mental disorder, and it at least becomes possible to discern the outlines of a model for mental disorder which takes into account of findings in both social psychiatry and molecular biology. However, we have not made corresponding progress in refining the administrative and architectural requirements for meeting the needs of the mentally ill and in most countries of the world services for the mentally ill survive on the crumbs left from the banquet of general health care.

Mental health professionals of the countries of the Region have both the challenge of addressing the vast mental health needs of the population and an opportunity to build these services using the community mental health principles. Economically rich countries have addressed the community psychiatry movement from the institutionalised care to community care building on the strengths of their social institutions. South Asian countries have begun this process more recently and have made significant progress utilising the strengths of their communities and building from the 'bottom of the pyramid' (Prahlad 2006). There is need to continue the process by widening the scope of the mental health interventions, increasing the involvement of all available community resources and rooting the interventions in the historical, social and cultural roots of the countries and systematic evaluation. This is a continuing challenge for professionals and people in the coming years. As the community mental health programs of the countries of the region, blossom and fulfil the mental health needs of the population, it will be an important contribution of the Region to the rest of the world (German 1975; Wig 1989; Susser et al. 1996; Tomlinson et al. 2008).

References

Agarwaal, S. P., D. S., Goel, R. L., Ichhpujani, R. N., Salhan, and S. Shrivatsava. 2004. *Mental health: An indian perspective (1946–2003)*. New Delhi: Directorate General of Health Services, Ministry of Health and Family Welfare.

Ahmad, I., K. Saeed, M. H. Mubbashar, et al. 2001. Minor psychiatric morbidity and socio-economic factors. *Medical forum monthly* 12:5–8

Ayoughil, S., I. Missmahl, R. Weierstall, T. Elbert. 2012. Provision of mental health services in resource-poor settings: a randomised trial comparing counselling with routine medical treatment in North Afghanistan (Mazar-e-Sharif). *BMC Psychiatry* 12:14.

Bhugra, D., G. Ranjith, and V. Patel. 2005 *Handbook of Psychiatry- a South Asian Perspective*, New Delhi: Byword Viva Publishers.
Carstairs, G. M. 1974. Community action for mental health care. WHO/SEARO/MENT/22.
Chandrashekar, C. R., M. K. Issac, R. L. Kapur, and R. Parthasarathy. 1981. Management of priority mental disorders in the community. *Indian Journal of Psychiatry* 23:174–178.
Chatterjee, S., V. Patel, A. Chatterjee, and H. A. Weiss. 2003. Evaluation of a community based rehabilitation model for chronic schizophrenia in India. *British Journal of Psychiatry* 182:57–62.
Chisholm, D., S. K. James, K. Sekar, M. H., Kumar, and V R. Mubasshar. 2000. Integration of mental health care into primary care: demonstration cost-outcome study in India and Pakistan, *British Journal of Psychiatry* 176:581–588.
Chisholm, D., A. J. Flisher, C. Lund, V. Patel, S. Saxena, G. Thornicroft, et al. 2007a. Scale up services for mental disorders: a call for action. *Lancet* 370:1241–53.
Chisholm, D., C. Lund, and S. Saxena. 2007b. Cost of scaling up mental health care in low and middle income countries. *British Journal of Psychiatry* 191:528–35.
Climent, C. E., B. S. M. Diop, T. W. Harding, H. H. A. Ibrahim, L. L. Ignacio, and N. N. Wig. 1980. Mental health and primary health care. *WHO Chronicle* 34:231–236.
Cohen, A. 2001. *The effectiveness of mental health care in primary health care: the view from the developing world.* Geneva: World Health Organization.
Cohen, A., A. Kleinmann, and B. Saraceno. 2002 *World mental health casebook. social and mental health programs in low income countries*. New York: Kluwer Academic.
de Silva Damini and S. Jayasinghe. 2003. Suicide in Sri Lanka, In Suicide prevention: Meeting the challenge together, ed. L. Vijayakumar, 178–190. Chennai: Orient Longman.
Desjarlais, R., L. Eisenberg, B. Good, and A. Kleinman. 1995. *World mental health: problems and priorities in low-income countries*. New York: Oxford University Press.
Diaz, J. O., R. Srinivasa Murthy, and R. Lakshminarayana, eds. 2006. *Advances in disaster mental health and psychosocial support*. New Delhi: American Red Cross.
Gadit, A. A. and N. Khalid. 2002. *State of mental health in Pakistan*. Karachi: Hamdard University.
Gater, R. and Rehman ChU. 2001. Mental health and service developments in the Rawalpindi district of Pakistan. *Journal of College of Physicians and Surgeons Pakistan* 11:210–214.
Gask L., H. Lester, T. Kendrick, and R. Peveler, eds. 2009. *Primary care mental health*. London: Royal College of Psychiatrists.
German, A. 1975. Trends in Psychiatry in Black Africa. In *New Dimensions in psychiatry—a world view*, eds. S. Arieti and G. Chrzanowski. New York: Wiley.
Ghodse, H., ed. 2011. *International Perspectives on Mental Health*. London: RCPsych Publications
Goldberg, D. 1987. Round the world: Pakistan- revolution in mental health care. *Lancet* i:736–737.
Government of India. 1982. *National mental health program for India*. New Delhi: Ministry of Health and Family Welfare.
Giel, R., M. V. de Arango, C. E. Climent, T. W. Harding, H. H. Ibrahim, L. Ladrido-Ignacio, et al. 1981a. Childhood mental disorders in primary health care: results of observations in four developing countries. A report from the WHO collaborative study on strategies for extending mental health care. *Pediatrics* 68:677–683.
Giel, R., E. d'Arrigo Busnello, C. E. Climent, A. S. Elhakim, H. H. Ibrahim, L. Ladrido-Ignacio, et al. 1981b. The classification of psychiatric disorder. A reliability study in the WHO collaborative study on strategies for extending mental health care. *Acta Psychiatrica Scandinavica* 63:61–74.
Giel, R., M. V. de Arango, A. Hafeiz Babikir, M. Bonifacio, C. E. Climent, T. W. Harding, et al. 1983. The burden of mental illness on the family. Results of observations in four developing countries. A report from the WHO collaborative study on strategies for extending mental health care. *Acta psychiatrica Scandinavica* 68:186–201.
Goldberg, D., and P. Huxley. 1992. *Common mental disorders: A bio-social model*, 162. London: Routledge.
Harding, T. W., C. E. Climent, M. Diop, R. Giel, H. H. A. Ibrahim, L. L. Ignacio, et al. 1983a. The WHO collaborative study on the strategies for extending mental health care, II: the development of new research methods. *American Journal of Psychiatry* 140:1474–1480.

Harding, T. W., E. d'Arrigo Busnello, C. E. Climent, M. Diop, A. El-Hakim, R. Giel, et al. 1983b. The sa collaborative study on the strategies for extending mental health care, iii: Evaluative design and illustrative results. *American Journal of Psychiatry* 140:1481–1485.
Horton, R. 2007. Launching a new movement for mental health. *Lancet* 370:806.
Husain, N., F. Creed, and B. Tomenson. 2000. Depression and social stress in Pakistan. *Psychological Medicine* 30:395–402.
IASC. 2007. *Inter agency standing committee guidelines on mental health and psychosocial support in emergency settings*. Geneva: IASC.
Indian Council of Medical Research—Department of Science and Technology (ICMR-DST). 1987. *A collaborative study of severe mental morbidity*. New Delhi: ICMR.
Indian Council of Medical Research. 2005. *Mental health research in India*. New Delhi: Indian Council of Medical Research.
Indian Council of Marketing Research. 2008. *Evakuation of the DMHP*. New Delhi.
Institute of Medicine (IOM). 2001. *Neurolological, psychiatric and developmental disorders: Meeting the challenges in the developing world*. Washington: National academy of Sciences.
Issac, M. K., and R. L. Kapur. 1980. A cost effectiveness of three different methods of psychiatric case finding in the general population. *British Journal of Psychiatry* 137:540–546.
Issac, M. K., R. L. Kapur, C. R., Chandrasekar, M., Kapur, and R. Parthasarathy. 1982. Mental health delivery in rural primary health care—development and evaluation of a pilot training program. *Indian Journal of Psychiatry* 24:131–138.
Issac, M. K., C. R. Chandrasekar, and R. Srinivasa Murthy. 1988. *Manual of mental health care for medical officers*. Bangalore: National Institute of Mental Health and Neurosciences.
Issac, M., and O. Guruje. 2009. Low and middle income countries. In *Primary care mental health* ed. L. Gask, H. Lester, T. Kendrick, and R. Peveler, 72–87. London: Royal college of Psychiatrists.
Jacob, K. S., P. Sharan, I. Mirza, M. Garrido-Cumbrera, S. Seedat, J. J. Mari, et al. 2007. Mental health systems in countries: Where are we now? *Lancet* 370:1061–1077.
Jadhav, S., and S. Jain. 2009. Pills that swallow policy: Clinical ethnography of a community mental health program in northern india. *Transcultural Psychiatry* 46:60–85.
James, S., D. Chisholm, R. Srinivasa Murthy, K. Kishore Kumar, K. Sekar, K. Saeed, et al. 2002. Demand for, access to and use of community mental health care: Lessons from a demonstration project in India and Pakistan. *International Journal of Social Psychiatry* 48:163–176.
de Jong JTVM. 1996. A Comparative public health program in Guinea-Bissau: a useful model for African, Asian and Latin American countries. *Psychological Medicine* 26:97–108.
Kapur, M. 1997. *Mental health in indian schools*. New Delhi: Sage.
Kapur, R. L. 1975. Mental health care in rural India: a study of existing patterns and their implications for future policy. *British Journal of Psychiatry* 127:286–93.
Kapur, R. L. 2004. The story of community mental health in India. In *Mental Health—An Indian perspective (1946–2003)*, eds. S. P. Agarwaal, D. S. Goel, R. L. Ichhpujani, R. N. Salhan, and S. Shrivatsava, 92–100. New Delhi: Directorate General of Health Services, Ministry of Health and Family Welfare.
Karim, M. R., F. Shaheed, S. Paul. 2011. Bangladesh. In *International perspectives on mental health*, ed. H. Ghodse, 83–87. London: RCPsych Publications.
Khan, M. M., and H. Reza. 1998. Gender differences in nonfatal suicidal behaviour in Pakistan: significance of sociocultural factors. *Suicide and life-threatening behavior*, 28:62–68.
Khan, M. M., and H. Reza. 2000. The pattern of suicide in Pakistan. Crisis. *Journal of Crisis Intervention and Suicide* 21:31–35.
Karim, S., K. Saeed, and M. H. Rana. 2004. Pakistan mental health profile. *International Review of Psychiatry* 16:83–92.
Kohn, R., S. Saxena, I. Laevav, and B. Saraceno 2004. The treatment gap in mental health care. *Bulletin of World Health Organization* 82:858–66.
Ladrido-Ignacio, L., M. V. de Arango, J. Baltazar, E. Busnello, C. E. Climent, B. S. Diop, et al. 1983. Knowledge and attitude of primary health care personnel concerning mental health problems in developing countries. *American Journal of Public Health* 73:1081–4.

Linhorst D. M. 2006. *Empowering people with severe mental illness: A practical guide*. Oxford University Press.

Mari J. D. J., D. Razzouk, R. Thara, J. Eaton, and G. Thornicroft. 2009. Packages of care for schizophrenia in low- and middle-income countries. *PLOS Medicine* 6: 1–8. Doi: e1000165.

Mendis, N. 2011. *International perspectives on mental health*, ed. Ghodse, 197–201. London: RCPsych Publications.

Mercier, C. 2007. *Expert opinion on barriers and facilitating factors for the implementation of existing mental health knowledge in mental health services*, p. 12. Geneva: World Health Organization, Geneva.

Miller, K. E., A. Rasmussen. 2010. War exposure, daily stressors, and mental health in conflict and post-conflict settings: bridging the divide between trauma-focused and psychosocial frameworks. *Social Science & Medicine* 70: 7–16.

Miller, K. E., P. Omidian, A. Rasmussen, A. Yaqubi, and H. Daudzai. 2008. Daily stressors, war experiences, and mental health in afghanistan. *Transcultural Psychiatry* 45: 611–638.

Minhas, H. 2007. Expert opinion on barriers and facilitating factors for the implementation of existing mental health knowledge in mental health services, Geneva: World Health Organization. p. 16.

Ministry of Health, Afghanistan. 2009. *National mental health strategy: 2009–2014*. Kabul: Islamic Republic of Afghanistan.

Mohit, A., K. Saeed, D. Shahmohammodi, et al. 1999. Mental health manpower development in Afghanistan: report of a training course for primary health care physicians. *Electronic Medical Records* 5: 215–219.

Mubbashar, M. H. 1989. Promotion of mental health through school health program. *EMR Health Services Journal* 6: 14–19.

Mubbashar, M. H. 1999. Mental health services in rural Pakistan. In *Common mental disorders in primary care*, eds. M. Tansella and G. Thornicroft, pp. 67–80. London: Routledge.

Mubbashar, M. H. 2011. Pakistan. In *International perspectives on mental health*, ed. H. Ghodse, 173–177. London: RCPsych Publications.

Mubbashar, M. H., S. J. Malik, J. R. Zar, and N. N. Wig. 1986. Community based mental health care program: Report of an experiment in Pakistan. *The Eastern Mediterranean Health* 1: 14–20.

Mubbashar, M. K., K. Saeed, and R. Gater. 2001. Reaching the unreached: evaluation of training of primary health care physicians. *Journal of the College of Physicians And Surgeons Pakistan* 11: 219–223.

Mumford, D. B. 1997. Stress and psychiatric disorder in rural Punjab—A community survey. *British Journal of Psychiatry* 170: 473–78.

Mumford, D. B., M. Nazir, F-U-M. Jilani, et al. 1996. Stress and psychiatric disorder in the Hindu Kush. A community survey of mountain villages in Chitral, Pakistan. *British Journal of Psychiatry* 168: 299–307.

Mumford, D. B., F. A. Minhas, I. Akhtar, et al. 2000. Stress and psychiatric disorder in urban Rawalpindi. Community survey. *British Journal of Psychiatry* 177: 557–562.

Mumford, D. B., K. Saeed, I. Ahmad, et al. 1997. Stress and psychiatric disorder in rural Punjab: a community survey. *British Journal of Psychiatry* 170: 473–478.

Nagaswami, V. 2012. The shrinking Universe, *The Hindu,* May 27, 2012.

National Human Rights Commission (NHRC). 1999. *Quality assurance in mental health*. New Delhi: NHRC.

National Human Rights Commission (NHRC). 2008. *Mental health care and human rights*. eds. D. Nagaraja and P. Murthy. New Delhi: NHRC–NIMHANS.

Pai, S. and R. L. Kapur. 1982. Impact on treatment intervention on the relationship between the dimensions of clinical psychopathology, social dysfunction and burden on families of schizophrenic patients. *Psychological Medicine*, 12:651–658.

Pai, S. and R. L. Kapur. 1983. Evaluation of home care treatment for schizophrenic patients. *Acta Psychiatrica Scandinavica* 67:80–88.

Pai, S., R. L. Kapur, and E. J. Roberts. 1983. Follow up study of schizophrenic patients initially treated with home care. *British Journal Psychiatry* 143:447–450.
Pai, S., S. M. Channabasavanna, and R. Raghuram. 1985. Home care for chronic mental illness in Bangalore: An experiment in the prevention of repeated hospitalization. *British Journal Psychiatry* 147:175–179.
Panter-Brick, C., M. Eggerman, A. Mojadidi, and T. McDade. 2008. Social stressors, mental health, and physiological stress in an urban elite of young Afghans in Kabul. *American Journal of Human Biology*. 20:627–641.
Panter-Brick, C., M. Eggerman, V. Gonzalez, and S. Safdar. 2009. Violence, suffering, and mental health in Afghanistan: A school-based survey. *The Lancet* 374:807–816.
Panter-Brick C, A. Goodman, W. Tol, and M. Eggerman. 2011. Mental health and childhood adversities: a longitudinal study in kabul, afghanistan, *Journal of the American Academy Child Adolescent Psychiatry*. 50:349–363.
Patel, V., and R. Thara eds. 2003. *Meeting mental health needs in developing countries: NGO innovations in India*. New Delhi: Sage (India).
Patel, V. and Saxena, S. 2011. India. In International perspectives on mental health, ed. H. Ghodse, 101–105. London: RCPsych Publications.
Patel, V., R. Araya, S. Chatterjee, D. Chisholm, A. Cohen, M. De Silva et al. 2007. Treatment and prevention of mental disorders in low-income and middle-income countries. *Lancet* 370:991–1005.
Prahlad, C. 2006. The future at the bottom of the Pyramid—eradicating poverty through profits, Upper Saddle River: Wharton Publishing House.
Prince, M., V. Patel, S. Saxena, M. Maj, J. Maselko, M. R. Phillips. et al. 2007. No health without mental health. *Lancet* 370:859–877.
Rahman, A., A. Malik, S. Sikander, C. Roberts, and F. Creed 2008. Cognitive behaviour therapy based intervention by community health workers for mothers with depression and their infants in rural Pakistan: A cluster-randomised controlled trial. *Lancet 13* 372(9642):902–909.
Ranganathan, S. 1996. *The empowered community: A paradigm shift in the treatment of Alcoholism*. Madras: TTR Clinical Research Foundation.
Ranganathan, S. 2011. *Treatment of alcoholism- Community treatment approach*. Chennai: TTR Foundation.
Regmi, S. K., A. Pokhrel, S. P. Ojha. et al. 2004. Nepal mental health country profile. *International Reviwe of Psychiatry* 16:142–149.
Roderigo E. K., D. Bhugra, and P. de Silva. 2005. Mental health planning in developing countries, In Handbook of psychiatry- a South Asian perspective. eds. D. Bhugra,, G. Ranjith, and V. Patel, 599–610. New Delhi: Byword Viva Publishers.
Saeed, K., R. Gater, A. Hussain, and M. K. Mubbashar. 2000. The prevalence, classification and treatment of mental disorder among attendees of native faith healers in rural Pakistan. *Social Psychiatry and Psychiatric Epidemiology* 35:480–485.
Saeed, K., S. Wirz, R. Gater et al. 1999. Detection of disabilities by school children: A pilot study in rural Pakistan. *Tropical Doctor* 29:151–155.
Saraceno B, M. van Ommeren, R. Batniji, A. Cohen, O. Gureje, J. Mahoney et al. 2007. Barriers to improvement of mental health services in low-income and middle-income countries. *Lancet* 370:1164–1174.
Sartorius, N. 1997. Psychiatry in the framework of primary health care: A threat or boost to psychiatry? *American Journal of Psychiatry* 154(Suppl. 6):67–72.
Sartorius, N., and T.T. Harding. 1983. The WHO collaborative study on strategies for extending mental health care, I: The genesis of the study. *American Journal of Psychiatry* 140:1470–1473.
Saxena S, G. M. Thornicroft, KnappM. Knapp, and H. Whiteford. 2007. Resources for mental health: scarcity, inequity, and inefficiency. *Lancet* 370:878–889.
Scholte W. F., M. Olff, P. Ventevogel et al. 2004. Mental health symptoms following war and repression in Eastern Afghanistan. *Journal of American Medical Association* 292:585–593
Sebastia B. 2009. *Restoring mental health in India-pluralistic therapies and concepts*. New Delhi: Oxford University Press.

Shankar R and K. Rao. 2005. From burden to empowerment: the journey of family caregivers in India, In *Families and Mental Disorders*, eds. N. Sartorius, J. Leff, J. J. Lopez-Ibor, M. Maj, and A. Okasha, Chichester: Wiley 259–290.

Shyangwa PM, A. Jha. 2011. Nepal, In: *International perspectives on mental health*, Ed. H. Ghodse, 161–166. London: RCPsych Publications.

Somasundaram D. J. 1994. War trauma in a civilian population. *British Journal of Psychiatry* 165:524–527.

Somasundaram D. J. 1997. Treatment of massive trauma due to war. *Advances in Psychiatric Treatment* 3:321–331.

Somasundaram D. J. 1998. *Scarred Minds*, New Delhi: Sage Publications.

Somasundaram D. J. 2001. War trauma and psychosocial problems: Patient attendees of Jaffna. *International Medical Journal* 8:193–197.

Somasundaram D., and C. S. Jamunanatha 2002. Psychosocial consequences of war: Northern Sri Lankan experience. In: *Trauma, war and violence: public mental health in socio-cultural context*. ed. JTVM. de Jong. New York: Plenum.

Somasundaram D. J. 2002. Using traditional relaxation techniques in minor mental disorders. *International Medical Journal* 9:191–198.

Somasundaram D. J. 2006. Disaster mental health care in Sri Lanka. In: *Advances in disaster mental health and psychosocial support*. eds. J. O. Diaz, R. Srinivasa Murthy, and R. Lakshminarayana, 36–49. New Delhi: American Red Cross.

Srinivasan N. 2006. Together we rise—Kshema family power. In *Mental health by the people*, ed. R. Srinivasa Murthy. India: Peoples Action For Mental Health.

Srinivasan N. 2008. *We are not alone—family care for persons with mental illness*. Bangalore: Action for Mental Illness (ACMI).

Srinivasa Murthy R. 1998. Psychiatry in the third world: Managing and rationalising mental health care. *Current Opinion in Psychiatry*. 11:197–199.

Srinivasa Murthy R. 2005. Perspectives on the stigma of mental illness. In: *Stigma of mental illness in the third world*, eds. A. Okasha and C. N. Stefanis, 112–30. Cairo: World Psychiatric Association.

Srinivasa Murthy R. 2004. The national mental health program: Progress and problems. In: *Mental Health—An Indian perspective*,(1946–2003). eds. S. P. Agarwaal, D. S. Goel, R. L. Ichhpujani, R. N. Salhan, and S. Shrivatsava, 75–91. New Delhi: Directorate General of Health Services, Ministry of Health and Family Welfare.

Srinivasa Murthy R. ed. 2006. *Mental health by the people*. Bangalore: Published by Peoples Action For Mental Health.

Srinivasa Murthy R. 2007a. Mental health program in the 11th five year plan. *Indian Journal of Medical Research*. 125:707–712.

Srinivasa Murthy R. 2007b. Mass violence and mental health—recent epidemiological findings. *International Review of Psychiatry* 19:186–192.

Srinivasa Murthy, R., and K. Kumar. 2006. Challenges of building community mental health care in developing countries. *World Psychiatry* 7:101–102.

Srinivasa Murthy R. and N. N. Wig. 1983. The WHO collaborative study on strategies for extending mental health care, IV: A training approach to enhancing the availability of mental health manpower in a developing country. *American Journal of Psychiatry* 140:1486–1490.

Srinivasa Murthy R, M. K. Issac, C. R. Chandrasekar and A. Bhide. 1987. *Manual of Mental Health for Medical Officers-Bhopal Disaster*. New Delhi: Indian Council of Medical Research.

Srinivasa Murthy, R., C. R. Chandrasekar, Nagarajaiah, M. K. Issac, R. Parthasathy, and A. Raghuram. 1988. *Manual of mental health care for multi-purpose workers*. Bangalore: National Institute of Mental Health and Neurosciences.

Srinivasa Murthy R., K. V. Kishore Kumar, D. Gautam, T. Thomas, K. Sekar, and C. R. Chandrasekar. 2004. Community outreach for untreated schizophrenia in rural India: a follow–up study of symptoms, disability, family burden and costs. *Psychological Medicine* 34:1–11.

Srinivasa Murthy R. 2007. Family and mental health care in India. *Alumni Journal of Christian Medical College* 41(2):19–30.

Srinivasa Murthy R. 2011. Mental health services in low and middle income countries. Chapter 39. In: *Oxford Textbook of Community Psychiatry,* eds. G. Thornicroft, and G. Szmukler. Oxford University Press, Oxford, 325–336.
Susser E, P. Collins, B. Schanzer, V. K. Varma, and M. Gittelman. 1996. Topic for our times: Can we learn from the care of persons with mental illness in developing countries? *American Journal of Public Health* 86:926–928.
Swift CR (1972) Mental Health Program in a developing country: Any reference elsewhere? *American Journal of Orthopsychiatry* 42: 517.
Szmukler G., G. Thornicrioft. 2001. What is community psychiatry? In: *Textbook of Community Psychiatry*. eds. G. Thornicroft, G. Szmukler. Oxford. New York. 1–12.
Thara, R. 2002. Community mental health in India: A vision beckoning fulfillment? *Canadian Journal of Community Mental Health* 21:131–137.
Thara, R., R. Padmavathi, and T. N. Srinivasan. 2004. Focus on psychiatry in India. *British Journal of Psychiatry* 184:366–373.
Thara, R. 2007. We must tackle stigma by offering proven treatments. *PLoS Medicine* 4:965–966.
Thara, R., R. Padmavati, R. A. Aynkran, and S. John. 2008. Community mental health in India: A rethink. *International Journal of Mental Health Systems* 2:11.
Thirthahalli, J., B. K. Venkatesh, M. N. Naveen, K. V. Kishore Kumar, and B. N. Gangadhar. 2006. Do antipsychotics limit disability in schizophrenia? A naturalistic comparative study in the community. *Indian Journal of Psychological Medicine* 28:14–19.
Thirthahalli, J., B. K. Venkatesh, K. V. Kishore kumar, U. Arunachala, G. Venkatasubramaniam, D. K. Subbukrishna, et al. 2009. Prospective comparison of course of disability in antipsychotic treated and untreated schizophrenia patients. *Acta Psychiatrica Scandinavica* 119:209–217.
Thirthahalli, J., B. K. Venkatesh, M. N. Naveen, G. Venkatasubramaniam, U. Arunachala, K. A. Kishore kumar, et al. 2010. Do antiopsychotics limit disability in schizophrenia? A naturalistic comparative study in community. *Indian Journal of Psychiatry* 52:37–41.
Thornicroft, G., and G. Szmukler. 2001. *Textbook of community psychiatry*. New York: Oxford University Press.
Thornicroft, T., M. Tansella, and A. Law. 2008. Steps, challenges and lessons in developing community mental health care. *World Psychiatry* 7:87–92.
Thornicroft, G., E. Brohan, D. Rose, N. Sartorius, and N. Leese. 2009. Global pattern of experienced and anticipated discrimination against people with schizophrenia-a cross-sectional study. *Lancet* 373:408–4015.
Thornicroft, G., G. Szmukler, K. T. Mueser, and R. E. Drake. 2011. *Oxford textbook of community mental health*. New York: Oxford University Press.
Tomlinson, M., I. Rudan, S. Saxena, L. Swartz, A. C. Tsai, and V. Patel. 2008. Setting priorities for global mental health research. *Bulleting of WHO* 87:438–446.
Trivedi, J. K., D. Goel, R. A. Kallivayil, M. Issac, D. M. Shreshta, and H. C. Gambheera. 2007. Regional cooperation in South Asia in the field of mental health. *World Psychiatry* 6:57–59.
Udomratn, P. 2011. Thailand. In *International perspectives on mental health*, ed. H. Ghodse, 213–218. London: RCPsych Publications.
United Nations Convention on the Rights of Persons with Disabilities (UNCRPD). 2006. Available from: www.un.org/disabilities/default.asp?id=150, last. Accessed 18 Oct 2012.
Ustun, T. B. and N. Sartorius. 1995. *Mental illness in general health care: an international study*. Chichester: Wiley.
Ventovogel, P., R. Nassery, S. Azimi, and H. Faiz. 2011. Afghanistan. In *International perspectives on mental health*, ed. H. Ghodse, 63–67. London: RCPsych Publications.
Verghen, P. J., Van H. M. Praag, Ibor J. Ibor Lopez, J. L Cox, and D. Moussaoui. 2010. *Religion and Psychiatry- beyond boundaries. Chichester*. Wiley-Blackwell.
Verghese, A., J. K. John, S. Rajkumar, J. Richard, B. B. Sethi, and K. Trivedi. 1989. Factors associated with the course and outcome of schizophrenia—a multicentric study-results of 2 years multicenter follow up study. *The British Journal of Psychiatry* 154: 499–503.
Vijayakumar, L. 2003. *Suicide prevention: Meeting the challenge together*. Chennai: Orient Longman.

Wang, P. S., M. Angermeyer, G. Borges, R. Bruffarts, W. T. Chiu, G. de Girolomo, et al. 2007. Delay and failure in treatment seeking after first onset of mental disorders in the World Health Organization's World Mental Health Survey Initiative. *World Psychiatry* 6:177–185.

WHO. 1975. *Organization of mental health services in developing countries Technical Report Series*. Geneva: World Health Organization 564.

World Health Organization. 2001. The World Health Report 2001. Mental Health: new understanding new hope. Geneva: World Health Organization.

WHO. 2003. WHO-AIMS Report on Mental Health system in Bhutan: a report of the assessment of the mental health system in Bhutan using the World Health Organization—Assessment instrument for mental health systems (WHO-AIMS). Kabul, Afghanistan, 2006, WHO, Bhutan Office, Thimphu, WHO, Regional Office for the South East Asia (SEARO), New Delhi, WHO, Department of Mental Health and Substance Abuse (MSD), Geneva.

WHO. 2005. *Mental Health Atlas*. Geneva: World Health Organization.

WHO. 2006a. WHO-AIMS Report on Mental Health system in Afghanistan: a report of the assessment of the mental health system in Afghanistan using the World Health Organization—Assessment Instrument for mental health systems (WHO-AIMS). Kabul, Afghanistan, 2006, WHO, Afghanistan Office, Kabul, WHO, Regional Office for the Eastern Mediterranean (EMRO), Cairo, WHO, Department of Mental Health and Substance Abuse (MSD), Geneva.

WHO. 2006b. WHO-AIMS Report on Mental Health system in Myanmar: a report of the assessment of the mental health system in Myanmar using the World Health Organization—Assessment Instrument for Mental Health Systems (WHO-AIMS). Yangon, Myanmar, 2006, WHO, Myanmar Office, Yangon, WHO, Regional Office for the South East Asia (SEARO), New Delhi, WHO, Department of Mental Health and Substance Abuse (MSD), Geneva.

WHO. 2006c. WHO-AIMS Report on Mental Health system in Maldives: a report of the assessment of the mental health system in Maldives using the World Health Organization—Assessment Instrument for Mental Health Systems (WHO-AIMS). Male, Maldives, 2006, WHO, Maldives Office, Male, WHO, Regional Office for the South East Asia (SEARO), New Delhi, WHO, Department of Mental Health and Substance Abuse (MSD), Geneva.

WHO. 2006d. WHO-AIMS Report on Mental Health system in Uttarkhand, India: a report of the assessment of the mental health system in Uttarakand, India using the World Health Organization—Assessment Instrument for Mental Health Systems (WHO-AIMS). New Delhi, India, 2006, WHO, India Office, New Delhi, WHO, Regional Office for the South East Asia (SEARO), New Delhi, WHO, Department of Mental Health and Substance Abuse (MSD), Geneva.

WHO. 2007a. Assessment Instrument of Mental health Systems. Geneva: World Health Organization.

WHO. 2007b. *WHO mental health policy and service guidance package*. Geneva: World Health Organization.

WHO. 2007c. WHO-AIMS Report on Mental Health system in Bangladesh: a report of the assessment of the mental health system in Bangladesh using the World Health Organization—Assessment Instrument for Mental Health Systems (WHO-AIMS). Kabul, Afghanistan, 2006, WHO, Bangladesh Office, Dhaka, WHO, Regional Office for the South East Asia(SEARO) New Delhi, WHO, Department of Mental Health and Substance Abuse (MSD), Geneva

WHO. 2007d. WHO-AIMS Report on Mental Health system in Thailand a report of the assessment of the mental health system in Thailand, using the World Health Organization—Assessment Instrument for Mental Health Systems (WHO-AIMS). Bangkok, Thailand, 2007, WHO, Thailand Office, Bangkok, WHO, Regional Office for the South East Asia (SEARO), New Delhi, WHO, Department of Mental Health and Substance Abuse (MSD), Geneva.

WHO 2009a. WHO-AIMS Report on Mental Health system in Pakistan: a report of the assessment of the mental health system in Afghanistan using the World Health Organization—Assessment Instrument for Mental Health Systems (WHO-AIMS). Kabul, Afghanistan, 2006, WHO, Pakistan Office, Islamabad, WHO, Regional Office for the Eastern Mediterranean (EMRO), Cairo, WHO, Department of Mental Health and Substance Abuse (MSD), Geneva.

WHO 2009b. Pharmacological treatment of mental disorders in primary health care, World Health Organization.
WHO-EMRO. 2006. *Mental health in the Eastern Mediterranean countries: Reached the unreached. WHO Regional Publications of Eastern Mediterranean, Publication No.29*. Cairo: World Health Organization.
WHO Mental health atlas. 2011. Available on http://www.who.int/mental_health/publications/mental_health_atlas_2011/en/index.html, last Accessed Oct 18 2012.
WHO World Mental Health Survey Consortium. 2004. Prevalence, severity and unmet need for treatment of mental disorders in the World Health Organization World mental health surveys, *Journal of American Medical Association* 291:2581–2590.
WHO-SEARO. 2005 Mental health and psychosocial relief efforts after the Tsunami in South East Asia, WHO-SEARO, New Delhi.
WHO-WONCA. 2008. *Integrating mental health in primary care-a global perspective*. Geneva: World Health Organization and World Organization of Family Doctors.
Wig, N. N. 1978. General hospital psychiatry-right time for evaluation. *Indian Journal of Psychiatry* 20:1–5.
Wig, N. N. 1989 The future of psychiatry in developing countrie—The need for national programs of mental health. *National Institute of Mental Health and Neuro Sciences Journal* 7:1–11.
Wig, N. N. 1997. Stigma against mental illness (Editorial). *Indian Journal of Psychiatry* 39: 187–189.
Wig N. N. 2006. *The joy of mental health, mental health forum*. Chandigarh: Servants of the People Society.
Wig, N. N., and R. Srinivasa Murthy. 1980. *Manual of mental disorders for peripheral health personnel*. Chandigarh: Department of Psychiatry, PGIMER (2nd printing 1993).
Wig N. N., M. A. Suleiman, R. Routledge, R. S. Murthy, L. Ladrido-Ignacio, H. H. Ibrahim, et al. 1980. Community reactions to mental disorders. A key informant study in three developing countries. *Acta Psychiatrica Scandinavica* 61:111–126.
Wig N. N., R. Srinivasa Murthy, and T. W. Harding. 1981. A model for rural psychiatric services–Raipur Rani experience. *Indian Journal of Psychiatry* 23:275–290.
World Federation of Mental Health (WFMH). 2009. World mental health Day, WFMH.
Wright C., M. K. Nepal, W. D. Bruce-Jones 1989. Mental health patients in primary health care services in Nepal, *Asia-Pacific Journal of Public Health* 3:224–230.
Yousaf, F. 1997. Psychiatry in pakistan. *International Journal of Social Psychiatry* 43:298–302.

Chapter 11
Mental Health Resources in South Asia

Anju Kuruvilla and K. S. Jacob

11.1 Introduction

South Asia comprises of several countries that are diverse in terms of ethnicity, religion and language. However, uniformly, the core countries fall into the low and lower-middle income levels as classified by the World Bank and the region is recognized to be the poorest on the earth after Sub-Saharan Africa. Seventy percent of the South Asian population and about 75 % of South Asia's poor live in rural areas and most rely on agriculture for their livelihood. Finances accorded to health in the region therefore are limited. Given the deficits in general health infrastructure, the huge burden of infectious and physical diseases and limited funding, mental health resources remain a lower priority and insufficient to meet the growing burden of neuropsychiatric disorders in the region. The countries included in this review are Bangladesh, Bhutan, India, Maldives, Nepal, Pakistan and Sri Lanka.

This chapter summarizes the available data on Mental Health resources in the South Asian region, gathered primarily from the Mental Health Atlases of 2005 and 2011 published by the World Health Organization and from the World Health Organization Assessment Instrument for Mental Health Systems (WHO–AIMS) reports (WHO 2005, 2006, 2009, 2011). The results are limited by data being unavailable in some cases, differences in coverage and quality, and quantitative data being reported only at the aggregate level possibly masking important regional differences. In addition, data presented in this summary has not taken into account changes that may have occurred since the publication of the above reports.

A. Kuruvilla (✉) · K. S. Jacob
Christian Medical College, Vellore 632002, India
e-mail: sanju@cmcvellore.ac.in

K. S. Jacob
e-mail: ksjacob@cmcvellore.ac.in

J. K. Trivedi, A. Tripathi (eds.), *Mental Health in South Asia: Ethics, Resources, Programs and Legislation,* International Library of Ethics, Law, and the New Medicine 58,
DOI 10.1007/978-94-017-9017-8_11

11.2 Mental Health Financing

A country's financial resources significantly influence per capita mental health expenditure. However, it is also apparent that the proportion of total health expenditures directed towards mental health is not solely accounted for by availability of financial resources but is also affected by several other factors and is an indication of the priority given to mental health within the health sector. Lower income countries spend a smaller percentage of their health budget on mental health. Mental health budget in South Asian countries ranges from less than 0.5 to 2.05 % of the total health budget. Table 11.1 highlights details of individual countries and the budgetary allocation for mental health.

The exhortation to increase the mental health budget is laudable. However, the very low current spending implies that even a doubling of the budget will have very limited impact on mental health care delivery (Jacob 2011). For example, the median mental health expenditure in South Asia is a very small fraction of the total health budget, which in itself is very low in actual terms (Jacob et al. 2007). The doubling of the current budgets would result in a very meager availability of funds for mental health care. The ring fencing of such monies for use in stand-alone mental health programs or its employment for integration with primary health care will have limited impact on mental health care delivery. This is particularly true due to the dysfunctional nature of existing primary care systems and directly empowering primary health care delivery systems will pay better dividends.

While the finances for mental health care have improved over the past few decades in many South Asian countries, their utilization leaves much to be desired (Saraceno et al. 2007). For example, the funding increased from Rs. 280 million (US$ 6.2 million) during the Ninth Plan to over Rs. 4000 million (US$ 100 million) in the Eleventh plan for India (Jacob et al. 2007). However, only a small fraction of the money allocated in the national 5-year plans have been utilized. The failure to have a clear and workable plan to integrate mental health into primary care and to move psychiatric services from mental hospital to the community has resulted in a return of the allocated funds (Goel 2010).

11.3 Governance

The issues related to mental health policy, plans, programs and legislation are mentioned in Table 11.2.

11.3.1 Mental Health Policy

Some nations have an official statement regarding principles, objectives, guidelines and plans for action specific to mental health. Others do not have a dedicated mental

Table 11.1 Mental health financing

Country	Income group (World Bank 2010)	Budget allocation for mental health	Details of mental health financing	Disability benefits for mental health
Bangladesh	Low	Present	Mental health is accorded less than 0.5 % of total health budget. Of this, 67 % is devoted to mental hospitals. Primary sources of finances for mental health are tax based, private expenditure, grants. No mental disorder is covered in social insurance schemes. For those that pay out of pocket, the cost of the cheapest antipsychotic medication is Taka 5.00 (US$ 0.07) per day and the cheapest antidepressant medication Taka 3.00 (US$ 0.04) per day	Provided for persons with mental disorders. Lifetime pension provided for mentally handicapped children after the death of parent receiving pension
Bhutan	Lower-middle	Present	0.17 % of the total health budget is spent on mental health. (According to the WHO-AIMS report of 2006 the figure is 1 %), Primary sources of mental health financing are tax based, grants and social insurance. Hundred percent of the population has free access to essential psychotropic medicines. All mental disorders are covered by social insurance schemes. For those who pay for their medicines out of pocket, the cost of either antipsychotic or antidepressant medication is Nu. 10 ($ 0.23 USD) per day	Provided for persons with mental disorders. Mentally ill patients are exempted from paying labour tax; some are given cash benefits
India	Lower-middle	Present	2.05 % of the total health budget is spent on mental health. The primary sources of mental health financing are tax based, private expenditure by the patient or family, private insurances and social insurance. State and Central Government provides funding for health services; services provided at Government health centres are free	Provided for persons with mental disorders in a limited way
Maldives	Lower-middle	Absent	Budget has allocations for provision of free medication to patients with mental illness. Government provides financing for the residential facility for People with Special Needs. Separate budgets allocated for National Narcotics Control Bureau and the Rehabilitation Centre also. Primary sources of mental health financing are out of pocket expenditure by patient and grants. Cost of antipsychotic medication is 3 Rf per day, antidepressant medication is 2 Rf per day. No social insurance schemes available. Government expenditures also directed by the Ministry of Health towards the outpatient and inpatient facilities within the general hospital	Provided for persons with mental disorders

Table 11.1 (continued)

Country	Income group (World Bank 2010)	Budget allocation for mental health	Details of mental health financing	Disability benefits for mental health
Nepal	Low	Present	0.08 % (0.17 % according to the WHO-AIMS report of 2006) of the total health budget is spent on mental health, largely on metal hospitals. Primary sources of mental health financing are out of pocket expenditure by the patient or family, tax based and grants. Other organizations (WHO) and NGOs (United Mission to Nepal etc.) are also important sources. A negligible portion of the population has free access to essential psychotropic medicines. For those that have to pay for their medicines out of pocket, the cost of antipsychotic medication is approximately 9 Nepalese Rupees (NRs) per day, and the cost of antidepressant medication is 9 NRs per day. There is no social insurance scheme available	Provided for persons with mental disorders. Chronic mental illness has been equated with other disabilities according to the Disability Act
Pakistan	Lower–middle	Present	0.4 % of the total health budget is spent on mental health of which 11 % is spent on mental hospitals. The primary sources of mental health financing are out of pocket expenditure by the patient or family, tax based, and private insurance. Five percent of the population has free access to essential psychotropic medicines. For those that pay out of pocket, the cost of antipsychotic medication is 2 dollars per day, and the cost of antidepressant medication is 5 dollars per day. None of the mental disorder is covered by social insurance schemes	Provided to persons with mental disorders who are unable to work due to mental illness
Sri Lanka	Lower-middle	Present	1.6 % of the total health budget is spent on mental health, mostly for mental hospitals. Drugs are provided free of charge to those receiving services from the Government sector. Primary sources of mental health financing are tax based, out of pocket expenditure by the patient and private insurances	No disability benefits are provided for persons with mental disorders

Table 11.2 Policy and Legislation

Country	Mental Health Policy	Substance Abuse Policy	National Mental Health Program	National Therapeutic Drug Policy/Essential List of Drugs	Mental Health Legislation
Bangladesh	Specific policy absent. Incorporated in policy, strategy and action plan for surveillance and prevention of Non-Communicable Diseases (NCD)-last revised in 2006	Present; formulated in 1990	Present; formulated in 1984	Present; formulated in 1983	Lunacy Act enacted in 1912. Draft version of Mental Health Act made in 1999, submitted 2002; formulation of final version of the act and its enactment pending. Latest legislation enacted in 1912. Disaster/ emergency preparedness plan for mental health submitted to government for approval
Bhutan	Present-initially formulated in 1997	Present (Narcotic Drugs and Psychotropic Substances Notification) -initially formulated in 1988	Present-formulated in 1997	Present; formulated in 1987	Nil Bhutan Penal Code, enacted in 2004, contains clauses to safeguard the interests of people with mental illness in the criminal justice system. Framework on disaster risk management including preparedness for mental health in Bhutan has been proposed to the government
India	Mental health policy absent. National Health Policy (2002) incorporates suggestions made in the draft of the National Mental Health Policy	Absent. National master plan for substance abuse discussed in 1994 guided by the Ministry of Health and Family Welfare and the Ministry of Social Justice and Empowerment. Smoking in public places banned by many Indian states recently	Present-formulated in 1982. National Mental Health Program (NMHP), initially formulated in 1982; restrategized in 2002. District Mental Health Program launched in 1995; covers 24 districts currently, with plans for expansion	Present; year of formulation unknown	Mental Health Act formulated in 1987 (latest legislation); Narcotic Drugs and Psychotropic Substances Act formulated in 1985 (amended in 2001); Other related acts include Persons with Disabilities Act,1995 (draft on Rights of Persons with Disabilities Bill made in 2011)

Table 11.2 (continued)

Country	Mental Health Policy	Substance Abuse Policy	National Mental Health Program	National Therapeutic Drug Policy/Essential List of Drugs	Mental Health Legislation
Maldives	Absent	Present-formulated in 1977	Absent	Present; year of formulation not known available	No mental health legislation Disaster preparedness plan for mental health and psychosocial issues established 2005
Nepal	Present; initially formulated in 1996	Present-formulated in 1994	Absent	Present; formulated in 1986	Some sections of the Civil Law have legal provisions concerning insanity. Mental health legislation has been drafted; awaiting official approval. Latest legislation enacted in 1964
Pakistan	Present; initially formulated in 1997	Present; initially formulated in 1997 and implemented by the Planning Commission of the Government of Pakistan	Present; formulated in 1986. It is a part of the general health policy of the country; fully implemented in 2001	Present-formulated in 1997	Mental health ordinance enacted in 2001
Sri Lanka	Absent (committee has been appointed to formulate draft policy and legislation)	Absent (policy has been developed but is still to be ratified by the parliament)	Present-formulated in 1966. Based on recommendations given in World Health Report, 2001 and the National Strategic Mental Health Plan, 2001	Present-formulated in 1985	Mental Disease Ordinance of 1873 (amended in 1956). Efforts to formulate a new Mental Health Act started in the year 2000

health policy but include mental health care in the general health policy. Still others have neither. Bhutan, Nepal and Pakistan are the countries that do have a dedicated mental health policy. India, the most populous country in the region, does not have a dedicated mental health policy. The failure of countries to have a dedicated and comprehensive mental health policy reflects the low priority for mental health in the region.

11.3.2 Mental Health Plan or Program

This is conceptualized as a detailed proposal that specifies the strategies and activities that will be implemented to realize the objectives of the mental health policy. It also specifies elements such as the budget, timeframe and specific targets that will be met. Most plans mention a focus on the promotion of mental health, the prevention of mental disorders, treatment and rehabilitation. Every country in the region other than Maldives and Nepal has a mental health program. However, the implementations of such programs are variable across counties and regions. Many countries have policies, which exist only on paper sans implementation, highlighting the problems related to mental health policies and programs in South Asia (Jacob 2011).

11.3.3 Mental Health Legislation

This covers a range of issues from access to mental health care, admission to mental health facilities, to provisions for legal mechanisms to promote and protect human rights. Most countries in the region have formulated mental health legislation with a few exceptions.

11.4 Health Care Delivery

Mental Health care delivery is detailed in Tables 11.3 and 11.4.

Primary health care provides the first point of entry into the health system. Factors that are important in the primary care system that cater to mental health needs include training of health professionals, medication availability and prescribing practices, referral protocols and availability of manuals. However, the primary health care delivery system in the public sector in the region is poor. It is not efficient even for managing the many physical health problems. This is particularly true in primary care and at the community level. Although recent inputs in some countries have increased the infrastructure, physical resources and personnel (e.g. India), decades of neglect, the overburdened system and the poor discipline and morale of the health professionals make the inclusion of mental health care provision into primary care difficult (Jacob 2011; Saraceno et al. 2007; Hanlon 2010).

Table 11.3 Mental health care delivery and facilities (a)

Country	Mental health as part of primary health care system	Regular training of primary care professionals	Community care facilities for patients with mental disorders/Community Mental Health Centers	Psychiatric rehabilitation centres
Bangladesh	MH is part of primary health care system; treatment of severe mental disorders available at primary level. Periodic extension services provided at the primary care level by the Institute of Mental Health Research, Dhaka. Assessment and treatment protocols available for key mental health clinics. Only doctors allowed to prescribe psychotropic medications in PHCs. Almost all physician-based primary health care clinics have at least one psychotropic medicine of each therapeutic category	Present. Provided by Ministry of Health and Family Welfare for primary care physicians, health workers and Medical administrators	Present. Thirty One community-based psychiatric inpatient units available (0.58 bed per 100,000 population)-2 % reserved for children and adolescents. Eleven community residential facilities available (0.92 beds/places per 100,000 population)-55 % reserved for children and adolescents Of available beds-20 % mental hospital, community based psychiatric inpatient units, 60 % other residential facilities, 20 % community residential facilities	No specific/organized rehabilitation programs available; attempts being made to implement day care facilities, sheltered workshops and rehabilitation programs for chronic schizophrenics
Bhutan	Mental health is part of primary health care system. Treatment of severe mental disorders is available at the primary level	Present	Present. Basic emergency and follow-up services done by health workers in the community. Sixty Three community-based psychiatric inpatient units available integrated with mental health outpatient facilities. All community-based psychiatric inpatient units had at least one psychotropic medicine of each therapeutic class and 30 Basic Health Units had all psychotropic drugs, with the exception of mood stabilizers. No community residential facilities available	Nil
India	Mental health is part of primary health care system is available in 22 of 600 districts. Treatment of severe mental disorders available at the primary level	Present but limited to some districts. Training materials have been developed and field tested. Workshops for sensitization/train personnel and mental health professionals have been undertaken	Present in some designated districts. NGOs provide different types of services in the community (telephone hotlines, residential rehabilitative services)	Provided by NGOs

Table 11.3 (continued)

Country	Mental health as part of primary health care system	Regular training of primary care professionals	Community care facilities for patients with mental disorders/Community Mental Health Centers	Psychiatric rehabilitation centres
Maldives	Mental health not part of primary health care system. Actual treatment of severe mental disorders is not available at the primary level No treatment protocols for key mental health conditions available. In primary health care only doctors are allowed to prescribe without any restrictions. 21–50 % of the physician-based PHC clinics have at least one psychotropic medicine of each therapeutic category	Present; is integrated in the community health workers training program	No community care facilities for patients with mental disorders. There is one community-based psychiatric inpatient unit (1.38 beds per 100,000 population) which had at least one psychotropic medicine of each therapeutic class No community residential facilities available	Nil
Nepal	Mental health not an integral part of primary health care; however treatment of severe mental health disorders available in ten districts where community health programs are run with the support of NGOs. Some physician-based and few non-physician-based primary health care clinics, have available assessment and treatment protocols for key mental health conditions. Primary health care nurses, non-doctor/non-nurse primary health care workers are allowed to prescribe but with restrictions. Primary health care doctors prescribe without restriction. Majority of physician-based PHC clinics have at least one psychotropic medicine of each therapeutic category	No regular training of primary care professionals in mental health but has occurred in some regions. Referral network has been set up including local faith healers	17 community-based psychiatric inpatient units available (1.00 beds per 100,000 population), none reserved for children and adolescents only. All had at least one psychotropic medicine of each therapeutic class available. No public community residential facilities available	Nil

Table 11.3 (continued)

Country	Mental health as part of primary health care system	Regular training of primary care professionals	Community care facilities for patients with mental disorders/Community Mental Health Centers	Psychiatric rehabilitation centres
Pakistan	Mental health is a part of primary health care system. Actual treatment of severe mental disorders is available at the primary level. Some physician-based primary health care clinics have assessment and treatment protocols for key mental health conditions. Only primary health care doctors are allowed to prescribe psychotropic medications. A few physician-based PHC clinics have at least one psychotropic medicine of each therapeutic category	Regular training of primary care professionals present. NGOs (e.g. National Rural Support Program (NRSP) are also being trained. No training facilities present for social workers. Mental health training included in the program of the District Health Development Centres with Institute of Psychiatry Rawalpindi Medical College as a resource centre. Training manuals, packages on counseling skills, and rehabilitation of mentally ill have been developed	Community care facilities for patients with mental disorders present. More than 78 psychiatrists have been trained in community mental health to act as resource persons in the development of programs in their areas Six hundred and twenty four community-based psychiatric inpatient units are available. One percent of these beds in community-based inpatient units are reserved for children and adolescents only. Thirty Four percent of community-based psychiatric inpatient units had at least one psychotropic medicine of each therapeutic class available in the facility. Community residential facilities are not available	There are many residential and day-care facilities, especially for people with learning disabilities providing social, vocational and educational activities
Sri Lanka	Mental health is a part of primary health care system but actual treatment of severe mental disorders is not available at the primary level. Primary care workers in some regions carry out mental health work including dispensing of medication	Regular training of primary care professionals is carried out in the field of mental health	No community care facilities for patients with mental disorders. Psychiatric services initially available only at tertiary care level are presently available at secondary care level also. Outreach, liaison, community mental health and school mental health services have been launched. With the assistance of the Nations for Mental Health Program (WHO), attempts have been made to resettle long-stay patients from mental hospitals in the community	Ten intermediate-stay units developed by Ministry of Health to provide rehabilitation services to patients discharged from the psychiatry units at tertiary care level. Day and community care services are also provided from these centres. NGOs (e.g. Sahanaya) also provide community residential facilities rehabilitation programs

Table 11.4 Mental Health Care Delivery and facilities (b)

Country	Human rights review body/training for staff on human rights	Organization of mental health services	Mental health outpatient facilities	Day treatment facilities	Mental Hospitals	Programs for Special Population
Bangladesh	Nil	No specific mental health authority or organization of mental health service areas	50 outpatient mental health facilities available, 4% for children and adolescents only. None provide follow-up care in the community. Two percent have mental health mobile teams. Fifty eight percent of outpatient facilities have at least one psychotropic medicine of each therapeutic class available	Nil	1 mental hospital (0.4 beds per 100,000 population) organizationally integrated with mental health outpatient facilities. Had at least one psychotropic medicine of each therapeutic class 15 beds available for persons with mental disorders in forensic inpatient units, 3900 beds in other residential facilities (homes for persons with mental retardation, detoxification inpatient facilities, homes for the destitute, etc.)	Special units in National Institute for child and adolescent and elderly population
Bhutan	Nil	National mental health authority present. Mental health services organized in terms of catchment/ service areas or districts and sub-districts	63 outpatient mental health facilities, fully integrated with mental health inpatient units. All provide follow-up care in the community; no mental health mobile clinic. teams. All have at least one psychotropic medicine of each therapeutic class	One (for adults)	No mental hospitals. No forensic inpatient units. Additional residential facilities: 50-bed facility for juvenile delinquents, 15-bed facility for drug detoxification rehabilitation, five-bed residential facility for elderly. Residential facility for victims of domestic violence and rape planned	Nil

Table 11.4 (continued)

Country	Human rights review body/training for staff on human rights	Organization of mental health services	Mental health outpatient facilities	Day treatment facilities	Mental Hospitals	Programs for Special Population
India	National Human Rights Commission established (1993) to facilitate Protection of Human Rights Act. Act also mandates setting up of State Human Rights Commission and Human Rights Courts. Supreme Court judgment passed (2002) for protection of rights of mentally ill persons	Central and State Mental Health Authority exist	Several present	Present	Present (37 psychiatric hospitals)	Specific programs for mental health for disaster affected population elderly, substance use disorders. Counseling and rehabilitation centres established in voluntary sector supported by Ministry of Welfare School health initiatives by NGOs. Residential facilities available for mentally challenged and mentally ill children
Maldives	Present	No mental health authority in the country. Mental health services are organized in terms of catchment/service areas	One outpatient mental health facility available; does not provide follow-up care in the community, but occasionally provides mental health mobile teams. None of mental health outpatient facilities had at last one psychotropic medicine of each therapeutic class. At the regional level, services provided by visiting psychiatrists. At the atoll and island levels, trained community health workers and nurses provide basic psychiatric services	One facility available	No mental hospitals available. 125 beds present in a home for persons with mental retardation and 500 beds in a residential facility specifically for people with substance abuse problems (children and adolescents are not admitted). No forensic inpatient units	Nil

Table 11.4 (continued)

Country	Human rights review body/training for staff on human rights	Organization of mental health services	Mental health outpatient facilities	Day treatment facilities	Mental Hospitals	Programs for Special Population
Nepal	Absent	Nil MH authority	18 outpatient mental health facilities available, none for children and adolescents only. All provide follow-up care in the community; no mental health mobile teams. All have at least one psychotropic medicine of each therapeutic class available	3 day treatment facilities available (none are for children/adolescents only)	One mental hospital exists (0.20 beds per 100,000 Population)-organizationally integrated with mental health outpatient facilities. No beds reserved for children/adolescents. At least one psychotropic medicine of each therapeutic class is available. No separate forensic inpatient units 145 beds in other residential facilities (homes for persons with mental retardation, detoxification inpatient facilities, homes for the destitute)	Special clinics for children, psychosexual disorders, headache, drug abuse treatment refugees Orientation programs organized for school teachers
Pakistan	National human rights review body exists Regular evaluation of quality of care done by National Steering Committee	National mental health authority exists. Mental health services are not organized in terms of catchment/service areas	3729 outpatient mental health facilities available in the country, 1 % is for children and adolescents only. Forty-six percent of outpatient facilities provide follow-up care in the community, 1 % has mental health mobile teams. Thirty three percent of mental health outpatient facilities had at last one psychotropic medicine of each therapeutic class available	Nil	Five mental hospitals present, organizationally integrated with mental health outpatient facilities. All had at least one psychotropic medicine of each therapeutic class available. 0.02 beds for persons with mental disorders in forensic inpatient units and 1620 in other residential facilities (homes for persons with mental retardation, detoxification inpatient facilities, homes for destitute, etc.)	For refugees, children, women and victims of torture. NGOs, international organizations also involved. Residential and day care facilities for people with learning disabilities, especially big cities. School mental health program present

Table 11.4 (continued)

Country	Human rights review body/training for staff on human rights	Organization of mental health services	Mental health outpatient facilities	Day treatment facilities	Mental Hospitals	Programs for Special Population
Sri Lanka	Details not available	Details not available	Present	Day, outreach, liaison, community mental health and school mental health services have been launched	Ministry of Health has recently initiated program to develop 10 intermediate-stay units at a regional level, provide rehabilitation services to patients with disability, discharged from the psychiatry units at tertiary care level. Day care services and community care services are also provided from these centres	Specific programs for mental health for children, trauma victims. Homes for children with severe learning and behavioural disorders are available. Counselling centres in the private and NGO sectors are present

The integration of mental health into primary health care delivery is only possible with well-established, functional and efficient systems.

Mental health facilities The region has wide variations with regard to the number of facilities, beds, admissions and follow-up contacts at outpatient facilities, day treatment facilities, psychiatric wards in general hospitals, community residential facilities and mental hospitals and facilities for special populations such as children and adolescents. The large majority of psychiatric beds in the region are in metal hospitals. Community residential and day treatment facilities are few.

It is also no secret that the majority of psychiatrists and their professional associations in South Asia are indifferent to empowering general physicians (Jacob 2011; Saraceno et al. 2007). The community psychiatry movement always had second-class status within the discipline. Psychiatrists prefer the safety of specialist institutions rather than moving out into the community. They favour referrals and consultations to transferring expertise to primary care professionals. For example, the community psychiatry movement was led in India, in the 1970s and 1980s, by many national institutes and centres of excellence across the country. However, the very ideas of decentralisation and empowerment gradually lost ground and are all but abandoned by these centres, resulting in a leadership vacuum. In addition, the concentration of trained mental health personnel in urban South Asian centres make the delivery of care and the supervision of community psychiatry programs deficient (Jacob 2011). Mental health professionals are also divided based on their disciplinary perspectives (e.g. psychiatry, psychology, nursing, social work) weakening the cause.

11.5 Human Resources

The number of trained mental health professionals is uniformly low in the South Asian countries. It is also evident that available facilities and personnel are largely concentrated in the urban areas with minimal resources available in the rural areas. Informal human resources such as family and user associations were scarce in the region. Tables 11.5 and 11.6 record the details across nations.

In addition, the complete lack of training to manage common psychiatric conditions seen in general medical settings is a major lacuna in the curriculum in many South Asian countries (Jacob 2011). Most training programs employ tertiary care and specialist perspectives and jargon, which are inappropriate for primary care settings. For example, most psychiatric classifications used in training physicians subclassify common mental disorders into depression, anxiety and somatisation, etc. (WHO 1996), despite the fact that the recognition of these syndromes is difficult in primary care practice in South Asia (Jacob 2006). The inability to recognize classical psychiatric syndromes by primary care physicians and nurses make them unable to choose from the many different management protocols. This often results in the dismissal of the patients complaints as unimportant or management using vitamins and benzodiazepines.

Table 11.5 Human resources

Country	Total psychiatric beds/ 10,000 pop	Psychiatric beds in mental hospitals/ 10,000 population	Psychiatric beds in general hospitals/ 10,000 population	Psychiatric beds in other settings/ 10,000 population	Number of psychiatrists/ 100,000 pop	Number of neurosurgeons/ 100,000 pop	Number of psychiatric nurses/ 100,000 pop	Number of neurologists/ 100,000 pop	Number of psychologists/ 100,000 pop	Number of social workers/ 100,000 pop	Other information
Bangladesh	0.065	0.03	0.009	0.024	0.05	0.01	0.06	0.02	0.002	0.001	One occupational therapist and 4011 medical assistants
Bhutan	0	0	0	0	0.3	0	.16	0	0	0	1 occupation therapist
India	0.25	0.2	0.05	0.01	0.2	0.06	0.05	0.05	0.03	0.03	200 mental health workers of other types present. One third of mental health beds are in a single state, several states have no mental hospitals. Some beds allocated to treatment of drug abuse and for child psychiatry. Few mental health professionals based in rural areas. Most states allow public sector psychiatrists to have private clinics. Many mental health professionals have emigrated Psychologists do not have prescription privileges, no formal system of licensing clinical psychologists
Maldives	0	0	0	0	0.36	0.36	0	0	1.2	0	Most personnel work in the capital and in tertiary centres. The psychiatrist visits other islands whenever needed

Table 11.5 (continued)

Country	Total psychiatric beds/10,000 pop	Psychiatric beds in mental hospitals/10,000 population	Psychiatric beds in general hospitals/10,000 population	Psychiatric beds in other settings/10,000 population	Number of psychiatrists/100,000 pop	Number of neurosurgeons/100,000 pop	Number of psychiatric nurses/100,000 pop	Number of neurologists/100,000 pop	Number of psychologists/100,000 pop	Number of social workers/100,000 pop	Other information
Nepal	0.08	0.02	0.02	0.04	0.12	0.04	0.08	0.08	0.08	0.04	40 beds (governmental and private sector) reserved for drug dependence treatment. All mental health professionals are located in urban and semi-urban areas. Eighty percent of the psychiatry beds in the country are located in or near the largest city
Pakistan	0.24	0.06	0.148	0.02	0.2	0.2	0.08	0.14	0.2	0.4	2000 other mental health personnel. Psychiatric units present in all medical colleges and allied hospitals in public and private sector. Some psychiatric care facilities are available at the district level. Beds for the treatment of drug abusers available at 232 centres. Forensic beds available at a few centres. There are two child psychiatrists in the country. Mental health professionals are concentrated in urban centres. Most psychiatrists have private clinics

Table 11.5 (continued)

Country	Total psychiatric beds/ 10,000 pop	Psychiatric beds in mental hospitals/ 10,000 population	Psychiatric beds in general hospitals/ 10,000 population	Psychiatric beds in other settings/ 10,000 population	Number of psychia-trists/ 100,000 pop	Number of neurosur-geons/ 100,000 pop	Number of psychiatric nurses/ 100,000 pop	Number of neu-rologists/ 100,000 pop	Number of psy-cholo-gists/ 100,000 pop	Number of social workers/ 100,000 pop	Other information
Sri Lanka	1.8	1.4	0.3	0	0.2	0.03	1.8	0.06	0.02	0.07	Most psychiatrists and other mental health profes-sionals are concentrated in urban areas; dirth in areas such as the northeast. Almost three fourth of the beds are in the capital. One private hospital provides inpatient services. Most psychiatrists work in the public sector, almost all have private clinics also

Table 11.6 NGO input, family and user associations, public education and links with other sectors

Country	Others input: NGO; family and users association	Public education/awareness campaigns on mental health	Legislative and financial provisions for persons with mental disorders	Links with other sectors	Monitoring
Bangladesh	10 NGOs involved mainly with treatment and rehabilitation No consumer association 1 family association, not supported by the government	Coordinating body (National Institute of Mental Health, Dhaka) oversees public education and awareness campaigns. Ministry of Health & Family Welfare, NGOs, professional associations, private trusts and foundations and international agencies have promoted public education and awareness campaigns in the last 5 years	None	Formal collaborations present between the government departments responsible for mental health, primary health care/community health, substance abuse, welfare and criminal justice. No mental health facility has access to programs outside the mental health facility that provide outside employment for users with severe mental disorders. Five percent of people who receive social welfare benefits do so for a mental disability	Formally defined list of data to be collected by mental health facilities exists (number of beds, admissions, length of stay and patient diagnoses in mental hospital). Initiatives have been taken to maintain continuous communication with four model sub districts around capital city to develop community mental health services
Bhutan	NGOs mainly involved in advocacy, promotion, prevention, treatment and rehabilitation; provide variety of services including counselling, suicide prevention, training of lay counsellors, rehabilitation programs, self-help groups	Information and Communication Bureau, Ministry of Health, is coordinating body that oversees public education and awareness campaigns on mental health and mental disorders. Other government agencies, NGO and professional associations also involved	None	Formal collaborations exist with departments/agencies responsible for primary health care/community health, HIV/AIDS, reproductive health, child and adolescent health, substance abuse, child protection and education. Many schools have school-based counselors and health educators. Three percent of people who receive social welfare benefits do so for a mental disability	Formally defined list of individual data items that ought to be collected by all mental health facilities exists Government health department receives data from community based psychiatric inpatient units, mental health outpatient facilities

Table 11.6 (continued)

Country	Others input: NGO; family and users association	Public education/awareness campaigns on mental health	Legislative and financial provisions for persons with mental disorders	Links with other sectors	Monitoring
India	NGOs mainly involved in advocacy, promotion, prevention, treatment and rehabilitation. NGOs involved in counselling, suicide prevention, training of lay counselors, provision of rehabilitation programs Family members of mentally ill persons have recently come together to form self-help groups	Government agencies and NGOs have promoted public education and awareness campaigns Nil official details regarding coordinating Body	Nil details	Nil official details available	Nil official details available
Maldives	NGOs mainly involved in treatment and rehabilitation for patients with severe mental illnesses and special needs, counseling for drug users, counseling or support groups. There are no consumer or family associations in the country	Government agencies and NGOs have promoted public education and awareness campaigns. No coordinating body to oversee them	None	Formal collaborations with the departments/agencies responsible for Primary health care/community health, HIV/AIDS, Reproductive Health, Child and Adolescent health, Substance Abuse, and Education and Welfare In terms of support for child and adolescent health, 3 % of the primary and secondary schools have either a part-time or full-time counselors	Defined data set is to be collected regularly from the residential facility Government health department receives data from community based psychiatric inpatient unit and mental health outpatient facility located within the general hospital (number of inpatient admissions, number of days spent in the hospital, diagnoses, number of users treated and number of user contacts)

Table 11.6 (continued)

Country	Others input: NGO; family and users association	Public education/awareness campaigns on mental health	Legislative and financial provisions for persons with mental disorders	Links with other sectors	Monitoring
Nepal	No consumer and family associations in Bhutan. Two NGOs (for with abused women and substance abusers) involved in individual assistance activities such as counseling, housing, or support groups	No coordinating body to oversee public education and awareness campaigns on mental health and mental disorders. Government agencies, NGOs, and professional organizations involved in promotion of public education and awareness campaigns	There is no mental health legislation protecting the right of patients with mental disorders. However, a draft has been prepared and is awaiting finalisation	Formal collaborations exist between the government department responsible for mental health and the departments/agencies responsible for Primary Health Care/Community Health, Child and Adolescent Health and Substance Abuse Less than 0.02 % of primary and secondary schools have either a part-time or full-time mental health professional, and a few (between 1 and 20 %) of primary and secondary schools have school-based activities to promote mental health and prevent mental disorders No mental health facilities have access to programs outside the mental health facility that provide outside employment for users with severe mental disorders	There is no formally defined minimum data set of items to be collected by mental health facilities

Table 11.6 (continued)

Country	Others input: NGO; family and users association	Public education/awareness campaigns on mental health	Legislative and financial provisions for persons with mental disorders	Links with other sectors	Monitoring
Pakistan	NGOs mainly involved in advocacy, promotion, prevention, treatment and rehabilitation, run mental health services for homeless psychotic patients, refugees and day care centres for drug users. No information on consumer associations for health issues. 5 NGOs in the country that are involved in individual assistance activities such as counseling, housing, or support groups	Coordinating body present to oversee public education and awareness campaigns on mental health and mental disorders. Government agencies, NGOs, professional associations, private trusts, international agencies involved in promotion of public education and awareness campaigns	The following legislative and financial provisions exist to protect and provide support for users: (1) provisions concerning a legal obligation for employers to hire a certain percentage of employees that are disabled, (2) provisions concerning protection from discrimination (dismissal, lower wages) solely on account of mental disorder, (3) provisions concerning priority in state housing and in subsidized housing schemes for people with severe mental disorders, and (4) provisions concerning protection from discrimination in allocation of housing for people with severe mental disorders. These provisions exist are not enforced	Formal collaborations exist between government department responsible for mental health and the departments/agencies responsible for primary health care/community health, HIV/AIDS, reproductive health, child and adolescent health, substance abuse, child protection, education, employment, housing, welfare, criminal justice, the elderly, and other departments/Agencies. Three percent of primary and secondary schools have either a part-time or full-time mental health professional. No mental health facilities have access to programs outside the mental health facility that provide outside employment for users with severe mental disorders	Formally defined list of individual data items that ought to be collected by all mental health facilities exists. Government health department receives data from all of the mental Hospitals, community based psychiatric inpatient units, and mental health outpatient facilities

Table 11.6 (continued)

Country	Others input: NGO; family and users association	Public education/awareness campaigns on mental health	Legislative and financial provisions for persons with mental disorders	Links with other sectors	Monitoring
Sri Lanka	NGOs are involved mainly in advocacy, promotion, prevention, treatment and rehabilitation Different NGOs provide specific services: day care services, training to mental health professionals, community mental health project, vocational rehabilitation, self-employment, community-based interventions in the areas of suicide prevention, disasters, mental retardation, alcohol related problems and mental disorders in elderly population	No official data available	No official data available	No official data available	No official data available

The lack of ability to diagnose and manage psychiatric presentations in primary care has spawned many short courses for physicians and nurses (Jacob 2011). However, these transfer knowledge, rather than skill and confidence making it difficult for physicians to translate the knowledge gained for use in their primary care practice.

11.6 Medicines for the Mental and Behavioural Disorders

Table 11.7 documents the details of the psychotropic medication available in the South Asian region. In comparison with the low and lower-middle income countries in the region, median expenditures on medicines in upper-middle and high income counties are approximately 340 times greater. Individuals in many of the countries in the south Asian region are largely dependent on out of pocket expenditure to finance their medical needs in the absence of widespread social security systems.

11.7 Information Systems

Data regarding numbers of admissions, contacts, days spent and interventions delivered at different types of mental health facilities, including mental hospitals, general hospitals, primary health care facilities, outpatient facilities and community residential facilities is shown in Table 11.7. Most countries collect mental health data on outpatient facilities from mental hospitals and general hospitals. Fewer countries collect data from other treatment facilities.

11.8 The Way Forward

Current attempts to revive community psychiatry programs at national and international levels are more about mental health advocacy and less about technical inputs and guidance (Jacob 2011). The technology to translate psychiatric research evidence into primary care practice, with its poor infrastructure, staffing and morale, does not exist in poor countries. The idealism of the original primary health care movement, without technical contribution for scaling-up, meant that implementation at the national level was problematic, patchy and unproductive. The situation has not changed since.

The suggested changes include restructuring basic medical and paramedical education, revamping psychiatric training, revitalizing primary care and renewing political pressure to improve mental health care resources and delivery (Jacob 2011).

Table 11.7 Training, Information systems, Medicines

Country	Psychiatric training: Post-graduate training in psychiatry/psychology	Information Gathering System		Therapeutic Drugs: availability	Other Information
		Mental health reporting system	Data collection system		
Bangladesh	Present. Passing per year per 100,000 (in year of data collection): 0.0036 psychiatrists, 0.18 psychologists with at least 1 year training in mental health care. 1–20 % of psychiatrists emigrate after completion of training. 4 and 2 % of training for medical doctors and nurses respectively, is devoted to mental health	Present	Hospital based service data collection present. Information also sent from community based psychiatric inpatient units (45 %), and mental health outpatient facilities (28 %). Some figures related to incidence and prevalence of mental disorders available from WHO supported survey	Drugs generally available at the primary health care level of the country: carbamazepine, phenobarbital, phenytoin sodium, sodium valproate, amitriptyline, chlorpromazine, diazepam, fluphenazine, haloperidol, lithium, levodopa	
Bhutan	Number of professionals graduated in year of data collection: 1 psychiatrist (0.15 per 100,000 population), 2 nurses with at least 1 year training in mental health care (0.30 per 100,000 population). No psychologists or occupational therapists	Present	Present. Patient treatment data available in one hospital since July 1999	Drugs generally available at the primary health care level: phenobarbital, diazepam	
India		Present. Limited to mental hospitals	No data collection system or national epidemiological study on mental health	Drugs available at the primary health care level in designated districts where special program is operational: carbamazepine, phenobarbital, phenytoin sodium, sodium valproate, amitriptyline, chlorpromazine, diazepam, fluphenazine, haloperidol, lithium, carbidopa, levodopa. A large, mostly indigenous, pharmaceutical industry ensures that most psychotropic drugs are available	The traditional Indian medical systems of Ayurved and Unani recognized mental illness and provided necessary treatment

Table 11.7 (continued)

Country	Psychiatric training: Postgraduate training in psychiatry/psychology	Information Gathering System		Therapeutic Drugs: availability	Other Information
		Mental health reporting system	Data collection system		
Maldives	Nil. One-year certificate course in counseling recently started with assistance from an NGO	Present. National registry maintained at the Ministry of Gender, Family Development and Social Security, National Narcotics Control Bureau	No data collection system or epidemiological study on mental health (prevalence study on mental disorders was under way in 2004)	Therapeutic drugs are generally available at the primary health care level of the country: carbamazepine, phenobarbital, phenytoin sodium, sodium valproate, amitriptyline, chlorpromazine, diazepam, fluphenazine, haloperidol, lithium, carbidopa, levodopa	
Nepal	Number of professionals who graduated last year in academic and educational institutions per 100,000: psychiatrists 0.016, psychologists with at least 1 year training in mental health care 0.004, nurses with at least 1 year training in mental health care 0.008; 21–50% of medical graduates immigrate to other countries after completion of training	Present. Morbidity form regarding outpatients filled by primary health centres	Present	Therapeutic drugs generally available at the primary health care level: carbamazepine, phenobarbital, phenytoin sodium, amitriptyline, chlorpromazine, diazepam. More psychotropic drugs are available in the district and primary health care level	National level Non-communicable Disease Prevention and Control Committee formed in the Ministry of Health. Among these, mental disorder has also been prioritized. Coordinator, Policy, strategies and activities have been formulated

Table 11.7 (continued)

Country	Psychiatric training: Post-graduate training in psychiatry/psychology	Information Gathering System		Therapeutic Drugs: availability	Other Information
		Mental health reporting system	Data collection system		
Pakistan	Number of professionals graduated in year of study: 0.002 psychiatrists; 0.07 psychologists with at least 1 year training in mental health care; 0.008 nurses with at least 1 year training in mental health care; 0.005 social workers with at least 1 year training in mental health care; 0.002 occupational therapists with at least 1 year training in mental health care. 1–20 % of psychiatrists emigrate after completion of their training	Mental health reporting system has been initiated in the National Health Management Information System	The country has data collection system or epidemiological study on mental health Information system in tertiary facilities developed at the WHO Collaborating Centre at Rawalpindi to collect information from primary care centres on depressive illness, substance abuse and epilepsy	Therapeutic drugs generally available at the primary health care level: carbamazepine, phenobarbital, chlorpromazine, diazepam, haloperidol, imipramine, procyclidine	Community mental health program includes faith healers. Print/electronic media utilized to spread mental health education. Collaboration with schools and NGOs (National Rural Support Program) established. Public educational material available. Guidelines for economic analysis of community mental health care program in low income countries being developed
Sri Lanka	Nil details	There is mental health reporting system in the country. Only hospital data is reported	The country has no data collection system or epidemiological study on mental health	Therapeutic drugs generally available at the primary health care level: carbamazepine, ethosuximide, phenobarbital, phenytoin sodium, sodium valproate, amitriptyline, chlorpromazine, diazepam, carbidopa, levodopa. Addition of newer anti-psychotics in the essential drug list being considered	WHO guided project initiated to encourage process of deinstitutionalization of psychiatric patients and reintegration into the community

Restructuring medical and nursing education Basic medical and nursing education needs to be skill-based to produce competent practitioners (Jacob 2011). The current specialty-based medical education has become a knowledge-based information transmission system. The specialist and western-based curriculum makes the need to master common regional diseases and national health priorities much less important. Consequently, basic doctors, without practical training, are forced to specialise in order to practice medicine. The curriculum needs to be restructured to focus on the essential competencies required.

Revamping psychiatric training The current elaborate psychiatric diagnostic systems and management, essentially watered down tertiary approaches, are irrelevant and impractical in primary care and general practice (Jacob 2011). The recognition and management of psychiatric presentations should be based on the reality of primary care rather than specialist perceptions. The identification of common clinical presentations (rather than specific diagnosis) and simple management protocols (rather than elaborate and separate routines) are mandatory (Jacob 2006).

Revitalising primary care The strengthening of the general health infrastructure, to improve primary health care delivery, is mandatory for the effective integration of mental health into primary care practice (Jacob 2011). The availability of effective and affordable treatments and improved national finances has not closed the gap between mental health need and services in most LMICs. The major challenge is to transform overburdened, underfunded and demoralised primary care systems. The focus of mental health care delivery should shift from attempting to incorporate mental health care into an impoverished system to strengthening and empowering primary health care in general. Only robust primary health care systems can meet the challenge of delivering mental health care to communities.

Renewing political pressure Leadership from politicians, administrators, health and mental health professionals is crucial (Jacob 2011). Educating the population about mental illness using the mass media will reduce stigma and increase the demand for services. A "HIV/AIDS model" of activism, where users, families, interest groups, health professionals and scientists come together with the single aim of service provision, is required for transformation. Slick documents, scintillating launches, stirring speeches and shallow programs, which repackage failed strategies, are no substitute for hard technical inputs for translating research evidence into primary care practice.

11.9 Conclusions

The South Asian countries have made progress over recent years in terms of mental health facilities and provision of care; however resources are still inadequate to cope with the burden of mental illness. Government spending on mental health is insufficient. Infrastructure needs to be improved—both in primary health care in general, as well as psychiatric settings. More training of personnel and greater accessibility to psychotropic medication is required. Co-ordination of the mental health sector with other relevant sectors needs to improve. Policy and legislation on mental health is still left wanting. Equity of access to mental health services for all individuals remains a challenging issue in the region. It is essential for users, families, interest groups, health professionals and scientists to come together with the single aim of service provision, to transform mental health care in South Asia.

References

Goel, D. S. 2010. The National Mental Health Program: Problems, perspectives, possibilities. In: *Public mental health: An evolving field.* New Delhi: Ministry of Health, Government of India.

Hanlon, C., D. Wondimagegn, and A. Alem. 2010. Lessons learned in developing community mental health care in Africa. *World Psychiatry* 9:185–189.

Jacob, K. S. 2006. The diagnosis and management of depression and anxiety in primary care: The need for a different framework. *Journal of Postgraduate Medicine* 82:836–839.

Jacob, K. S. 2011. Repackaging mental health programs in low-and middle-income countries Indian Journal of Psychiatry [serial online] [cited 2012 Jan 28] 53:195–198. http://www.indianjpsychiatry.org/text.asp?2011/53/3/195/86798.

Jacob, K. S., P. Sharan, I. Mirza, M. Garrido-Cumbrera, S. Seedat, J. J. Mari, V. Sreenivas, and S. Saxena. 2007. Mental health systems in countries: Where are we now? *Lancet* 370:1061–1077.

Saraceno, B., M. van Ommeren, R. Batniji, A. Cohen, O. Gurejee, J. Mahoney, D. Sridhar, and C. Underhill. 2007. Barriers to improvement of mental health services in low-income and middle-income countries. *Lancet* 370:1061–1077.

World Health Organization. 1996. *The diagnostic and management guidelines of the international classification of diseases 10 primary care version.* Geneva: World Health Organization.

World Health Organization, Mental Health Atlas. 2005. http://www.who.int/mental_health/evidence/mhatlas05/en/index.html. Accessed 28 Jan 2012.

World Health Organization, Mental Health Atlas. 2011. http://whqlibdoc.who.int/publications/2011/9799241564359_eng.pdf. Accessed 28 Jan 2012.

World Health Organization, WHO-AIMS report on Mental Health System in Bangladesh. 2006. http://www.searo.who.int/LinkFiles/Mental_Health_Resources_WHO-AIMS_Report_MHS_Ban.pdf. Accessed 28 Jan 2012.

World Health Organization, WHO-AIMS report on Mental Health System in Bhutan. 2006. http://www.searo.who.int/LinkFiles/Mental_Health_Resources_WHO-AIMS_Report_MHS_Bhu.pdf. Accessed 28 Jan 2012.

World Health Organization, WHO-AIMS report on Mental Health System in Maldives. 2006. http://www.searo.who.int/LinkFiles/Mental_Health_Resources_WHO-AIMS_Report_MHS_Mal.pdf. Accessed 28 Jan 2012.

World Health Organization, WHO-AIMS report on Mental Health System in Nepal. 2006. http://www.searo.who.int/LinkFiles/Mental_Health_Resources_WHO-AIMS_Report_MHS_Nep.pdf. Accessed 28 Jan 2012.

World Health Organization, WHO-AIMS report on Mental Health System in Pakistan, World Health Organization. 2009. http://www.who.int/mental_health/pakistan_who_aims_report.pdf. Accessed 28 Jan 2012.

Chapter 12
Allied Mental Health Professionals: Clinical Psychologists, Psychiatric Nurses and Psychiatric Social Workers: Availability and Competency

R. K. Chadda and R. Prashanth

12.1 Introduction

Management of mental disorders requires a team approach as the disorders have a multifactorial etiology. Both assessment as well as treatment involves interventions by a number of members of the mental health care team. The mental health care team has both medical as well as non-medical mental health professionals. Psychiatrist is the leader of the multidisciplinary mental health team and is a mental health professional. The non-medical mental health professionals, also called allied mental health professionals, include clinical psychologists, psychiatric social workers and psychiatric nurses.

Broadly, mental disorders may be grouped under severe mental disorders, common mental disorders, personality disorders, alcohol and substance use disorders, and psychiatric disorders in special age groups or populations. Severe mental disorders include illnesses like schizophrenia, bipolar affective disorder, severe depression and other psychotic disorders. Common mental disorders include mild to moderate depression, anxiety disorders, somatoform disorders and dissociative disorders. Since etiology of most psychiatric disorders is multifactorial, management is also on multimodal lines. The allied mental health professionals have an important role to play in management.

This chapter discusses the role of allied mental health professionals in mental health care with a focus on South Asia, especially covering the issues of their limited availability and competency issues in South Asian countries. The chapter also focuses on their availability, training and barriers to their incorporation in services.

R. K. Chadda (✉) · R. Prashanth
Department of Psychiatry, All India Institute of Medical Sciences, New Delhi 110029, India
e-mail: drrakeshchadda@gmail.com

J. K. Trivedi, A. Tripathi (eds.), *Mental Health in South Asia: Ethics, Resources, Programs and Legislation,* International Library of Ethics, Law, and the New Medicine 58,
DOI 10.1007/978-94-017-9017-8_12

12.2 Role of Allied Mental Health Professionals in Mental Health Care Delivery

All the mental health professionals have an important role to play in the delivery of mental health services. Psychiatrist is the leader of the mental health team and is a medical professional, who makes a clinical diagnosis and finalises the management plan in discussion with the multidisciplinary team consisting of clinical psychologists, psychiatric social workers, and psychiatric nurses. The management plan should also be discussed with the patient and the family members, and suitably modified if required. This helps in better adherence to treatment.

This section discusses the role of various allied mental health professionals.

12.2.1 Clinical Psychologist

Clinical psychology is a branch of psychology, which deals with psychological assessment and management of mental disorders. A clinical psychologist is a health professional who has completed formal training in clinical psychology at a recognized university-level school for a diploma or degree in clinical psychology. Clinical psychology is a specialised branch of psychology. So to become a clinical psychologist, first one has to study psychology and then specialise in clinical psychology.

The professional roles of a clinical psychologist in a mental health team are varied. In the past, the focus used to be primarily on psychodynamic understanding and counselling. Nowadays, the clinical psychologists carry out a number of activities covering different areas of the psychodiagnostics and psychotheraputics.

Clinical psychologists work in a range of settings including mental hospitals, in various departments of general hospitals like psychiatry, paediatrics, neurology, otorhinolaryngology, rehabilitation and others, rehabilitation homes, homes for mentally retarded, non-government organisations, office based practice, academic institutions and many other kinds of settings. Some clinical psychologists work as a member of the multidisciplinary mental health team, while others work in independent practice.

Clinical psychologists do a wide range of work, which includes diagnostic assessment, disability assessment, and therapeutic work. Diagnostic work may include assessment of intelligence, personality, neuropsychological functioning, disability, specific abilities and other areas. The assessments are made using clinical interviews as well as standardized instruments. The assessment further helps in diagnostic clarification, identifying the intra-psychic and interpersonal factors contributing to the development and maintenance of psychological problems (Prabhu and Shankar 2004).

The therapeutic work may include different kinds of psychological treatments like supportive psychotherapy, behaviour therapy, cognitive behaviour therapy, interpersonal therapy, brief dynamic therapy, counselling, family therapy, marital therapy, rehabilitation and cognitive training. The clinical psychologist develops treatment plan after a detailed assessment, tailored as per the individual needs.

Clinical psychologists are an important constituent of almost all settings of mental health services including child and adolescent mental health services, geriatric mental health services, deaddiction services, disaster management and others. Some clinical psychologists work as school counsellors, taking care of the mental health needs of the school children. Child guidance, behaviour modification, play therapy and parental counselling are some important techniques used in child and adolescent psychiatry (Prabhu and Shankar 2004).

Clinical psychologists are also an important constituent of the community awareness programs on mental health. Stress management, promotion of mental health and well being are some other areas, where clinical psychologists have special expertise.

12.2.2 Psychiatric Social Worker

Psychiatric social work refers to application of various methods of social work like case work, group work, community organisation, social welfare and social action in the field of mental health for the purposes of psycho-social intervention, promotion of mental health and prevention of mental health problems.

Psychiatric social worker is a health professional who has completed formal training in social work at a recognized, university-level school with a diploma or degree in social work followed by a specialised degree or diploma in psychiatric social work.

Psychiatric social workers have a wide range of responsibilities besides the psychosocial interventions, which may include mental health promotion, disability assessment and rehabilitation, psychoeducation, group therapy, family therapy, vocational guidance, etc. They have an important role in providing vocational and occupational skills training to the patients who have developed marked deficits due to severe mental disorders. Psychiatric social workers may carry out daily therapeutic interventions in form of group therapy or group work to improve the social skills or group behaviour skills of the mentally ill patients (Parthasarathy and Ranganathan 2004).

Psychiatric social workers also help the affected families using different types of family therapy targeting at the family pathology. Working with families also includes strategies at assessing and reducing the caregiver or family burden due to the mental illness of their patient and strengthening the coping skills of the families to cope with the stress of caregiving. Development of self help groups is an important method used by the psychiatric social workers. This may be done in institutional as well as community settings. The groups may be of the improved mentally ill patients, family members, persons with drug or alcohol related problems, parents of children with mental retardation, etc (Parthasarathy and Ranganathan 2004).

Psychiatric social worker mostly works as a part of the multi-disciplinary mental health service team, working in different settings like in-patient and out-patient services, child guidance clinics, family psychiatric centres, deaddiction centres, rehabilitation and after-care centres and community mental health programs. They are also an important part of disaster mental health services (Department of Psychiatric Social Work, NIMHANS—online).

Some psychiatric social workers may work exclusively in the rehabilitation and after care or welfare settings.

12.2.3 Psychiatric Nurse (PN)/ Mental Health Nurse (MHN)

Psychiatric or mental health nursing is a specialized area of nursing practice providing specialised nursing care to the mentally ill patients. Psychiatric nurse is a health professional who has completed formal training in psychiatric nursing at a recognized, university-level school followed by a diploma or degree in psychiatric nursing.

The psychiatric nurse is an integral part of the mental health team, working in outpatient, inpatient and community settings. In the outpatient services, they provide psychoeducation and crisis intervention and may assist in therapeutic procedures like electroconvulsive therapy or behaviour therapy, conduct group therapy, etc. In inpatient settings, the psychiatric nurse is often the first contact in emergency or crisis and handles the crisis providing immediate intervention, which may include administration of some medications or counselling. They also provide basic nursing care, administer medications, observe and record the behaviour of the admitted patients, and conducts group sessions. In the community settings, the psychiatric nurse may be involved in mental health assessment and delivery of services, providing counselling, psychoeducation, behaviour therapy, group therapy, etc (Department of Psychiatric Nursing, NIMHANS—online).

12.3 Available Manpower and Their Competency

Broadly, the allied MHPs in South Asia could be divided into 2 groups: one, who have received a formal training in their respective field from a recognised institution or university in form of a diploma or degree, and the other are those professionals who have a professional qualification in the field of psychology, social work or general nursing, but not specifically in the field of mental health. The latter group have been working in various mental health care settings and have learnt the skills of mental health care while on job. Some of the latter group got an opportunity of short term training programs lasting few weeks to months towards refinement of their mental health care skills.

A global survey of mental health resources was conducted by the World Health Organization (WHO) in 2005 and 2011 respectively (WHO Mental Health Atlas 2005, 2011—online). A comparison of strength of various mental health professionals, as available in different South Asian countries in 2005, is shown in Table 12.1. It is apparent that the availability of allied MHPs is very low and in some countries like Bhutan and Maldives, some cadres are not present at all. Amongst various

Table 12.1 A comparison of mental health manpower resources in different SEAR countries.[a] (Source—World Mental Health Atlas 2005)

Resource	Bangladesh	Bhutan	India	Maldives	Nepal	Pakistan	Sri Lanka
Psychiatrists	0.05	0.3	0.2	0.36	0.12	0.2	0.2
Psychiatric nurses	0.06	0.16	0.05	0	0.08	0.08	1.8
Psychologists	0.002	0	0.03	1.2	0.08	0.2	0.02
Psychiat-ric social workers	0.001	0	0.03	0	0.04	0.4	0.07
Total	0.113	0.46	0.31	1.23	0.32	0.88	2.09

[a] All figure are in number per 10,000

Table 12.2 A comparison of mental health manpower resources in different parts of the world in 2005 and 2011 (All figures in the table are in number per 100,000)

WHO region	Psychiatrist (2005/2011)	Nurses (2005/2011)	Psychologist (2005/2011)	Social worker (2005/2011)
America	2.0/1.57	2.60/3.92	2.8/1.29	1.0/0.39
Africa	0.04/0.05	0.20/0.61	0.05/0.04	0.05/0.03
Europe	9.8/8.59	24.8/21.93	3.10/2.58	1.50/1.12
SEAR	*0.20/0.23*	*0.10/0.77*	*0.03/0.03*	*0.04/0.01*

South Asian countries, Pakistan has the highest number of allied MHPs currently followed by India and Sri Lanka. Bangladesh had the minimum number of psychiatrists amongst various South Asian countries and also very less psychologists and psychiatric social workers, whereas Maldives did not have any psychiatric nurses and psychiatric social workers. Bhutan, a small Himalayan country, also does not have any psychologists and psychiatric social workers.

During the period of 2005–2011, there have been only minor changes in the manpower in SEAR countries (Table 12.2). A comparison of the data of 2005 and 2011 reveals a slight increase in the number of psychiatrists from 0.2 to 0.23 per 100,000 of population and a marked increase in the number of psychiatric nurses from 0.10 to 0.17 per 100,000 of population. The definition of nurses and social workers was broadened in the latest mental health atlas to include professionals, who though not formally qualified in the mental health sector had received training while on job. The change in definition brought a marked change in the strength of nurses, but there was rather a fall in the number of social workers. Implication of the finding is that all the countries need to make a systematic effort to increase manpower in this field to provide mental health care access to a majority of the population.

Another important matter of concern is that in most of the South Asian countries, MHPS are often concentrated in the urban areas compared to rural areas. This inequity in resource distribution leads to non-availability of services to a vast majority of population living in rural areas in these countries. For example in Bangladesh and Nepal the concentration of nurses is 5–8 times greater in urban areas than in rural areas (WHO AIMS Country Reports, Bangladesh, Nepal—online). The situation is similar in India where upto 70 % of doctors and probably even a much higher number of all the MHPs are concentrated in urban areas (Khandelwal et al. 2004). One ray of hope could be that many voluntary non govern-mental organisations (NGOs) make periodic visits to rural areas and the local governments have taken many active efforts in recent years by active efforts like the National Mental Health Program of India.

No reliable data is available about various allied MHPs without a formal degree or diploma, who are working in the mental health sector. But this number, though many times that of the formally trained manpower, is still grossly inadequate. A large percentage of this group works in state run or private psychiatric hospitals or nursing homes, and general hospital psychiatric units and rehabilitation centres in public as well as private sector.

12.4 Training of Allied Mental Health Professionals (MHPs)

There have been very few training facilities in various South Asian countries available for the allied MHPs including clinical psychologists, and psychiatric nurses and psychiatric social workers. In fact this was the reason that most psychiatric institutions and sets ups in South Asia employed psychologists, social workers and nurses without any formal mental health qualifications. The course of development of the training facilities for various allied MHPs and their current status in South Asia is discussed as below.

12.4.1 Clinical Psychologist

Teaching in clinical psychology first started in the 1940s in India as a study of abnormal psychology. The teaching did not have a focus on clinical application. Following the establishment of All India Institute of Mental Health in Bangalore in 1954, Diploma in Medical Psychology (DMP) was started as a full-time residential training program. The training was based in mental hospital setting and had a number of the medical model. The nomenclature has undergone a number of changes over the period like Diploma in Medical and Social Psychology (DM & SP) in 1960, MPhil Medical and Social Psychology in 1978, and later MPhil Clinical Psychology. In parallel, DM & SP was started at the Central Institute of Psychiatry, Ranchi in 1960s, which was later redesignated as MPhil. A number of institutions have started MPhil clinical psychology in the last 10 years, the number currently may be about 15 (Prabhu and Shankar 2004).

Certain other courses are also available in clinical psychology in India, which include Professional Diploma in Clinical Psychology (P.D. Cl.Psy), Doctor of Psychology (Psy.D.) and Doctor of Philosophy (PhD) in clinical psychology. These courses are not residential programs as the MPhil. The basic requirement for M.Phil and PhD is MA/MSc in psychology, which is provided by many universities in

India. The Rehabilitation Council of India (RCI) is the accreditation authority for all 3 courses except PhD (Rehabilitation Council of India—online).

MPhil in clinical psychology is a 2 years residential course in a hospital setting, which makes the trainee competent in various roles as a clinical psychologist like diagnosing mental health problems and therapeutic psychological interventions for mental illnesses. A person with MPhil in clinical psychology can also undertake teaching and research in the areas of clinical psychology and mental health.

PhD is more a research based program as in other fields. The basic requirement for enrolling in PhD course is more rigorous with more emphasis on research preparing the individual for a career in academia. The Psy.D program lays emphasis on applied clinical knowledge and provides training in a clinical setting. The trainee is exposed to the wide range of psychiatric disorders (Rehabilitation Council of India—online). The course duration is 4 years for candidates holding M.A/M.Sc in psychology and 2 years for candidates holding M.Phil in psychology. The course also allows students to select sub-specializations like clinical child and adolescent psychology, behaviour therapy, cognitive-behaviour therapies, community mental health, marital and family therapy and rehabilitation of mentally ill etc. Availability of these new courses in the recent yeas is likely to reduce the deficit in mental health manpower in the coming years.

In other south Asian countries like Bangladesh, there has been a beginning in form of availability of training opportunities in clinical psychology. No formal services in clinical psychology existed till 1996, when it was first setup in the University of Dhaka. Two degrees (M.S. and M. Phil) are now awarded in Clinical Psychology each year, comprising three components, namely academic, clinical and research work. PhD course is expected to be started soon (Department of Clinical Psychology, University of Dhaka—online). Similarly in Sri Lanka the Department of Educational Psychology, University of Colombo offers M.Phil and post graduate diploma courses in clinical psychology (Department of Educational Psychology, University of Colombo—online).

In Pakistan, the number of psychologists increased from 62 in 1980 to about 450 in the year 2000. But many of them are not qualified clinical psychologists, though they may be practising clinical psychology. Some of the post graduate programs available in Pakistan include 2 year MS course in clinical psychology, one year top up MS in clinical psychology and Diploma in Child Guidance and Counselling and PhD programs in clinical psychology.. The clinical training courses are skill based and emphasize on developing an empirically derived indigenous knowledge base (Karim et al. 2004).

In Nepal the first postgraduate training program in form of MPhil in clinical psychology was started from the Institute of Medicine, Kathmandu in 1997 and at the BP Koirala Institute of Health Sciences, Dharan from 2000 (Regmi et al. 2004).

More recently under the National Mental Health Program in India, short term training programs of 3 months duration have been started for psychologists with a Masters degree but without specialization in mental health. This has been started with the objective of enabling the psychologists to function as an effective member of a multidisciplinary mental health team. The training consists of various teaching methods including didactic teaching, role play, demonstration and clinical rotation in adult psychiatry, child and adolescent unit, deaddiction and rehabilitation settings (District Mental Health Program—online).

The current data suggests two major issues in the field of clinical psychology training. First, there are very few accredited institutions available for training and secondly, inequitable distribution of resources between the government and private sectors.

12.4.2 Psychiatric Social Worker

Psychiatric social work is also a specialised filed of social work, requiring a formal training in the mental health sector. However, there are very limited facilities available in South Asia for specialised training in the field of psychiatric social work. At most of the places, persons qualified in social work have been working in the mental health services.

Facilities of training in psychiatric social work were till recently available only at two places, in Bangalore and Ranchi, in India, though in recent years it has been expanded to few more institutions. Two types of courses are available in India, MPhil and PhD in psychiatric social work. For the both, the essential requirement is Masters in Social Work. There is no central registration body for social workers and hence no clear estimate of the number of trained psychiatric social workers is available (Parthasarthy and Ranganathan 2004; Department of Psychiatric Social Work, NIMHANS—online).

In recent years, three month short-term training of social workers in mental health has also been underway under the National Mental Health Program.. This program prepares social workers to carry out the role of a psychiatric social worker in the community health team. The training is divided in three main parts viz theoretical, community and clinical orientation. At the end of the course the social worker is expected to carry out activities like education of public about mental illness in the community, create awareness, carry out mental health education, adolescent education and life skills education, facilitate community involvement in mental health programs and provide social case work and management at the community level (District Mental Health Program—online).

In Pakistan, Masters in Social Work is offered from approximately 7 universities and includes optional training in medical and psychiatric social work (Karim et al. 2004). The training of psychiatric social workers in countries like Nepal, Bangladesh and Sri Lanka also doesn't meet the requirement of needed personnel in the community, though exact estimates of number of centres or personnel trained per year are available.

In Sri Lanka, to cope up with the after effects of the 2004 tsunami, the government created a new team of mental health workers called 'community support officer'. They provide social support and psychological first aid to the affected people and also help in detection and referral of the affected individuals in the community (Kakuma et al. 2011).

Among various allied MHPs, training and resource generation in the field of social work is still at a low level. It shows that the potential of social workers in the mental health has not been adequately realized. Besides creating facilities for training, there is also a need to create adequate employment opportunities in the mental health sector for the qualified psychiatric social workers.

12.4.3 Psychiatric Nurse

As in the field of psychiatric social work, most of the nurses working in the psychiatric hospitals or psychiatric wards in the general hospitals in various South Asian countries are trained in general nursing with no formal training in psychiatric nursing. Very few facilities are available for training in psychiatric nursing.

Specializing in mental health nursing involves developing specific skills in mental health care and would involve direct clinical care. In India, two kinds of courses are available, one year diploma in psychiatric nursing and two years MSc course in psychiatric nursing. The diploma course, which is a purely service oriented training program, is available only at two places in India, Bangalore and Ranchi. The MSc courses are not targeted at preparing the clinical psychiatric nurses, but are more academically oriented, preparing the candidates to take up faculty positions in the nursing schools and colleges. For MSc psychiatric nursing also, there are only a few institutions offering the course (Redamma and Nagarajaiah 2004).

In India as well as in other south Asian countries like Bangladesh and Nepal, the respective nursing councils have laid out guidelines and standards for training in psychiatric nursing (Indian Nursing Council—online). It is important to mention here that a 2 month clinical rotation in psychiatry is also part of all diploma and graduate nursing programs.

In Sri Lanka, no documented programs are available for post graduate training in the field of psychiatric nursing. The country still has around 800 nurses working in the mental health sector with some degree of training. The new National Mental Health Policy of Sri Lanka, approved in 2005, aims to create a new cadre of psychiatric nurses for placement in mental health settings (Integration of mental health into primary care in Sri Lanka—online).

Pakistan has postgraduate training in psychiatric nursing, available in form of MSc in psychiatric nursing. The course is available at a number of places, but has more of academic orientation rather than a clinical focus like in India.

In India, short term training courses of nurses in mental health have been introduced as a part of the district mental health program. This is intended at training nurses to work in community mental health and delivery of nursing care at primary care level. The training program is conducted at psychiatric hospitals or general hospital psychiatric units for a duration of four weeks. The main objectives of the program are to impart skills which will enable nurses to identify, refer and follow up mentally ill patients in the community, make home visits, provide first aid in emergencies, carry out IEC activities in the community and coordinate rehabilitation services in the community (District Mental Health Program—online).

Implementation of such policies may lead to a strong workforce at the community level and enhance the capabilities of mental health teams to provide much needed relief and care to the affected population. Also, provision of adequate employment opportunities in clinical and research settings with provision of good salaries will encourage more candidates to take up training in this field.

The existing scenario shows that nurses are the most widely available work force in the field of mental health. The current focus in most of the low and middle income countries of south Asia is focussing on training of existing nursing staff in the field of mental health using short term training courses.

12.5 Barriers to Availability and Training

According to the WHO, the estimated total number of mental health care workers needed in the 58 countries of low and middle income in 2005 was 362,000, coming to 22.3 workers per 100,000 population in low-income group countries and 26.7 workers per 100,000 in middle-income countries, comprising 6 % psychiatrists, 54 % nurses in mental health settings, and 41 % psychosocial care providers. The shortage amounts to 17.3 workers per 100,000 population in low income countries and 14.9 in middle-income countries (Kakuma et al. 2011). This also applies to the various countries in South Asia.

Lack of financial resources to mobilise resources in the mental health sector remains the biggest challenge faced currently. Both India and Pakistan spend around 1 % of their GDP on health, out of which a negligible amount is allocated to mental health (Khandelwal et al. 2004). There is also a shortage of training institutions to cater to the growing need in most of the south Asian countries. Provision of training relevant to the community health care needs is important for smooth transition of personnel to the field, where they can implement their skills. This is often lacking. There is also the problem of ill-defined roles of professionally trained mental health workers.

The majority of the population in most of the South Asian countries live in rural areas, but most of the mental services are situated in big cities, leave aside smaller cities. Though the local governments have taken some initiatives at starting community mental health programs, but they are still grossly inadequate and under staffed in the absence of various MHPs.

The profession of mental health care also suffers from negative attitude of health care workers who are hesitant to take up this field. There are a number of misconceptions about mental illness, career in mental health, poor scope for training and employment, all of which contribute to the problem.

There is also a lack of identification of other professional health workers who may be trained to work in the field of mental health to expand the resource base. It has been shown that the non-specialist health workers can take up various roles like detection and referral, rehabilitation, providing psychoeducation follow up care and ensuring adherence to treatment (Saraceno et al. 2007).

12.6 Future Directions

The chapter has discussed the role of various of MHPs in mental health care, their availability and the steps taken by the respective governments in tackling the shortages. The barriers to the resources have been under the scanner of most nations for quite some time and some remedial measures are on the way.

In India under the NMHP, the total amount of funding was increased to 472 crores in the eleventh five year plan (2007–2012) and manpower development has been one of the major stated objectives of this program, which is proposed to be substantially increased in the next five year plan of 2012–2017 (Murthy 2011). Measures are being undertaken to expand the facilities of training and also in-service training for nurses, social workers and psychologists in mental health to increase the manpower.

The other issues including inequitable distribution of manpower still pose a challenge. Provision of good work conditions, career growth and development and financial incentives will help in addressing these problems.

The improvement of human resource in the field of allied mental health is a continuing process. We need to appreciate the efforts underway in many of the South Asian countries but also remember that it requires strong political and administrative will, innovative solutions to overcome the gross shortage in manpower and strong coordination between different stakeholders including government, non-governmental organisations, health workers, affected people and their care-givers in the community, if the situation is to be changed.

References

Department of Clinical Psychology, University of Dhaka. http://www.univdhaka.edu/department/common/home.php?bodyid=CPS. Accessed 25 Dec 2011.

Department of Educational Psychology, University of Columbo. http://www.cmb.ac.lk/academic/edu/edupsy/index.htm. Accessed 25 Dec 2011.

Department of Nursing, NIMHANS. http://www.nimhans.kar.nic.in/nursing/default.htm. Accessed 25 Dec 2011.

Department of Psychiatric Social Work, NIMHANS. http://www.nimhans.kar.nic.in/psw/default.htm. Accessed 25 Dec 2011.

District Mental Health Program. http://www.nimhans.kar.nic.in/dmhp/resources3.htm. Accessed 25 Dec 2011.

Indian Nursing Council. http://www.indiannursingcouncil.org/types-nursing-programs.asp. Accessed 25 Dec 2011.

Integration of mental health into primary care in Sri Lanka, World psychiatric association. http://www.wpanet.org/detail.php?section_id =24 & content_id=927. Accessed 25 Dec 2011.

Kakuma, R., H. Minas, N. V. Ginneken, M. R. Dal Poz, K. Desiraju, J. E. Morris, S. Saxena, and R. M. Scheffle. 2011. Human resources for mental health care: Current situation. *Lancet* 378:1654–1663.

Karim, S., S. Khalid, and M. H. Rana. 2004. Pakistan mental health country profile. *International Review of Psychiatry* 16:83–92.

Khandelwal, S. K., H. P. Jhingan, S. Ramesh, R. K. Gupta, and V. K. Srivastava. 2004. India mental health country profile. *International Review of Psychiatry* 16:126–141.

Murthy, R. S. 2011. Mental health initiatives in India (1947–2010). *The National Medical Journal of India* 24:98–107.

Parthasarathy, R., and M. Ranganathan. 2004. Psychiatric social work: Past, present and future. In *Mental Health: An Indian Perspective 1946–2003,* ed. S. P. Agarwal, D. S. Goel, R. L. Ichhpujani, R. N. Salhan, and S. Shrivastava. India: Elsevier.

Prabhu, G. G., and R. Shankar. 2004. Clinical psychology: Coming of age. In *Mental health: An indian perspective, 1946–2003,* ed. S. P. Agarwal, D. S. Goel, R. L. Ichhpujani, R. N. Salhan, and S. Shrivastava. India: Elsevier.

Reddemma, K., and S. Nagarajaiah. 2004. Psychiatric nursing: The perpetual cinderella. In *Mental health: An indian perspective, 1946–2003,* ed. S. P. Agarwal, D. S. Goel, R. L. Ichhpujani, R. N. Salhan, and S. Shrivastava. India: Elsevier.

Regmi, S. K., A. Pokharel, O. P. Ojha, S. N. Pradhan, and G. Chapagain. 2004. Nepal mental health country profile. *International Review of Psychiatry* 16:142–149

Rehabilitation Council of India. RCI. http://rehabcouncil.nic.in/programs/crregular.htm. Accessed 25 Dec 2011.

Saraceno, B., M. V. Ommeren, R. Batniji, A. Cohen, O. Gureje, J. Mahoney, D. Sridhar, and C. Underhill. 2007. Barriers to improvement of mental health services in low-income and middle-income countries. *Lancet* 370:1164–1174

WHO-AIMS Country Reports, Bangladesh. http://www.who.int/mental_health/who_aims_country_reports/en/index.html. Accessed 25 Dec 2011.

WHO-AIMS Country Reports, Nepal. http://www.who.int/mental_health/who_aims_country_reports/en/index.html. Accessed 25 Dec 2011.

WHO Mental Health Atlas. 2005. http://apps.who.int/globalatlas/default.asp. Accessed 25 Dec 2011.

WHO Mental Health Atlas. 2011. http://www.who.int/mental_health/publications/mental_health_atlas_2011/en/index.html. Accessed 25 Dec 2011.

Chapter 13
Private Sector Psychiatry: What it has to Offer to Mentally Ill Persons in South Asia?

Charles Pinto and Malay Dave

13.1 Introduction

A variety of psychiatric disorders are prevalent in South Asian region; including *Serious Mental Disorders (SMD)* like Schizophrenia and Bipolar Mood Disorders, *Common Mental Disorders (CMD)* like Depression and Anxiety Disorders (Patel and Thara 2001), and Specific Culture Bound Syndromes like the *Dhat Syndrome. Prameha*, a culture bound syndrome similar to *Dhat Syndrome* is found in countries like Nepal & Sri Lanka (Wen-Shing Tseng 2006).

This kaleidoscope of populace, psychiatric disorders and scarcity of resources make the management of these problems challenging as well as interesting.

13.2 Prevalence of Psychiatric Disorders

The prevalence rates of Psychiatric disorders, whether serious common, seem to be following much the same pattern as the rest of the world, developed or otherwise. The prevalence rate of Schizophrenia seems to be more or less fixed at about 1 % of the total population (Versola-Russo 2006). Prevalence of Depression is around 26 % (Patel and Shidhaye 2010). Suicide rates are reported to be about 11–37 per 100,000 population in India and Sri Lanka respectively (Desai and Isaac 2001).

C. Pinto (✉)
Topiwala National Medical College & BYL Nair Charitable Hospital, Mumbai, India
e-mail: charlespinto@yahoo.com

M. Dave
Gayatri Polyclinic, Ghatkopar East, Mumbai, India

J. K. Trivedi, A. Tripathi (eds.), *Mental Health in South Asia: Ethics, Resources, Programs and Legislation,* International Library of Ethics, Law, and the New Medicine 58,
DOI 10.1007/978-94-017-9017-8_13

The population structure of the region is fast changing with increase in survival rates and life expectancy. The effects of globalization are also seen in this region of the world. The effects of this are also seen in the perceived increase in the prevalence of psychiatric disorders like Alzheimer's disease (Ferri et al. 2005), and Eating disorders (Anorexia Nervosa—0.002 to 0.9%; Bulimia Nervosa—0.46 to 3.2%) (Muazzam and Khalid 2008).

13.3 Need for Trained Personnel

In spite of the huge number of people making South Asia their home, the number of trained psychiatric professionals, Psychiatrists as well as Psychiatric Nurses available, are not very encouraging. For example, in India with a population of nearly 1.5 billion, only 3500–4000 trained psychiatrists are available in the country. Similar figures are found for the paramedical staff as well. This makes managing psychiatric disorders difficult, if not impossible (WHO-AIMS Report 2006). In Bangladesh the number of Psychiatrists per 100,000 population is 0.05, and Psychiatric Nurses is 0.06 (Mental Health Atlas 2005a, countries a, b); and figures for Pakistan are 0.2 Psychiatrists per 100,000 population, and 0.08 Psychiatric Nurses per 100,000 population (Mental Health Atlas 2005c, countries n-r).

The distribution of services in this region is not uniform, and often not well regulated. Services may be available in the urban setting, but would be totally lacking in many other parts, especially in the rural or hard to access areas. The existing infrastructure is equally short staffed, with many posts remaining vacant. In many countries of the region, the number of practicing Psychiatric physicians is much more in the private set up than in the various Departments or Institutes. And the same pattern is observed for the allied and para-professionals also (Patel 2007).

13.4 System of Medical Education

Medical education is very long and exhaustive universally. The goal of each and every medical college or institute or university is to produce physicians of the highest calibre who can cater to the ever growing population, and who can contribute not only to the society, but also to the field of medicine. Psychiatry, though understood to be of immense importance to any physician as well as to all medical specialties, remains a Postgraduate specialty till date (Patel and Shidhaye 2010).

13.5 Undergraduate and Postgraduate Training

Almost universally, the average duration for obtaining a medical degree is 6 years or more. The duration includes 5 years of medical training and teaching, and 1 year of compulsory internship before the graduate degree is conferred onto an individual. In this period of teaching, the time allotted for Psychiatric teaching is very less, not more than 2 weeks of clinical posting and a few lectures. The internship in Psychiatry ranges from a period of 15–30 days in a period of 1 year. Hence, an average medical graduate is not well aware of, or confident enough to treat an individual suffering from a psychiatric disorder (Badrakalimuthu and Sathyavathy 2009).

Post graduation in Psychiatry is of 2 years (for a *Diploma*) or 3 years (for a *Degree*). The post graduate training is comprehensive and imparts necessary skills and expertise in the management of all psychiatric disorders (Yamaguchi 2011; Das et al. 2002). However, in most countries of South Asia, there is a lack of specialty training in various branches of Psychiatry (Trivedi and Dhyani 2007). So, a practicing Psychiatrist in any region will deal with each and every kind of disorder with little or no backup.

13.6 Role of General Physicians in Treating Psychiatric Disorders

As mentioned earlier, the number of trained professionals and the distribution of available services leave a lot to be desired. The General Practitioner or Physician is the first point of contact for most people suffering from psychiatric problems. A large number of patients, thus, are identified and managed by a *General Physician (Family Physician)*. A General Physician is a *Medical Graduate with basic medical qualifications*. With available resources, efforts are made to provide the best possible treatment under the given circumstances, and a referral to a qualified Psychiatrist or an Institute may be made. Treatment by a General Physician is not always appropriate or adequate. A study by Mykletun et al. (2010) showed that General Practitioners needed more time for assessment of disorders, more knowledge and competence, and more collaboration with secondary care services (Mykletun et al. 2010).

13.7 Psychiatrists in Primary Health Centres (PHC)

Primary Health Centres (PHC), which are the backbone of any government medical service, are mostly managed by Physicians, who have had little or no psychiatric training. Referrals to institutes for the management of difficult cases tend to be cumbersome and time consuming. In addition to lack of qualified staff, PHC also face a shortage of essential psychotropic medication (Mental Health Atlas 2005b, countries e-i). In India, essential Psychotropic medication is made available in certain districts only, where a National Program is operational (Mental Health Atlas 2005b, countries e-i). Due to lack or shortage of qualified personnel, regular and continuous therapies may not be possible.

13.8 Lack of Opportunities for Full Time Practice in Institutes

The number of academic positions that are available in Institutes and Departments, after qualifying, are very less. Majority of the postgraduates, therefore, opt out of academics and into private practice. Most Departments have inadequate research programs or facilities. Those keen or research or academics, often, leave their native lands and travel to western countries to further their ambitions (Patel 2007).

All of the above mentioned factors prompt the freshly qualified doctor and the professional, as well as the patient to approach the private sector for their problems.

13.9 Perception of Services in the Public and Private Sectors

13.9.1 Social Stigma

Stigma is associated with psychiatric disorders since time immemorial. In today's modern world too, stigma continues to play a part in the understanding and management of these. Stigma prevents individuals and families from seeking help and understanding their disorder (Chandra 2001). Stigma is also associated with poor outcomes for most disorders (Patel and Thara 2001). This phenomenon is evident in both private & public setups. Awareness programs can help in reduction of stigma. The private sector with its available resources can contribute to reducing stigma to a large extent. An expert clinician can have a good rapport with the patient and their families thereby helping in reducing stigma and making them active partners in the treatment process. Commendable work is done by various support groups and NGOs in creating awareness and reducing stigma.

13.9.2 Confidentiality

Patients do not want others to know of their disorders, this is the ethical basis of patient confidentiality. The common fear that many patients have that the more the number of people they come in contact with the less is the confidentiality. Hence, the perception is that a private consultation that involves fewer people will be able to maintain the necessary confidentiality (Marks and Thornicroft 1990). Moreover, many a times the patients come with a referral letter from their Family Physician, which makes it easier for them to trust the treating Psychiatrist. A General Practitioner—Psychiatrist collaboration is necessary for effective patient management (Younès et al. 2005).

13.9.3 Quality of Care in Institutions and Departments

In general, the care provided by public institutions is perceived as inadequate by patients. This can be attributed to a number of reasons like huge patient load, inadequate facilities, and a long waiting time for consultation and hospitalization (Galletly et al. 2011). Another factor would be a number of doctors that a patient encounters in typical department. This is due to the rotation that the resident doctors undergo during the period of residency. This may be misconstrued as discontinuity of care and a breakdown of communication. In a private setup, since the treating doctor remains the same, this issue is taken care of.

13.9.4 Approachability and Accessibility

For most patients, the private practitioner is easily accessible and mostly approachable (Marks and Thornicroft 1990). The institutes and PHC are often located out of reach of people. In towns and smaller places the private set up or the private psychiatrist is often the only mental health service available. Hence, the private sector in such areas is the initial and the most important source of care provision. A common difficulty experienced by patients visiting institutes is the sense of bewilderment because of new and intimidating surrounding, coupled with the inability to get a senior doctor's opinion (Chandra 2001).

13.9.5 Costs and Affordability

Treatment in the private sector is costly. Each consultation is charged, and each investigation has a considerable expense. Therapy sessions are charged by the hour. Chronic psychiatric disorders often entail a lifetime of psychotropic medication and rehabilitation, all of which are constant and recurrent. In countries like India, where the medical insurance does not cover psychiatric disorders, the cost of all treatment is borne by the patient and his family members. Yet, there is a willingness to spend money for good services in the private sector. Free or subsidized treatment is considered to be not of good standard. For example, an episode of hospitalization which costs *INR 320–385* in the public sector in India, would cost anywhere from *INR 735–1206* depending on the setting *(rural or urban)* (Uton and Sein 2001).

13.9.6 Time Consuming Services

Due to some of the factors mentioned above, viz. problems in accessing a PHC, large number of patients and a long waiting period for consultations and appointments, it is considered that treatment in public sector setups is time consuming. A great deal of paperwork is involved in these setups, which is considered undesirable. In Bangladesh, for example, *only one* district hospital is equipped to provide mental health services. This gives an idea of the time involved in obtaining services for the common people (Mental Health Atlas 2005a, countries a, b).

13.9.7 Satisfaction with Services

The perceived satisfaction with services offered by the private practitioners is often more than the institutes. This may be due to the easy accessibility and approachability, and the level of rapport and confidentiality maintained in the treatment process (Collaboration between Public and Private Sector Psychiatry 2000).

13.9.8 Social Class Disparity

It is perceived that a public institution or a PHC caters to only the lower socio-economic class of society. A preference is therefore given to a private consultation.

13.9.9 Ground Reality

There is a disparity in the distribution of mental health professionals–psychiatrists, psychologists, and trained psychiatric nursing staff—in any given region. Most of the professionals are concentrated in the more developed nations, and more in urban centres than in rural settings. Many PHC, in a country like India, do not have a dedicated mental health professional. However, most of the psychiatric patients are treated by the practitioners in the private sector (Uton and Sein 2001). Private facilities are increasing as compared to the settings in the public sector (Dorwart and Schlesinger 1988). Private setups include clinics, small nursing homes or hospitals, large multi-specialty hospitals, rehabilitation centres and day care facilities.

13.9.10 Poor Allocation of Governmental Budget

There is little budgetary allocation for mental health, in general, and no incentive for setting up a private service. The National Mental Health Program, 1986, a government initiative for the development of services at grassroots level, has failed to deliver the goals (Patel and Shidhaye 2010). It offers assistance for developing public sector units, but does not in anyway, contribute to developing services in the private sector. Majority of psychiatric patients are treated by private practitioners, and so are out of any government financial benefits. In India, 2.05 % of the total health budget is spent on Mental Health (Mental Health Atlas 2005b, countries e-i), and in Sri Lanka, 1.6 % of the total health budget is spent on Mental Health (Mental Health Atlas 2005d, countries s).

13.9.11 Lack of Public–Private Collaboration

There is poor collaboration between the private and institute based practitioners (Yung et al. 2005). It is difficult for most practitioners to provide admission facilities for their needy patients in a public setup. There is a fear of losing a patient to the public setup. The necessary feedback about patient management to the private practitioner is often lacking (Yung et al. 2005).

13.9.12 Costs Involved in Setting Up and Maintaining Services in the Private Sector

Setting up and maintaining any kind of service in the private sector is a costly affair, whether it is a simple clinic for consultation or a multi bedded service providing additional services like counseling-therapy or electroconvulsive therapy. The nature of cost varies from country to country and region to region. Suffice it to say that all the cost of setting up and maintaining services is borne by the individual. Private setups are generally small as compared to the public setups (Uton and Sein 2001).

13.9.13 Costs of Medicines, Investigations and Other Services and Lack of Medical Insurance

It is unfortunate that in a developing, densely populated country like India, and in many parts of South Asia, the medical or health insurance services do not include any psychiatric disorder in their schedule of medical disorders for which reimbursement could be claimed (Shah 1997). In the private sector, therefore, all the cost of consultation, medication, counseling–therapy is borne by the patient and his or her family members. More often than not, patients suffering from chronic mental disorders have no permanent or sustained employment which makes meeting expenses very difficult. One of the common reasons for poor adherence with medication is an inability to afford medication over a period of time. Many patients with poor financial support shift over to psychiatric care in the public sector (Marks and Thornicroft 1990; Tang 1997).

13.9.14 Lack of Advocacy

There is poor advocacy about private practitioners when it comes to decision making at the highest levels of power. So, there is no law or policy that is formulated which has the private sector services in consideration.

13.9.15 Consumerism

With the widely prevalent consumer movement, the private sector faces considerable difficulties. There is an increase in the number of litigation against the psychiatrist. Many psychiatrists have lawsuits against them for alleged violation of human rights and, inhuman and involuntary treatment. Psychiatrists frequently seek help of medico-legal cells more than any other specialty group (Bradley 1997).

13.10 Functioning of Services in the Private Sector

13.10.1 Legal Provisions and Regulation of Services

Every nation has laws and legal provisions to care for their mentally ill. The *National Mental Health Act, 1987* of India is one such document that governs the care for mentally ill in both public and private setups. There are provisions in the *Act* regarding setting up licensed services in accordance with the norms laid down, the procedures for admission and discharge, periodic inspection of premises by authorities, and care of human rights of the mentally ill. *Licensing* is for a period of 5 years, which is then renewed. Such legislation provides for uniformity and regulation of services (Das et al. 2002). *Mental Health Legislation* is drafted, but awaiting approval in Nepal (Mental Health Atlas 2005c, countries n-r), Pakistan has a *Mental Health Ordinance* (Mental Health Atlas 2005c, countries n-r).

13.10.2 Dearth of Trained Staff, Especially for Inpatient Services

Most private setups, be it a clinic or a hospital face shortage or unavailability of trained personnel for management of patients, *especially inpatients*. In private hospitals, where there is a psychiatry service there will be regular nursing staff who would be caring for psychiatric patients. This can often lead to conflicts between the psychiatrists, staff and patients' relatives (Patel 2007).

13.10.3 All Services and Investigations Not Available Under One Roof, and are Often Not Integrated

With high costs involved it is difficult to integrate all psychiatric and psychological services under one roof. It is also seen that these services *are not well coordinated.*

13.10.4 Private Hospital Settings, Clinics and Nursing Homes

Private psychiatric setups range from consultation clinics to small few bedded nursing homes to large multi specialty hospitals with psychiatric services. In the first two types of setups the liaison services with other specialties are not satisfactory. Liaison with other specialties is often an individual endeavour, and not necessarily a result of a large setup.

13.10.5 Poor Rehabilitation Facilities

Private psychiatric setups are generally designed for short to medium term care, for a few months at the most. There are very few centres in the private sector which are designed for long term care. Few half way homes or residences with supervised living are found.

13.10.6 Lack of Subspecialty Centres—Child and Adolescent Psychiatry, Geriatric Psychiatry and Substance Use Disorders

Like rehabilitation services, there is a lack of dedicated centres for the various subspecialties of psychiatry in the private sector. Though post PG courses in Child & Adolescent Psychiatry, and Geriatric Psychiatry are available in a few institutes, this has not translated into better of targeted services for these subpopulations in the private sector.

13.10.7 Cost and Reimbursement

As mentioned before, psychiatric disorders are not reimbursed under various health insurance schemes. It, therefore, is the individual patient's responsibility to meet the expenses of treatment (Marks and Thornicroft 1990). Assisting patients in this aspect are many trusts, community associations and NGOs that meet a part of the cost of treatment. Many employers in their individual capacity meet the expenses of their employees' treatment. There is no managed care available in South Asian countries. A practitioner can prescribe medication and other therapies as he deems fit for the patient. This has its pros and cons (Kalman and Goldstein 1998).

13.11 Positives Despite the Drawbacks

A private psychiatrist is often the first point of contact for most people, especially in cities & towns. This psychiatrist is then the sole treating doctor (Collaboration between Public and Private Sector Psychiatry 2000).

Services of the private sector are prompt. The waiting period is short and personal attention is the norm. The services are reasonably good often bordering on the excellent, and mostly conscientious (Marks and Thornicroft 1990). Evidence Based Medicine is followed keeping in mind the practicalities of the patient's circumstances and organizational limitations (Aarons et al. 2009; Badrakalimuthu and Sathyavathy 2009).

Accessibility is a hallmark of a private practitioner. For a service to endure and to be meaningful, doctors and other mental health professionals have to be available and accessible (Collaboration between Public and Private Sector Psychiatry 2000).

The local and national legislation ensures that private sector services are reasonably well regulated. Most doctors address all types of psychiatric disorders. With considerable clinical experience and liaison with colleagues of other medical specialties some practitioners have excelled in psychiatric subspecialties like Child and Adolescent Psychiatry, Geriatric Psychiatry and Substance Use Disorders (Collaboration between Public and Private Sector Psychiatry 2000). Specialty centres are needed for well planned treatment of specific populations which may not be handled well in general hospital settings (Draper and Koschera 2001; Reif et al. 2010).

Advocacy is much needed for private services, and people are waking up to this reality. With associations of private mental health professionals providing a platform for exchange of ideas and experiences, and a voice advocacy is making foray into the fabric of society. In India, the *Indian Association of Private Psychiatrists* is a body of private psychiatrists. This association is at the forefront of updating the skill and knowledge base of private psychiatrists. It holds *regular CMEs and workshops* on clinical and psychotherapeutic issues.

Nonprofessionals are an important component of the private sector. They are responsible for the various support groups that cater to increasing the awareness about mental disorders, and provide an edge to advocacy. *Individuals*, in their own capacity, have established centres for management of disorders like Autism and Specific Learning Disorders.

13.12 What is Needed?

Integration and collaboration is of utmost importance if services have to meaningful and beneficial in the long run. Recognition of the efforts of private practitioners, and of the sector also, will encourage more people to venture into private practice and provide quality service to the needy.

Collaborative models of service could be *Collaborative Care, Collocated Suites, Telepsychiatry* for consultations (Glueck 2011), *Linkage Units*, and *Individual Initiatives*. Collaboration will lead to better access, improved and continuous care (Collaboration between Public and Private Sector Psychiatry 2000; Pirkis et al. 2004). *Case conferencing* can also be considered (Pirkis et al. 2005).

Collaboration and Transition between private and Public sector [Fig. 3b.5.1] The Victorian Public–Private model has shown that a collaborative approach is feasible and can work in realistic settings (Collaboration between Public and Private Sector Psychiatry 2000). Each component of the model has its own pros and cons. Collectively, these components can effectively negate the cons of the other. For example, in collaborative care costly treatment in the private sector can be managed by the same consultant in a public setting (Collocated Suites or Linkage Units). For highly specialized services the patient can be transferred to the required care without a break in the continuity of services, if communication between the psychiatrists in both the components is good (Fig. 13.1).

Innovation, along with a patient centred approach, is necessary in increasing the longevity of a private facility (Geller 2006).

Fellowship and association are needed, and need to be strengthened to garner an opinion on matters of policy and decision making (Trivedi et al. 2007). *Development*

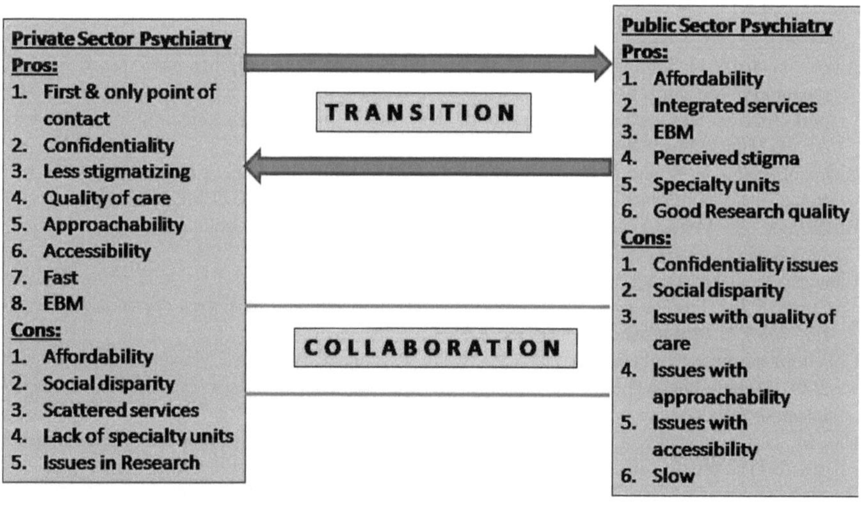

Fig. 13.1 Private Sector Psychiatry—Transition and Collaboration of services

of guidelines keeping in mind the private clinician can be a possibility when associations are a strong lobbying force. *Corporations* involved in setting up private services can also help in shaping policy (Hutchins et al. 2011).

Health insurance is another challenge that needs to be addressed. The ever increasing number of the mentally ill and the costs of prolonged treatment, coupled with limited resources mean that health insurance should include mental disorders as valid disease entities.

13.13 Research in the Private Sector

Conducting research in the private sector–clinic or a hospital setting is difficult. Facilities required to conduct basic scientific research and large multicentre randomized drug trials are not easily available. The necessary manpower and resources like funding are also not easy to procure. Another difficulty encountered is the setting up of an Institutional Review Board which can oversee the ethical conduct and quality of a study. However, many private hospitals have their own academic and research wings, and can look into the necessities required to conduct research studies of good quality (Uton and Sein 2001).

Small clinical studies can be carried out in such setups. Another kind of research work that can be done is Post Marketing Studies for psychotropic medication that has been made available in the market. Much of this research is sponsored by Pharmaceutical Companies. This research collaboration has its own plethora of issues (Singh and Singh 2007).

References

Aarons, G. A., D. H. Sommerfeld, and C. M. Walrath-Greene. 2009. Evidence-based practice implementation: The impact of public versus private sector organization type on organizational support, provider attitudes, and adoption of evidence-based practice. *Implementation Science* 4:83. doi:10.1186/1748-5908-4-83.

Badrakalimuthu, V. R, and V. R. Sathyavathy. 2009. Mental health practice in private care in rural India: A survey of practitioners. *World Psychiatry* 8:124–125. PMCID:PMC2691161.

Bradley, J. J. 1997. Litigation against psychiatrists psychiatric bulletin. *The Psychiatrist* 21:321–322. doi:10.1192/pb.21.6.321.

Chandra, V. 2001. Agenda for the mental health programme in South-East Asia regional health forum. 5:1–2. http://www.searo.who.int/LinkFiles/Regional_Health_Forum__Volume_5_No_1_rhf–Vol-5_No-1.pdf. Accessed 11 Dec 2011.

Collaboration between Public and Private Sector Psychiatry: Victoria's Mental Health Services. 2000. Department of Human Services, State Government Victoria. http://health.vic.gov.au/mentalhealth/archive/private/public-private.pdf. Accessed 11 Dec 2011.

Das, M., N. Gupta, and K. Dutta. 2002. Psychiatric training in India. *Psychiatric Bulletin* 26:70–72. doi:10.1192/pb.26.2.70.

Desai, N. G., and M. Isaac. 2001. Mental health in South-East Asia: Reaching out to the community. *Regional Health Forum* 5:6–13. http://www.searo.who.int/LinkFiles/Regional_Health_Forum__Volume_5_No._1_rhf-Vol-5_No-1.pdf. Accessed 11 Dec 2011.

Dorwart R. A., and M. Schlesinger. 1988. Privatization of psychiatric services. *American Journal of Psychiatry* 145:543–53. PMID 3282448.

Draper, B. M., and A. Koschera. 2001. Do older people receive equitable private psychiatric service provision under Medicare? *Australian and New Zealand Journal of Psychiatry* 35:626–630. doi:10.1046/j.1440-1614.2001.00930.x.

Ferri, C. P., M. Prince., C. Brayne, et al. 2005. Global prevalence of dementia: A delphi consensus study. *The Lancet* 366:2112–2117. doi:10.1016/S0140-6736(05)67889-0.

Galletly, C., S. Bell, C. Burton, et al. 2011. Bed accessibility in a private psychiatric hospital. *Australasian Psychiatry* 19:160–162. Epub 21 Feb 2011. doi:10.3109/10398562.2010.540246.

Geller J. L. 2006. A history of private psychiatric hospitals in the USA: From start to almost finished. *Psychiatric Quarterly* 77:1–41. doi:10.1007/s11126-006-7959-5.

Glueck, D. A. 2011. Telepsychiatry in private practice. *Child and Adolescent Psychiatric Clinics of North America* 20:1–11. doi 10.1016/j.chc.2010.08.006.

Hutchins, E. C., R. G. Frank, and S. A. Glied. 2011. The evolving private psychiatric inpatient market. *Journal of Behavioral Health Services and Research* 38:122–131. Epub 26 Nov 2009. doi:10.1007/s11414-009-9199-6.

Kalman, T. P., and M. A. Goldstein. 1998. Satisfaction of Manhattan psychiatrists with private practice—Assessing the impact of managed care. *Journal of Psychotherapy Practice and Research* 7:250–258.

Marks, I., and G. Thornicroft. 1990. Private inpatient psychiatric care. *British Medical Journal* 300:1272. http://www.ncbi.nlm.nih.gov/pmc/articles/PMC1662840/pdf/bmj00178-0064b.pdf. Accessed 11 Dec 2011.

Mental Health Atlas. 2005a. http://www.who.int/mental_health/evidence/atlas/profiles_countries_a_b.pdf. pp. 80–81. Accessed 17 Jan 2012.

Mental Health Atlas. 2005b. http://www.who.int/mental_health/evidence/atlas/profiles_countries_e_i.pdf. pp. 232–235. Accessed 17 Jan 2012.

Mental Health Atlas. 2005c. http://www.who.int/mental_health/evidence/atlas/profiles_countries_n_r1.pdf. pp. 357–360. Accessed 17 Jan 2012.

Mental Health Atlas. 2005d. http://www.who.int/mental_health/evidence/atlas/profiles_countries_s1.pdf. pp. 438–440. Accessed 17 Jan 2012.

Muazzam, A., and R. Khalid. 2008. Disordered eating behaviors: An overview of Asian cultures. *Journal of Pakistan Psychiatric Society* 5:76.

Mykletun, A., A. K. Knudsen, and T. Tangen, et al. 2010. General practitioners' opinions on how to improve treatment of mental disorders in primary health care. Interviews with one hundred Norwegian general practitioners. *BMC Health Services Research* 10:35. doi:10.1186/1472-6963-10-35.

Patel, V. 2007. Redefining the role of psychiatry in south asia. *Journal of Pakistan Psychiatric Society* 4:4.

Patel, V., and R. Shidhaye. 2010. Depression fact sheet—South Asia network for chronic disease. http://sancd.org/uploads/pdf/Depression_fact_sheet.pdf. Accessed 11 Dec 2011.

Patel, V., and R. Thara. 2001. Mental health policies in developing countries: A radical rethink. *Regional Health Forum* 5:3–5. http://www.searo.who.int/LinkFiles/Regional_Health_Forum__Volume_5_No._1_rhf-Vol-5_No-1.pdf. Accessed 11 Dec 2011.

Pirkis, J., J. Livingston, H. Herrman, et al. 2004. Improving collaboration between private psychiatrists, the public mental health sector and general practitioners: Evaluation of the partnership project. *Austystraukab and New Zealand Jourbal of Psychiatry* 38:125–34. PMID 14961930.

Pirkis, J., A. N. Headey, P. M. Burgess, et al. 2005. Remunerating private psychiatrists for participating in case conferences. *Australia New Zealand Health Policy* 2:33. doi:10.1186/1743-8462-2-33.

Reif, S., C. Horgan, M. Torres, and E. Merrick. 2010. Economic grand rounds: Types of practitioners and outpatient visits in a private managed behavioral health plan. *Psychiatric Services* 61:1066–1068. doi: 10.1176/appi.ps.61.11.1066.

Shah, A. 1997. Private private sector psychiatry—Can it be improved?. *Journal of The Royal Society of Medicine* 90:237–238. PMCID: PMC1296105

Singh, A. R., and S. A. Singh. 2007. Guidelines, editors, pharma and the biological paradigm shift. *Mens Sana Monographs* 5:27–30. doi: 10.4103/0973-1229.32176.

Tang, W. 1997. Previous private psychiatric treatment among public mental patients: A preliminary local survey. *Hong kong Medical Journal* 3:321–324. PMID 11847380.

Trivedi, J. K., and M. Dhyani. 2007. Undergraduate psychiatric education in South Asian countries. *Indian Journal of Psychiatry* 49:163–165. doi:10.4103/0019-5545.37314

Trivedi, J. K., D. Goel, R. A. Kallivayalil, et al. 2007. Regional cooperation in South Asia in the field of mental health—Mental Health Policy Paper. *World Psychiatry* 6:57–59. PMCID:PMC1805715

Uton, M. R., and U. T. Sein. 2001. Role of private hospitals in health care. *Regional Health Forum* 5:41–48. http://www.searo.who.int/LinkFiles/Regional_Health_Forum__Volume_5_No._1_rhf-Vol-5_No-1.pdf. Accessed 11 Dec 2011.

Versola-Russo, J. 2006. Cultural and demographic factors of schizophrenia. *International Journal of Psychosocial Rehabilitation* 10:89–103.

Wen-Shing, Tseng. 2006. From peculiar psychiatric disorders through culture-bound syndromes to culture-related specific syndromes. *Transcultural Psychiatry* 43:554–576. doi:10.1177/1363461506070781.

WHO-AIMS report on mental health system in Uttarkhand, India. 2006. World Health Organization & Ministry of Health, Uttarkhand. Accessed 11 Dec 2011.

Yamaguchi, N. 2011. Considering residency programs in psychiatry in a private hospital. *Seishin Shinkeigaku Zasshi* 113:267–70. PMID: 21591402.

Younès, N., M. C. Hardy-Bayle, B. Falissard, et al. 2005. Differing mental health practice among general practitioners, private psychiatrists and public psychiatrists. *BMC Public Health* 5:104. doi:10.1186/1471-2458-5-104.

Yung, A., L. Gill, E. Sommerville, et al. 2005. Public and private psychiatry: Can they work together and is it worth the effort? *Australian and New Zealand Journal of Psychiatry* 39:67–73. doi:10.1080/j.1440-1614.2005.01511.x.

Chapter 14
NGOs and Mental Health: Initiatives and Progress

R. Thara

This chapter provides an overview of the NGO movement in mental health in the South Asian countries, describes the various activities taken up by them such as care, treatment, rehabilitation, education, awareness, community outreach, advocacy and human rights. Some of them have gained international recognition for their work and have acted as good role models for others to emulate. Intense and focused work, credibility, closeness to the communities they serve, and capacity to innovate and network are the some of the positive aspects of these NGOs. Limitations such as lack of sustainability and transparency, high staff turnover and burn out are also discussed. The recent, rather disconcerting trend of mushrooming of small, unregistered and unmonitored facilities delivering sub human levels of care for the mentally ill is an area of concern.

14.1 Introduction

Today, mental health is much more than just a branch of medicine. It has widespread psychological, social, behavioural, spiritual and even physical health ramifications. The behaviours of children ranging from drug and mobile phone abuse to suicide for what seems trivial reasons, is of great concern not just for the family, but for the entire society. Increasing farmer suicides all over the country are due to socio-economic-psychological reasons. Older people, by living longer are now more susceptible to age related disorders such as dementias. Families have not much time to invest on ill and disabled people. In this scenario, prevention of psychological morbidity and managing mental health problems requires a multi-pronged, multi-sectoral and multidisciplinary effort.

R. Thara (✉)
Schizophrenia Research Foundation, Chennai, India
e-mail: scarf@vsnl.com

J. K. Trivedi, A. Tripathi (eds.), *Mental Health in South Asia: Ethics, Resources, Programs and Legislation,* International Library of Ethics, Law, and the New Medicine 58,
DOI 10.1007/978-94-017-9017-8_14

At the level of the Government, the departments of health, social welfare, education, labour and law are all players in the mental health field. But a concerted effort of these agencies and the local community is desirable. It is in this context that NGOs have a major role to play.

14.2 India

The NGO movement in the country has seen a steady upswing in the last 2 decades. NGOs are driven by a passion towards a certain cause and back it up with commitment and drive. While the reach of their work cannot parallel that of Govt. agencies, the quality of care and their efforts in reaching out to the various stakeholders gives them a distinct advantage. By working with limited resources in a limited area, NGOs can also develop models of care which can be replicated in other parts (Pachauri 1994).

This chapter seeks to provide an overview of the contributions of Mental Health NGOs (MHNGOs). A brief profile of some NGOs working in key and distinct areas should enable the readers to understand better the ways in which NGOs can innovate and replicate and complement state run services.

14.2.1 The Diversity of Mental Health NGOs (MHNGOs)

Despite the considerable challenges faced in developing mental health programs, it is gratifying to note the achievements of many MHNGOs (Patel and Thara 2003). They are distributed throughout the country, although there are a greater number in urban areas, and in the southern states. Although MHNGOs are predominantly urban in location, many have begun to extend services into rural areas. Most MHNGOs serve a defined community; however, the work of some has spread to more than one centre or geographical region. Examples of such NGOs are the Alzheimer and Related Disorders Society of India (ARDSI), which was started in Cochin, and has now spread to more than a dozen centres in India. And the Richmond Fellowship Society which has three centres. The oldest MHNGOs in India are probably those working in the field of child mental health, and in particular, mental retardation. This may not be surprising given the close nature of the relationship between mental retardation and the concept of childhood disabilities which has been one of the bedrocks of the MHNGO movement for several decades. The concept of child mental health has broadened from its earlier focus on mental retardation to include the far commoner mental health problems seen in children, such as autism, hyperactivity and conduct disorders. MHNGOs such as Sangath Society (Goa), The Research Society (Mumbai) and Samadhan (New Delhi) provide outpatient and school based services for such problems.

The other large group of NGOs are those working with people with severe mental disorders. Many of these MHNGOs such as the Schizophrenia Research Foundation (SCARF) in Chennai, the Medico-Pastoral Association (MPA) in Bangalore and

Shristi in Madurai were started by psychiatrists who already held full-time faculty positions in the local medical schools. These MHNGOs were started as a means to fulfil the need for a broader, holistic approach to the management of severe mental disorders. Thus, activities ranging from family counseling to vocational rehabilitation which were rarely provided in psychiatric out-patient clinics were given greater attention. Another area of mental health which attracted considerable interest and attention was substance abuse. Alcohol abuse and, in particular, drug abuse captured the public imagination and received considerable media interest in the 1970s and 1980s. This public attention and the obvious need for community based rehabilitation services for persons affected by substance abuse led to the development of numerous MHNGOs working in this area. The TTK Hospital in Chennai, the TRADA in Kerala and Karnataka, Alcoholics Anonymous and the Samaritans in many parts of the country and the National Addiction Research Centre in Mumbai are examples of MHNGOs focusing on substance abuse problems.

More recently, the scope of activities of MHNGOs has broadened further. Perhaps, this has to do with the expansion of our understanding of the range and nature of mental health problems. MHNGOs providing community based counseling and suicide prevention activities have mushroomed. Reports highlighting the rising rates of suicide in India, in particular amongst young people, have alerted health professionals and the community about this serious mental health problem. Sneha (Chennai), MPA (Bangalore), Saarthak (Delhi) and Prerana (Mumbai) work on suicide prevention activities. Some MHNGOs focus on women's mental health; common mental disorders, which are often linked to stress and oppression, are not surprisingly, more frequent in women.

Some MHNGOs, such as ACMI (Bangalore) and Aasha (Chennai) are entirely run by and focus on, families of those affected by severe mental disorders. ARDSI works with families who have a member affected by dementia. The growth of these non-professional, family oriented MHNGO sector is to be welcomed for it is very likely that the needs of the mentally ill may be expressed and met in different ways by families and by mental health professionals.

There are some MHNGOs like the Ashagram in Madhya Pradesh whose primary focus was initially physical disabilities, especially persons affected by leprosy but later expanded its community based rehabilitation program to include severe mental disorders which also produce a profound disability in some persons. Other examples of broad-based NGOs which are integrating mental health in their agenda include the Voluntary Health Associations of India and the Community Health Cell (Bangalore). These are healthy trends facilitating the view of mental health as an integral component of the broader rubric of public health.

14.2.2 What Motivates NGOs?

The perceived need of the community appears to have been a major catalysing factor for the sustainability of all the MHNGOs. In some cases, personal tragedies and first hand experiences have been inspirational factors. Skepticism and cynicism, especially of the medical community, non-cooperation and lack of sensitivity of government officials have been uniform experiences especially in the founding years. Not unexpectedly, a high premium is placed on involvement of families and other stakeholders in the activities and programs of all the MHNGOs. For many MHNGOs, government funding support is minimal and most are dependant on general public or donor agencies for financial resources. A few have been able to mobilise research funds, by virtue of having established research credentials. Many MHNGOs charge fees for services. Let us consider the kinds of activities which MHNGOs are engaged in working towards their objectives towards improving the health of those affected by mental disorders.

14.2.3 MHNGOS: Activities and Programs

The activities of the MHNGOs can be grouped as below:

14.2.3.1 Treatment: Care and Rehabilitation

It was natural for many MHGNOs to identify treatment and rehabilitation as their priorities, based on the felt needs of the populations they wished to serve. Models of care and rehabilitation have been developed, with a focus on psycho social rehabilitation (PSR), a sadly neglected area. The lack of trained staff to carry out PSR activities and difficulty in raising resources are other reasons for PSR not being main streamed into psychiatric services. Hence, many NGOs have taken it upon themselves to develop modules of PSR in both urban and rural areas. These programs include a spectrum of activities such as individual and group counseling, life skills training, vocational rehabilitation, cognitive retraining, family support and counseling, recreation and leisure activities. Out-patient clinics, in patient care, day care programmes, long term residential care and community outreach services form the spectrum of services provided by NGOs, specially the ones dealing with chronic psychotic conditions.

Drug and psychological treatments including individual counseling, marital and family therapies and group therapies, specific interventions targeted at children, their parents and sometimes even class room interventions are offered (Isaac 1998).

14.2.3.2 Community Based Activities

Although the National Mental Health Program was initiated in 1983 to ensure minimum standards of mental health care even in rural areas by integration with existing primary health care services, this still remains an utopian dream in many states. On the other hand, NGOs specially in the south and eastern parts of the country have initiated a number of community based mental health programmes.

These programs have ranged from primary prevention activities to provision of treatment in community clinics, increasing awareness and working on CBR models. NGOs are better poised to approach local communities, establish ties with them and locate their programs in and for the community.

Community based rehabilitation is an essential ingredient of community care programmes. SCARF, as part of vocational support activities has distributed livestock, cows and helped expansion of petty shops in rural areas to help persons with schizophrenia. This is not just a means of livelihood, but has improved their functioning and involvement in many ways. Empowerment of the local community is equally important and involvement of key and influential persons in the community such as teachers, religious heads, local administrators has yielded good results.

Community programs also gained a lot of significance when the Tsunami left in its wake a number of psycho social problems which required intensive counseling, support and sometimes medication to allay anxiety and depression (Vijaykumar et al. 2006).

In keeping with latest advances in technology and communications, telemedicine has been used in community mental health. SCARF started conducting community clinics through telemedicine in 2005 wherein a psychiatrist from Chennai connected up to the outreach clinic in Nagapattinam. Subsequently a network of seven telemedicine outreach clinics was set up. Gaining from this experience and leveraging the advantages that the technology provided in terms of expanding the reach of services, SCARF in 2010 initiated the delivery of mobile telepsychiatry in Pudukottai district. The successful demonstration of the model has resulted in several other NGOs from India as well as from other developing countries now looking to replicate it.

Homelessness and destitute mentally ill have also received attention in the last decade or so. Here again, the state has contributed little to address the issue in a comprehensive manner. NGOs such as Banyan and Anbagam in Chennai have developed comprehensive services for the wandering mentally ill. However, in order to sustain these programmes, a national plan is required for the provision of care to the homeless and wandering mentally ill, shrouded as it is in social, economic and human rights issues.

14.2.3.3 Substance Abuse

Alcoholism and drug abuse are serious public health problems with an impact not just on mental, but on physical health as well. IV drug use and unsafe sex in the context of alcoholism are associated with increased risk of HIV infections. With a 7–8 % of the work force in India being addicted to alcohol (Thirumagal and Ranganathan 2003), absenteeism, reduced work output and increased crime rates are also issues to contend with.

In the 1970s and 1980s, there was a mushrooming of NGOs to deal with drug abuse, but few of these organizations have sustained themselves. Again, there are not many NGOs which deal with alcoholism in a scientific manner. The TTK hospital/TTRanganathan clinical Research Foundation in Chennai started in 1980 is an organization which has expanded itself in various activities and become a referral centre for training and awareness building. This NGO has been active in the various fields of out-patient and in-patient care, extended care to those who need it, community outreach programs and CBR. Education, awareness and interventions at work place are other activities.

The Indian Alcohol Policy Alliance, a network of centres and individuals working in de-addiction has released an "Alcohol atlas of India" which is a good reference guide for policy makers and professionals.

One of the nodal drug abuse centres is the National Addiction Research Centre (NARC) established in 1985. Like NARC, there are other NGOs that advocate for policy change, besides creating awareness. A lot of IV drug use also resulted in the launch of the Needle Exchange Program (NEP) in Chennai.

Both alcoholism and drug abuse have far reaching consequences and implications, much beyond the medical boundaries. It is therefore critical that national and state level policies and programs be drawn up in consultation with NGOs committed to this cause so that a comprehensive strategy can be put together.

14.2.3.4 Suicide Prevention

Working with suicide has been Chennai based Sneha's prerogative and this NGO has definitely made a positive impact by increasing public access to the crisis centre and thereby decreasing the suicidal behaviour in the community. Formation of support groups for suicide survivors has been their other significant contribution. There are other centres dealing with suicide prevention such as Samaritans Sahara and networks such as the Befrienders India, a national association with 10 centres/members in India.

14.2.3.5 Women and Elderly

A noteworthy contribution by an NGO towards women's mental health especially with a history of abuse, violence and trauma has been IFSHA (Interventions for Support, Healing and Awareness) in Delhi. Later they made inroads into areas related to sexuality such as HIV/AIDS, adolescent health, lesbian and gay issues and reproductive health by conducting workshops. These workshops emphasize the participation of men and cover a range of subjects such as morality, gender, violence, pleasure, childhood conditioning, masculinity, femininity and sexual stereotyping. Ultimately the participants understood the concept and language of sexuality that is not sexist but gender sensitive and empowering. The impact of this new learning is most evident in the balance that it creates in a mixed group, helping men and women understand sexuality and vulnerability from a gender neutral point of view.

While there are several homes and centres for the elderly, most of them are wary to take care of those with mental disorders. The ARDSI, an off shoot of the

International ADI has been singularly effective in training workers to deal with persons with dementia. They are also engaged in a host of other activities such as awareness building, day care programmes, home based interventions and have a number of chapters all over the country.

14.2.3.6 Support Groups

As part of the broad perspective on health care, many MHNGOs are adopting methods to enhance the effectiveness of treatments provided to individuals. Support groups are widely used as a way to ensure that persons recovering from substance abuse can remain sober. The globally recognised organisation, Alcoholics Anonymous, is an example of the kind of support group philosophy which becomes the core to the process of treatment of alcohol dependence. Support groups are also evident in the residential and day care facilities geared to those with severe mental disorders. Some MHNGOs run support groups not for those actually affected by a particular disorder, but for their families. Here, families of elders with Alzheimer's disease, adults with schizophrenia and children with autism, meet regularly to discuss common problems, support each other and provide practical solutions to everyday difficulties.

The All India Federation for Mental Health is an umbrella organization of many NGOs working in the field of mental health and has been trying to lobby with Govt. on a few issues.

The Nodal Association of the Mentally Ill (NAMI- India) is another platform started by a consumer of mental health services in Mumbai to bring together consumers, mentally challenged persons, mental health professionals and the general public (www.namiindia.in). The FACEMI is a more recent addition to family support groups.

14.2.3.7 Research and Training

Until recently, MHNGOs were primarily concerned with service provision and advocacy related activities. Research was considered as an academic exercise, best reserved for the ivory towers of universities and teaching hospitals. This has changed so much in recent years that today, MHNGOs are at the forefront of ground-breaking research in India. Major research programs in health areas as diverse as infectious diseases to nutrition are now conducted under the aegis of NGOs. MHNGOs are no exception to this trend. The SCARF studies on schizophrenia are the most widely-cited research on the subject from any developing countries (Thara and McCreadie 1998). Many published studies of dementia in the community in India are from work done by MHNGOs who were part of the 10/66 dementia group of collaborators (Prince et al. 2011; 10/66 Dementia Research Group 2009; Dias et al. 2004). Sangath's studies on the treatment of depression are amongst the largest such studies from India (Patel et al. 2002, 2003) Ashagram's community program for schizophrenia has generated the first scientific evidence of the use of the CBR approach for rehabilitation of a mental disorder (Chatterjee et al. 2003). These are just some examples of innovative, action-oriented research emanating from MHNGOs.

The first community based Randomised Control Trial (RCT) for CBR in schizophrenia was also initiated by MHNGOs such as Sangath, SCARF, Parivartan' and 'Nirmittee'. This trial proved that trained lay health workers can deliver effective community based interventions for severe mental disorders. This can be used to scale up services in low resource settings (Chatterjee et al. 2014).

Many of the MHNGOs actively invest in the development of skills of their staff. Participation in workshops, conferences and seminars, and formal training in courses such as rehabilitation are often offered as opportunities for career development. Most of the MHNGOs provide opportunities for training other professionals and health workers in specific areas of mental health, such as counseling skills. Many colleges, for example, send their students to MHNGOs for field placements. Workshops with health workers, teachers and other key groups are a standard feature of the activities of many MHNGOs. The Richmond Fellowship has successfully established a full 2-year MSc program in psychosocial rehabilitation. Many of these organisations regularly organise local, national or international conferences, seminars, workshops or symposia to discuss current issues in the field (Kalyanasundaram and Varghese 2000). SCARF trains students in the post graduate course offering DNB in psychiatry and is also a PhD centre.

Sangath has been running a Leadership in Mental Health program in Goa in collaboration with SCARF, the London School of Hygiene and Tropical Medicine and the University of Melbourne. The course has been designed to equip participants in the methods to develop and scale up interventions for people with mental disorders in low resource settings, based on a population model to meet the goals of the Movement for Global Mental Health (www.globalmentalhealth.org).

14.2.3.8 Advocacy and Awareness Building

Advocating for the needs of under-served and underprivileged sections of the population has been the *raison d'etre* for most MHNGOs. At present, there is very low awareness of the considerable advances in our knowledge of the causes and treatment of mental disorders in India. This low awareness, coupled with the enormous stigma attached to mental illness, means that the needs and rights of mentally ill persons are largely ignored. MHNGOs have made raising awareness in different sectors of the community, such as health workers, teachers and lay persons, a priority area. Documentation and dissemination of relevant facts and research, and lobbying policy makers for changes in the law are vital instruments for improving mental health care. Prominent examples of the success of efforts of MHNGOs are the inclusion of mental disabilities in the disability legislation of the country.

The film festival organised by SCARF called the "Frame of Mind" which features films portraying mental illness and an international competition for short films on mental health and stigma is a huge success and has had four editions so far. Similar festivals have since been held in other cities like Calcutta and New Delhi. Many NGOs use short films to spread awareness about their work/cause. Many publish regular newsletters and host web sites marking the close affinity of MHNGOs with contemporary technological advances.

14.2.4 Other Service Providers

In the last few decades, several private centres which offer treatment, rehabilitation and long term, even life time care have sprung up all over the country, especially in the south. These centres do not have any qualms in calling themselves commercially run organizations. Nevertheless, they also seem to fulfill a need in the community. Affluent and NRI families who do not have the human resources to take care of the chronic mentally ill and disabled seem to be benefiting from such centres.

What is however not a welcome trend is the mushrooming of very small to medium sized outfits, many of them in a small portion of the homes of those who run it. People with MI, MR, dementia and other ailments are taken in and kept in a cramped place with frugal food and poor housing. Many of them sleep on mats even during winters and are given a standard dose of medication, usually to put them to sleep at nights, such as chlorpromazine 100 mgs. These places do not have any professional inputs and don't register themselves under the MHA. The quality of care is abysmal and the poor residents are unable to complain about it.

14.2.5 Licensing and Quality Control

While most of the well established NGOs are registered in their respective states under the MHA and are liable to be monitored and scrutinized, this is not the case with the kind of centres referred to in the earlier paragraph. In fact, many old age homes take in the mentally ill since they do not require any license. This has to be seriously looked into and violation of human rights of the mentally ill in these non professional, small centres cannot be allowed to continue.

Apart from India, other south Asian countries also have significantly contributed to mental health in their respective countries. A brief description of some of these NGOs is presented below:

14.3 Sri Lanka

Although voluntary organizations have existed in Sri Lanka for over 2 decades, the post tsunami period witnessed an upsurge in community based activities, both by the Government and in the NGO sector.

On such program was by the UK based organization—**Voluntary Services Overseas.**

In this, VSO volunteers worked within government and non-governmental organisations (NGOs) to strengthen the delivery of community-based mental health services. This included working on training curricula for new mental health staff, strengthening the management of government departments and supporting the expansion of successful rehabilitative projects, such as a gardening project run by VSO partner Basic Needs.

The other well known NGO is **Sahanaya or The National Council for Mental Health.** Sahanaya is a non profit Non Governmental Organization, working towards the development of mental health and mental health care in Sri Lanka. Established in 1982, it was incorporated by an Act of Parliament in 1986 and is a registered charity with the Ministry of Finance. Sahanaya maintains the **Borella Resource Centre** and the **Gorakana Residential Facility. Its objective** is to develop a centre of excellence with expertise and the necessary competence to respond to emerging mental health needs/issues of the individual, communities and special groups of people.

Navajeevana, meaning "new life", is a non-governmental organisation based in the Hambantota district of southern Sri Lanka. While beginning to work with persons with disabilities, since 2002 Navajeevana has provided community mental health services in partnership with Basic Needs. In 2008 Basic Needs moved away from the project area and Navajeevana officially became the implementing body on mental health in collaboration with the Health Ministry. The program has several components such as Community Mental Health, sustainable livelihood, Capacity Building.

Currently there is an individually tailored residential life skills development program in operation for the long-term mentally affected individuals especially for those who are suffering from schizophrenia to improve their daily living skills. Day care facilities for mentally affected individuals and Respite care facilities for the individuals who are mentally affected and need a short term stay for a period of 1–6 months are also offered to the local community.

14.4 Pakistan

Since its inception in 1971, **Fountain House** has been providing state of the art psycho social rehabilitation facilities to persons with mental disorders not only from Pakistan but also for Pakistani families living abroad.

A recent bulletin of October 2008 published by the Pacific Rim College of Psychiatrist Japan, spoke of Fountain House as a model even for developed countries. Over 7000 **psychiatric patients have been rehabilitated and sent back to the community**,

The Royal Kingston Rehab Centre was inaugurated in 2007 and now caters to the needs of private psychiatric patients as well as the patients from Kashmir Earth Quake zones.

Fountain House also regularly hosts international conferences. A WPA approved One Year Certificate Course in Psychiatry for Family Physicians is also being held.

Pakistan Association for Mental Health was established in 1965 under the convenership of Dr. S. Haroon Ahmed. PAMH is a non-profit, non-commercial organization for public service and for creating awareness about mental health, with a mission to: (1) Evolve a community-oriented (family involvement), cost effective (affordable) and culturally relevant (acceptable) program of treatment and aftercare of mentally ill patients and (2) develop local need-based programs in the field of behavioral sciences for: teaching, training, research and psycho-education.

The current activities of PAMH are its outpatient department/MIND Centre, help line of trained psychiatrists and Community Mental Health Workers who are available for help and information over the phone number six days a week and The Community Mental Health Program which is run with a concept of public service, providing basic mental health care at the doorstep of patients involving local resources and primary health providers in the community.

14.5 Nepal

The small country of Nepal does have some NGOs working in mental health.

The Centre for Mental Health and Counseling—Nepal (CMC-Nepal) is a national level non-governmental organization, registered on 1st May 2003 in Kathmandu District Administration Office and affiliated to the Social Development Council. It is working on various levels with preventive, promotive and curative aspects of mental health, aiming to provide mental health services in the community. It is also supporting other organizations in their psychosocial programs.

The main aim of CMC is to build the capacity of government health staff in mental health and of I/NGO staff in psychosocial approaches, in order to increase access to mental health services and psychosocial support at community level. Some of the projects they run are a community mental health and psychosocial support programme, child mental health programme, Support for Effective Empowerment (SEE) for the Swiss Agency for Development and Cooperation (SDC) Projects, Social Responsiveness Programin Brick Kiln Industries Psychosocial Support to the Children and Women Affected by HIV/AIDS (CABA Project).

They are also involved in general capacity building and run a National Mental Health Resource Centre.

Transcultural Psychosocial Organization is one of Nepal's leading psychosocial organizations. TPO Nepal's primary aim is to promote psychosocial well-being and mental health of children and families in conflict-affected and other vulnerable communities, through development of sustainable, culturally appropriate, community-based psychosocial support systems. TPO Nepal is affiliated with Health Net TPO, an Amsterdam based international organization that works in conflict and disaster settings, with the aim of re-establishing and improving public health and mental health care systems.

Maryknoll Nepal, established in 1991, is a non-profit, non-governmental voluntary social organization, duly registered with His Majesty's Government and Social Welfare Council of Nepal. It was established with the main aim of releasing all the chronically mentally ill patients locked in different jails like Central Jail in Kathmandu and Dhulikhel jail in Kavre. Those patients were imprisoned for many years, solely for being mentally ill. There were no hospital facilities to accommodate them and the families did not want them back due to the chronic and relapsing nature of their illness. Psychiatric treatment within the jail does not exist. The other two aims are to provide treatment and then, to rehabilitate the mentally ill within their own families and communities.

Koshish is a non-profitable, national Non-governmental organization (NGO) working to improve the quality of mental health policies and programs, and at the same time challenge existing discriminating attitudes towards people affected by mental illness. Koshish is functioning as a "self help" organization where the mentally ill persons and their families are themselves given a voice—instead of just other people "speaking on their behalf". In fact the very existence of Koshish evolved from the first hand experiences of people with mental health problems, along with the support and solidarity of their families and related professionals at strategic levels. Koshish has been working informally and voluntarily on mental health issues since 2004. During these first years Koshish became involved in the rehabilitation of dozens of people affected by mental illness.

14.6 Bangladesh

Founded in 1992, the Bangladeshi Mental Health Association (BMHA) is a company limited by guarantee, now seeking registration as a charity. The Association works in both the UK and Bangladesh; The Association in Bangladesh is now registered as a Non-Governmental Organisation with the NGO Affairs Bureau. Its main objectives are to promote and, where necessary, to provide a culturally appropriate community-based mental health service to the people of Bangladesh.

To work to counter all kinds of violence and, when it occurs, to contribute to the amelioration of its consequences, to advance and encourage the conditions for self-support by way of mental health education, education in consequences, and in any other way are other aims.

14.7 Afghanistan

Dr. Peter Ventevogel, an Amsterdam-based psychiatrist has worked to improve Afghanistan's mental health care through the nonprofit Healthnet TPO. "It's a new way of thinking that a patient with a mental disorder is someone you have to consult," Ventevogel said. "Those are notions that are not yet ingrained in the health-care system in Afghanistan. The whole idea of client participation is a difficult one."

Yarzada and other Afghan therapists at the hospital have been training with a medical anthropologist to practice more transcultural therapy, which incorporates patients' traditional beliefs about mental illness, treating them for what they believe ails them.

It is still common in Afghanistan for families to chain the severely mentally ill at home or at Sufi shrines where they are left to subsist for weeks on water and peppered bread.

14.8 MHNGOS: Strengths and Limitations

Why is it that the MHNGO movement has continued to survive despite the lack of resources and other problems? This is probably because MHNGOs have some inherent and intrinsic advantages. We can consider the advantages of MHGNOs under three broad categories: Working in Partnership, Innovations in Practice and Transparency in Administration.

- Working in Partnerships: One of the great strengths of MHNGOs is their ability to strike up collaborations and partnerships with other agencies or individuals with ease, unlike the public health sector where layers of permissions stifle the scope for collaboration and unlike the private health sector where collaborations may be perceived as a threat to the practice. Most MHGNO activities are provided by multidisciplinary teams of doctors, therapists, health workers, other professionals and volunteers. Partnerships are built not only between medical and non-medical professionals, but also between professionals and families. The close collaboration between academics, clinicians, social workers, rehabilitation workers, remedial teachers, clinical and educational psychologists are a distinct feature which marks MHGNOs as being a very different breed of animal from traditional psychiatric clinics in hospitals or private psychiatry.
- Innovations in Practice: MHGNOs are typically, closer to the community they serve and hence in a better position to be more sensitive to changing needs and perceptions. Furthermore, MHNGO services may be attached with much less stigma than formal psychiatric services, and may thus attract a much wider range of clients. Clinical support, involving diagnosis and treatment of specific mental disorders, is the key to many MHNGO activities. The success of MHNGOs lies in providing services which are accessible, such as through outreach camps, and which rely on available human resources, such as the community participatory model of rehabilitation. Many MHGNOs provide a wide range of services which are especially suited for severe and childhood mental health problems. By taking up the process of promoting attitudinal changes in the community and amongst policy makers, MHGNOs also play a key role in advocacy for changes which can benefit all persons with mental illness.
- Transparency in Administration: The activities of MHNGOs are driven not by profit but by the desire to achieve a basic quality of care for all clients, irrespective of their ability to pay. They are governed by a relative flexible set of

regulations. Employment and promotional avenues can be based on merit as opposed to the traditional governmental holy grail of seniority. Because they are dependent on external funding, MHGNOs are constantly kept on their toes in achieving program objectives and ensuring fiscal accountability. MHNGOs can explore, with remarkable entrepreneurial dynamism, collaborations with any other organisation or individual to achieve their objectives.

However, MHGNOs have their fair share of limitations and problems. We can consider these under the broad themes of Sustainability, Accountability and Scope.

- Sustainability: A key problem facing most MHNGOs is the source of their funding which is largely project based. The periodic fund raising required to augment resources can take up a good deal of time and energy. Staff has no guarantee of employment beyond a defined project period. Some MHNGOs suffer a high turn over of staff. This is partly because staff is appointed on specific funded projects and their continuity depends on the funding available. There might be a temptation to dilute goals and objectives as a response to availability of funding. Donor funding is notoriously fickle; priorities change over time, and MHNGOs often reinvent their objectives to keep afloat. The current trend for massive investment in HIV/AIDS related work, though important in its objectives, is concentrating the bulk of donor money to this one-disease issue. Many MHNGOs and, indeed, some MHGNOs are adding HIV/AIDS as core priorities to secure these funds. While this may broaden the scope of MHNGOs by enabling an integration of existing priorities with new ones, there is equally a need not to allow the focus on mental health to be diluted to the point that it becomes irrelevant.
- Accountability: Some MHGNOs have poorly established mechanisms for evaluation and monitoring. Although networking is actively sought for project collaboration, there is no similar zeal for review and monitoring from external assessors. There has been considerable public concern regarding the misuse of funds and lack of financial accountability of NGOs in general. Although this may not be as significant an issue in the context of MHGNOs where funds are scarce, MHGNOs would be well advised to ensure transparency in accounting for their funds. As MHNGOs become larger and more professionalised, there is a danger of increasing bureaucratization with increasing administrative costs. MHNGOs should be wary of this from the beginning since it could well dampen creativity and flexibility, two elements which give MHNGOs their unique flavour.
- Scope: Finally, and perhaps the most important limitation is the limited scope of individual MHNGOs. The world of most MHNGOs is confined to a city or a few villages. There is however a need to transplant the wide experience of these onto a larger canvas, ideally through influencing policies and programs for the entire state and country. For changes to occur on this wider canvas, there is little doubt that the public or government health sector must play a key and leading role. MHGNOs can, in this context, be seen as innovators who develop locally relevant models which can then be implemented on a national scale.

14.9 Conclusion

In conclusion, it can be said that MHNGOS are increasingly becoming a critical resource in the mental health scenario of the South Asian region. However, given the unmet needs in the area of mental health services, it would be unrealistic to expect NGOs to cater to all those in need of intervention. The strength of NGOs does not lie in numbers, but in evolving and perfecting quality programs and models which can be replicated. With the increasing emphasis on community participation and the recognition of the role of user and family groups, NGOs are bound to play a leading role in the years to come, not just as a sporadic movement but become a more enduring and unified force.

References

Chatterjee, S., V. Patel, A. Chatterjee, and H. Weiss. 2003. Evaluation of a community based rehabilitation model for chronic schizophrenia in a rural region of India. *British Journal Psychiatry* 182:57–62.

Chatterjee S., S. Naik, S. John, H. Dabholka, R. Thara, V. Patel, and Thornicroft. 2014. Effectiveness of a community-based intervention for people with schizophrenia and their caregivers in India (COPSI): a randomised controlled trial. *Lancet*. 383(9926):1385–1394.

Dias, A., R. Samuel, V. Patel, M. Prince, R. Parameshwaran, and E. S. Krishnamoorthy. 2004. The impact associated with caring for a person with dementia: A report from the 10/66 Dementia Research Group's Indian network. *International Journal Geriatric Psychiatry* 19 (2): 182–184.

Isaac, M. K., ed. 1998. *Psychosocial rehabilitation: A handbook*. Bangalore: Medico Pastoral Association.

Kalyanasundaram, S., and M. Varghese, eds. 2000. '*Innovations in Psychiatric Rehabilitation', Proceedings of the RF–ASPAC International Symposium (1995) at Bangalore*. Bangalore: Richmond Fellowship Society.

Pachauri, S., ed. 1994. *Reaching India's Poor: Non-governmental approaches to community health*. New Delhi: Sage.

Patel, V., and R. Thara, eds. 2003. *Meeting mental health needs in developing countries: NGO innovations in India*. New Delhi: Sage.

Patel, V., M. Rodrigues, and N. De Souza. 2002. Gender, poverty and post-natal depression: A cohort study from Goa, India. *American Journal Psychiatry* 159:43–47.

Patel, V., D. Chisholm, S. Rabe-Hesketh, F. Dias-Saxena, G. Andrew, and A. Mann. 2003. The efficacy and cost-effectiveness of a drug and psychological treatment for common mental disorders in general health care in Goa, India: A randomised controlled trial. *Lancet* 361:33–39.

Prince, M., D. Acosta, C. P. Ferri, M. Guerra, Y. Huang, K. S. Jacob, J. J. Llibre Rodriguez, A. Salas, A. L. Sosa, J. D. Williams, and K. S. Hall. 2011. A brief dementia screener suitable for use by non-specialists in resource poorsettings–the cross-cultural derivation and validation of the brief CommunityScreening instrument for dementia. *International Journal Geriatric Psychiatry* 26 (9): 899–907.

Thara, R., and R. G. McCreadie. 1998. Research in India: Success through collaboration. *Advances in Psychiatric Treatment* 5:221–224.

Thirumagal, V., and S. Ranganathan. 2003. *Meeting challenges in the field of addiction (289–308) in meeting mental health needs in developing countries: NGO Innovations in India*. New Delhi: Sage.

Vijaykumar, L., R. Thara, S. John, and S. Chellappa. 2006. Psychosocial interventions after tsunami in Tamil Nadu, India. *International Review Psychiatry* 18 (3): 225–231.

10/66 Dementia Research Group. 2009. Subjective memory deficits in people with and without dementia: Findings from the 10/66 dementia research group pilot studies in low- and middle-income countries. *Journal American Geriatrics Society* 57 (11): 2118–2124.

Chapter 15
Roadmap for the Future: What is Needed in the Region?

Pratap Sharan and T. Sivakumar

15.1 Introduction

Those who cannot remember the past are condemned to repeat it- George Santayna.

World health organization (WHO) estimates that 12 % of global burden of disease is accounted by mental and behavioural disorders (WHO 2001). This may be an underestimation considering the inter-connectedness between mental illness and other health conditions. Almost three quarters of the global burden of mental and behavioural disorders are in low- and middle-income countries (LAMICs) (WHO 2008). A WHO survey showed that 35–50 % of serious cases in high-income countries and 76–85 % in LAMICs had received no treatment in the previous 12 months (WHO 2008). Mental disorders are associated with poverty, marginalisation, and social disadvantage; and directly affect progress toward achievement of many of the Millennium Development Goals (MDGs) (Lancet Global Mental Health Group 2007). To break the vicious circle, patients with mental disorders should receive effective interventions which are as cost-effective as interventions for other chronic, non-communicable conditions (Saxena et al. 2007). Investment for mental health care should be considered a public health, human rights and development priority (Lund et al. 2011).

The Lancet global mental health group proposed the presence of official policy, programmes or plans for mental health; specified budget for mental health; strength of mental health workforce; availability of essential psychotropic medications in primary care; increased treatment coverage; reduced suicide rates and protection of human rights of mentally ill as some indicators of progress (Lancet Global Mental Health Group 2007).

P. Sharan (✉)
Department of Psychiatry, AIIMS, New Delhi 110029, India
e-mail: pratapsharan@gmail.com

T. Sivakumar
Department of Psychiatry, NIMHANS, Bengaluru 560029, India

J. K. Trivedi, A. Tripathi (eds.), *Mental Health in South Asia: Ethics, Resources, Programs and Legislation,* International Library of Ethics, Law, and the New Medicine 58,
DOI 10.1007/978-94-017-9017-8_15

15.2 Advocacy

Unified and concerted advocacy from international organizations, influential national opinion makers and users are necessary to create political will to develop mental health services (Saraceno et al. 2007). Global Initiatives like 'Mental Health Gap Action Programme' (mhGAP), 'grand challenges in mental health,' 'movement for global mental health,' and 'The World Psychiatry Association (WPA) 2008–2011 Action Plan' have raised the profile of mental health on the global health and development agenda (Eaton et al. 2011); and highlighted the need for more funds and political support for mental health services reform (Saraceno et al. 2007).

In the future, advocacy efforts must involve users of mental health services and their families as equal partners (Lancet Global Mental Health Group 2007).

15.3 Policy, Plan and Legislation

The most important step towards providing well-considered and comprehensive mental health care is the drafting of a policy and a plan that will guide mental health system and services development (WHO 2009a). At country level, mental health and overall health system planning and development must be integrated to the greatest extent possible (WHO 2009a).

In the South Asian region, Nepal and Sri Lanka have mental health policies (United Nations Statistics Division 2011) (Mental health policy group 2011). In India, the Mental Health Policy Group was formed in May 2011 to formulate the nation's mental health policy (Mental health policy group 2011; Dhar 2011). It is expected that the group will conduct a situation analysis, assess population needs, blend views of various stakeholders and learn from experience of other countries, before formulating the policy (WHO 2009a). The vision, values, principles and objectives for mental health should be described in the policy. Specific areas for action and responsibilities of different sectors need to be identified (WHO 2009a). Special focus should be kept for the groups most at risk—women, children and the elderly people (Kapur 2004). Similarly, policies and legislation should be put in place to prevent the inappropriate imprisonment of the mentally ill and provisions for their treatment in hospitals preferably or in the prison itself if unavoidable should be made (WHO 2001).

The three main obstacles to better mental health especially in LAMICs are scarcity of available resources, inequities in their distribution, and inefficiencies in their use (Saxena et al. 2007). Proper planning and implementation is required to overcome these obstacles. A mental health plan should outline the tangible details that will allow the implementation of the policy (WHO 2009a). Steps in developing a plan include: determining the strategies and timeframes; setting indicators and targets; determining the major activities; establishing the costs and available resources; and budgeting accordingly (WHO 2009a). The presence of nodal agency should ensure its proper implementation.

In India, National Mental Health Programme (NMHP) formulated in 1982 was relatively successfully implemented in Kerala and Gujarat while it wasn't implemented at all in Bihar (Goel 2011b). Problems in implementation of the NMHP have been attributed to inefficient administration, failure to develop indicators to address objectives, inadequate emphasis in creating awareness among users and uncoordinated, fragmented efforts by various stakeholders among other reasons (Murthy 2011). The Indian experience has shown that it is important to continuously assess performance at the ground level with independent audits and periodic review to identify problems at an early stage so that necessary corrective measures can be initiated (Goel et al. 2004). In contrast, a relatively well-developed national plan for mental health developed in a participatory way by the government and all key stakeholders has been shown to be more successful in Sri Lanka (Saraceno et al. 2007).

15.4 Legislation

Dedicated mental health legislation legally reinforces the goals of mental health policies and plans (WHO 2011a). Only 36% of low-income countries are covered by mental health legislation (WHO 2011a). Countries have to decide about following consolidated or dispersed mental health legislation or a combination of both (WHO 2009a). It is expected that existing laws in LAMICs will be changed to bring them in harmony with Convention on Rights of Persons with Disabilities (UNCRPD) 2006 which has given a new dimension to the rights of the mentally ill.

In India, the process for amending the mental health Act of 1987 has been initiated to meet the international and national obligations of the State towards its citizens. It is hoped that the new legislation will improve the lives of people with mental health issues and their caregivers at the grassroots by addressing issues like human rights, access to care, family burden, stigma, etc. that impact them. The draft of 'mental health care bill' has been introduced in the Parliament in 2013. Some of its proposals, viz. definition of mental illness, advance directives, banning ECT for children and adolescents, considering general hospital psychiatry units in the same way as mental health institutions and putting onus of transferring patients on the hospital, have elicited active debate; while others like extending insurance coverage for mental illness, have been universally welcomed.

Mental health legislation has to engage and work in tandem with legislation for people with disability and mental health policy. In LAMICs where the majority live in poor socioeconomic conditions, greater benefits and welfare measures like job reservation and housing schemes for patients with mental illness is needed (WHO 2001).

15.5 Financing of Mental Health

More than a third of low-income countries rely primarily on out-of-pocket payments for their treatment as compared to just 3 % in high-income countries (Saxena et al. 2007). LAMICs need to promote prepayment financing mechanisms such as social insurance, voluntary health insurance, and tax-based arrangements so that the needy can afford treatment (Saxena et al. 2007).

Mental health budget should be specified in the nation's health budget (Jacob et al. 2007) so that regular source of money is allocated to achieve mental health objectives (Lancet Global Mental Health Group 2007). Efforts should be made to improve financing at each level of governance rather than only at the federal level.

With increasing priority for mental health in LAMICs, federal financing of mental health are less of a constraint than before (Goel 2011b). However, many countries (including India) have not been able to utilize the allocated funds optimally (Goel 2011a). Budget for mental health should be ring fenced at each level of governance to ensure its proper utilization and limit chances of diversion in other directions.

15.6 Delivery of Mental Health Services

Currently, mental health services in LAMICs are largely centralised in big institutions which is more expensive than community care; isolates patients from their family and community support systems; and is associated with undignified life conditions, human-rights violations, and stigma (Saraceno et al. 2007). There is a long-standing public-health and public mental health recommendation that mental health care be deinstitutionalized and mental health resources be geographically decentralised so that care is available and accessible in the community (Saraceno et al. 2007).

Deinstitutionalization is a complex process in which reduction of beds in stand-alone mental hospitals is associated with implementation of a network of community alternatives that can avoid the institutionalization of individuals with mental illness. Neither closure of mental hospitals without provision of extramural alternatives (which can lead to homelessness and transinstitutionalization in prisons, nursing homes, etc.), nor creation of extramural alternatives without restriction on admission to mental hospitals (adding new services eventually recruits new patients but leaves the mental hospital unaffected) is optimal (Saxena and Sharan 2010). In LAMICs, existing mental hospitals should not be closed as they will be required for care of patients with severe mental illness, insufficient social support and for medico-legal cases (Wig 2004). Rather, resources should be focussed on development of acute psychiatric units in general hospitals, accessible psychiatric outpatient clinics, integration of mental health into primary health care, and community-based residential care and day services (Lancet Global Mental Health Group 2007).

During the transition to predominantly community based services, dual funding will be required for meeting concurrent running costs for mental hospitals while community-based services are established (Lancet Global Mental Health Group

2007). Decentralisation of resources is technically complex and LAMICs might be able to learn a lot from other countries that have successfully implemented community care (Saraceno et al. 2007).

A good model for LAMICs would be 'WHO Service Organization Pyramid for an Optimal Mix of Services for Mental Health' (WHO 2007) which proposes the integration of mental health services with general health care. The WHO model is based on the principle that no single service setting can meet all population mental health needs (WHO 2009a). Support, supervision, collaboration, information-sharing and education across the different levels of care will promote good use of resources and lend a human rights and community-based orientation (WHO 2009a). This is in line with the 1978 Declaration of Alma-Ata, which promoted a primary care model "sustained by integrated, functional and mutually supportive referral systems" as an integral part of the country's health system (Saraceno et al. 2007). The WHO model also assumes that people with mental disorders need to be involved, albeit to differing degrees, in their own recovery from mental disorders (WHO 2009a).

LAMICs should aim to procure the best possible mix of services from all levels of the pyramid and regularly evaluate what is available, with the aim of gradually improving the range of available services (WHO 2009a). Mental health services should be designed on a needs-led basis and the services offered should be comprehensive, evidence based, accessible, affordable, adapted to local realities and acceptable to the users (WHO 2009a).

In the South Asian region (United Nations Statistics Division 2011), WHO has cited the mental health services provided at general hospitals of Nepal (WHO 2010a) and community mental health services provided at Sri Lanka (WHO 2010a) as best practice examples. These can serve as role models for other countries in the region.

15.7 Human Resources for Mental Health

15.7.1 The Mental Health Workforce

The mental health workforce comprises of

1. Mental health professionals: psychiatrists, neurologists, psychiatric nurses, psychologists, mental health social workers, and occupational therapists (Kakuma et al. 2011).
2. Non-specialist health workers: doctors, nurses and lay health workers, affected individuals, and caregivers (Kakuma et al. 2011).
3. Other professionals: teachers, police and judiciary, and community-level workers (Kakuma et al. 2011).

According to the nation's population needs, mental health service delivery systems, and resources of the country, modifications need to be made to the mental health workforce (Kakuma et al. 2011).

15.7.2 Policy Framework for Mental Health Workforce

A national human resources policy for mental health workforce should define the overall values and goals and provide a coherent framework to plan, train, and develop the workforce (WHO 2009a). The Human Resources for Health Action Framework consisting of six interconnected components necessary in human resource development (policy, health workforce management, finance, education, partnerships, and leadership) can be helpful to address shortages in human resources for mental health (Dal Poz et al. 2006). In LAMICs where the role of mental health workforce differs across settings, skill-mix is more important than the staff-mix (Kakuma et al. 2011). For optimal outcome, mutual interaction and support between the primary health workforce and specialist community mental health services is essential (Saraceno et al. 2007).

15.7.3 Dealing with the Shortage of Mental Health Professionals

The shortage of mental health professionals in LAMICs is complicated by its predominantly urban concentration, preference for private practice and brain drain

(WHO 2001) (Saraceno et al. 2007). Lancet "call for action" proposes concurrent and systematic training of more specialist professionals and expansion of the non-specialist professional workforce (Lancet Global Mental Health Group 2007). The greater challenge is in retaining and ensuring equitable distribution of human resources for mental health (Kakuma et al. 2011). To minimize brain drain and attract talent to public sector, it is essential to develop financial incentives, provide career development opportunities and favourable workplace conditions (Kakuma et al. 2011).

Task shifting (also known as task sharing), defined as "delegating tasks to existing or new cadres with either less training or narrowly tailored training" has been suggested as a strategy to handle shortage in mental health professionals (Kakuma et al. 2011). Task sharing has been shown to be effective in immunisation uptake and management of tuberculosis and HIV (Eaton et al. 2011). For mental health services, task shifting entails: employment of mental health care professionals in different sectors; inter-sectoral collaborations with other professionals, to strengthen mental health awareness, detection of mental disorders, referrals, and service delivery; or both of these (Kakuma et al. 2011).

15.7.4 Implications of Task Shifting for the Mental Health Workforce

If task shifting is implemented extensively, the role of psychiatrists and neurologists will be that of public mental health practitioners to influence policy makers, manage complex psychiatric cases, and train, supervise, and mentor non-specialist workers (Kakuma et al. 2011). Shortage of these public health skills among mental health leaders has been identified as a major barrier to progress in mental health service reform (Eaton et al. 2011). This deficit in leadership and public health skills among mental health professionals is being addressed by emerging training options like 'International Diploma in Mental Health Law and Human Rights' run by the Indian Law Society and 'Sangath Leadership in Mental Health' Course in South Asian region (Eaton et al. 2011) (United Nations Statistics Division 2011). Universities in LAMICs must establish public mental health courses that cover policy, legislation, organisation of services, prevention, epidemiology of mental disorders and their risk factors (Lancet Global Mental Health Group 2007).

While innovative programmes are necessary for taking mental healthcare to all in need; continued efforts to improve psychiatry education in the undergraduate medical courses must be made. The inability to diagnose and manage psychiatric presentations in primary care has led to many short courses for physicians and nurses which have been criticized for transferring knowledge rather than skill and confidence and is of little practical use (Jacob 2011). If psychiatry is taught well in MBBS courses, there would be little need of manuals for the PHC doctors (Kapur 2004). At All India Institute of Medical Sciences, undergraduates are trained in psychiatry as part of residential posting in community medicine (Sood and Sharan 2011). The training programme is being modified to enable them to diagnose and treat common mental disorders in the community setting on their own (Sood and Sharan 2011).

Non-specialist health workers have contributed to services across the 'WHO Service Organization Pyramid for an Optimal Mix of Services for Mental Health' (WHO 2007) and have played a part in detection, diagnosis, treatment, and prevention of common and severe mental disorders as part of a complex stepped-care intervention or single intervention (Kakuma et al. 2011). More of them should be provided in-service training and effective supervision to effectively meet the mental health needs of the population (Kakuma et al. 2011).

A review on training of non-mental healthcare providers in mental health in LAMICs concluded that despite the wide variation in training programmes, para-professional training in mental health may improve mental health outcomes in LAMICs (Painuly and Sharan 2008). In the South Asian region (United Nations Statistics Division 2011), Pakistan has an effective national programme for community health workers (Oxford Policy Management 2002). Additional evidence is needed of the effectiveness and cost-effectiveness for identification and management of mental disorders by non-specialist health workers (Kakuma et al. 2011).

15.7.5 Family and Community Resources

Participatory action methods involving people with mental disorders, their family members, and other informal resources in the community can ensure a high level of motivation with least chances of attrition (Saraceno et al. 2007). Family caregivers can contribute to detection, treatment seeking, and management of family members with mental disorders (Kakuma et al. 2011). Patients can provide similar support to others, share personal experiences, and participate in self-help and mutual aid initiatives (Kakuma et al. 2011). Though this model holds a lot of promise, evidence is needed for the same (Kakuma et al. 2011).

In LAMICs, patients with mental disorders seek treatment from faith healing, religious places and alternative systems of medicine before seeking psychiatric care from psychiatrists (Trivedi and Jilani 2011). There is a need for both mental health professionals and non-psychiatric care providers to learn from each other in providing better services (Kapur 2004).

15.8 Integration of Mental Health Interventions in Primary Care

Though integration of mental health services into primary care has been shown to be the most viable way of closing the mental health treatment gap (WHO 2008), it is far from practical reality in most LAMICs (Jacob 2011). Some notable exceptions in South Asian region which have been appreciated by WHO are the nationwide integration of mental health care in Islamic republic of Iran and the integrated primary care for mental health in Thiruvanathapuram district of Kerala state in India (United Nations Statistics Division 2011; WHO, WONCA 2008).

The various barriers for integration of mental health into primary health care in LAMICs are

1. Overburdened primary health-care systems with multiple tasks and patient loads (Saraceno et al. 2007).
2. Insufficient training and supervision of primary health care workers with inadequate support from specialised services (Saraceno et al. 2007).
3. Problems in availability of essential psychotropic medicines in primary health care (Saraceno et al. 2007) (WHO 2011b). Ideally, they should be available within the context of a well-functioning mental health system, at all times, in adequate amounts, in the appropriate dosage forms with assured quality and adequate information, and at a price the individual and the community can afford (WHO 2009a).

It should be ensured that the primary care system is adequately well trained and staffed and that a functioning secondary referral system exists before incorporating mental health care in it.

The feasibility of alternative vehicles of programme delivery needs to be explored. In India, other more robust alternatives like the District Rural Development Agency (DRDA) and the National Rural Health Mission (NRHM) which are well-funded and well managed need to be explored (Goel 2011b). Promising models using local community resources in a cost-effective manner need to be considered (Chatterjee et al. 2003). Other possible solutions are integrating mental health care with services for people with chronic conditions or with other systems, such as social care and education (Eaton et al. 2011).

With weak health systems and short timeframes, vertical implementation of interventions like HIV antiretroviral services have achieved success at scale within a short period of time (Walley et al. 2008). A similar approach can be considered for mental health care and over time interventions should be integrated and delivered in a coordinated fashion with participation from communities (Rohde et al. 2008; Lawn et al. 2008).

15.9 Diagnosing and Managing Mental Illness in Primary Care

Detection and diagnosis of common mental disorders (CMD) can be done by brief screening questionnaires (Beaglehole et al. 2008). A two-stage case-finding procedure using community case-finding strategies followed by a diagnostic interview by an appropriately trained health worker can be used for less common disorders (Beaglehole et al. 2008).

There are several resources for mental health interventions in LAMICs, e.g.:

a) The Inter-agency Standing Committee guidelines on emergency interventions (Inter-agency Standing Committee 2007)

b) WHO Community-Based Rehabilitation Guidelines (WHO/UNESCO/ILO/IDDC 2010)
c) PLoS Medicine series on "packages of care" for mental, neurological, and substance-use disorders in LAMICs (Patel and Thornicroft 2009) covering treatment of depression (Patel et al. 2009), epilepsy (Mbuba and Newton 2009), schizophrenia (Mari et al. 2009), alcohol use disorders (Benegal et al. 2009), dementia (Prince et al. 2009) and attention-deficit hyperactivity disorder (Flisher et al. 2010)
d) The mhGAP Intervention Guide for eight priority mental, neurological, and substance misuse disorders in non-specialised health settings (WHO 2010b).

The proposed packages recommend stepped-care models where treatments are tailored to the needs of each individual (Beaglehole et al. 2008).

The guidelines for managing common mental disorders (CMD) in primary care have been criticized by some experts as impractical for routine use because they propose elaborate and separate protocols for many traditional psychiatric categories (Jacob 2010). Critics suggest that the recognition and management of psychiatric presentations should be based on the reality of primary care rather than specialist perceptions; and point out that the identification of common clinical presentations (rather than specific diagnosis) and simple management protocols (rather than elaborate and separate routines) would be more useful at the primary care level (Jacob 2006). However, the above guidelines are evidence based and have been developed following much consultation with experts from LAMICs, it would be inappropriate to let go the good, while waiting for the perfect. It should also be remembered that primary care professionals need support for executing primary care protocols. LAMICs do not have a supporting network of primary care mental health professionals that is available in high-income countries; hence, if the network support has to come from secondary care professionals there may be a need to follow protocols that they are familiar with. It goes without saying, that these protocols need to be tested in actual field conditions prior to uncritical acceptance.

LAMICs should aspire to reintegrate patients in the society by providing coordinated community care with the help of the community resources, utilizing innovative methods (e.g. mobile phones), and voluntary organizations (Murthy 2011). In the future, the focus will eventually shift to prevention of mental disorders and promotion of mental health (Murthy 2011).

15.10 Scaling Up

According to WHO, "scaling up" is defined as deliberate efforts to increase the impact of health service innovations successfully tested in pilot or experimental projects so as to benefit more people and to foster policy and programme development on a lasting basis (Eaton et al. 2011). Although there are examples of services being scaled up, few have been evaluated and shown to deliver consistent standard of care to the masses (Eaton et al. 2011). In India, the Bellary model (on which the NMHP was based) worked under resource-intensive experimental conditions over a limited time-frame but had limited success in real-life field conditions (Goel 2011b).

To scale up mental health services in LAMICs, It has been recommended to gather information about felt needs of the population, identify available resources, prioritize conditions for service delivery, identify barriers, design service delivery that can be integrated with existing mental health care system, involve all stakeholders, monitor the interventions and modify strategy accordingly (Eaton et al. 2011).

15.11 Mental Health Information Systems

It is important to obtain valid and reliable information about a country's mental health system to understand its functioning (WHO 2009a), have a baseline for monitoring the change (WHO 2009b) and provide insights to improve it (WHO 2009b). Privacy, confidentiality, access to information, and informed consent are especially important when implementing an information system for mental health (WHO 2009a).

WHO has given the World Health Organization Assessment Instrument for Mental Health Systems (WHO-AIMS): a tool primarily intended for assessing mental health systems in LAMICs (WHO 2009b). WHO-AIMS 2.2 consists of 6 domains: policy and legislative framework, mental health services, mental health in primary care, human resources, public information and links with other sectors, monitoring and research (WHO 2009b). These domains address ten recommendations of the 'World Health Report 2001' through 28 facets and 155 items (WHO 2009b). In the South Asian region (United Nations Statistics Division 2011), Bangladesh, Nepal, Bhutan, Maldives, Pakistan, Afghanistan, Islamic Republic of Iran, Uttarakhand and Gujarat states of India have used WHO-AIMS to make initial assessments of their mental health care systems (WHO 2009b). Further refinement and repeated use of WHO-AIMS would aid us in measuring progress (Eaton et al. 2011). The information derived from WHO-AIMS can be used to develop plans for strengthening community care and scaling up services for people with mental disorders.

15.12 Mental Health Research

The Mental Health: Global Action Programme of the WHO envisioned an active role for research in the multidimensional efforts required to change the current mental health situation in LAMICs (Saxena et al. 2004). However, only 10% of the world's medical research addresses the health needs of the 90% of the global population who live in low-income and middle-income countries (Lancet Global Mental Health Group 2007). This '10/90 divide' in health research output also holds true for mental health field and has not diminished over time (Saxena et al. 2006).

Further, research often does not influence practice and policy because of an absence of dialogue between researchers and the community (Saxena and Sharan 2003). A systematic approach including attention to the relevance of research questions; use of appropriate methodology; early involvement of practitioners, policymakers, advocacy groups and family members/consumers; effort at fund and resource generation, and manpower development; and judicious use of the media and forums for dissemination; should be formulated to enhance the utilization of research for policy and practice (Saxena and Sharan 2003; Saxena and Sharan 2004; Sharan et al. 2009).

In the context of global mental health and with a time frame of the next 10 years, it would be best to fill critical knowledge gaps by investing in research into health policy and systems, epidemiology and improved delivery of cost-effective interventions (Tomlinson et al. 2009).

15.13 Anti-Stigma Initiatives

Lack of awareness about mental disorders in the community, stigma towards and discrimination against people with mental health problems are important barriers to identify and treat mental disorders worldwide (Thornicroft et al. 2010). Protest by stigmatized individuals, education interventions for public to replace myths and negative stereotypes about mental disorders and personal social contact with people with mental disorders are the main strategies to reduce public stigma and discrimination (Thornicroft et al. 2010). Anti-stigma initiative like 'Time to change' in England, 'See me' in Scotland, 'Like me, See me' in New Zealand, WPA anti-stigma initiative and similar programmes in Japan, Brazil, Egypt and Nigeria have reported positive outcomes (Thornicroft et al. 2010).

15.14 Conclusion

All levels—individual, family, community, facility, district, national, and global—have a role and responsibility if health for all is to be achieved (Walley et al. 2008). Despite the challenges and restrictions in implementation so far, the ideals expressed at Alma-Ata and the primary health-care approach are as valid now as ever for LAMICs (Walley et al. 2008). Lessons learned during the last 30 years from all countries include the need for a nationally agreed package of prioritised and phased mental health care that all stakeholders are committed to implementing, attention to management systems at all levels of governance, and consistent investment in manpower linked to the (mental) health system (Rohde et al. 2008).

The major hurdle in providing mental healthcare to the community seems to lie not in the purse but in the mind (Goel et al. 2004).

References

Beaglehole, R., et al. 2008. Improving the prevention and management of chronic disease in low-income and middle-income countries: A priority for primary health care. *The Lancet* 372:940–949.

Benegal, V., P. K. Chand, and I. S. Obot. 2009. Packages of care for alcohol use disorders in low- and middle-income countries. *PLoS Medicine* 6 (10): e1000170.

Chatterjee, S., et al. 2003. Evaluation of a community-based rehabilitation model for chronic schizophrenia in rural India. *The British Journal of Psychiatry* 182:57–62.

Dal Poz, M., et al. 2006. Addressing the health workforce crisis: Towards a common approach. *Hum Resour Health* 4 (1): 21.

Dhar, A. 2011. At India's insistence, mental health included as non-communicable disease. *The Hindu*. http://www.hindu.com/2011/05/05/stories/2011050557552000.htm. Accessed 4 Jan 2012.

Eaton, J., et al. 2011. Scale up of services for mental health in low-income and middle-income countries. *The Lancet* 378:1592–1603.

Flisher, A. J., et al. 2010. Packages of care for attention-deficit hyperactivity disorder in low- and middle-income countries. *PLoS Medicine* 7 (2): e1000235.

Goel, D. 2011a. The national mental health program: problems, perspectives, possibilities. In *Public mental health: An evolving field,* eds. L. Swartz and L. Richter et al. New Delhi: Ministry of Health, Government of India.

Goel, D. 2011b. Why mental health services in low- and middle-income countries are under-resourced, underperforming: An Indian perspective. *The National Medical Journal of India* 24 (2): 94–97.

Goel, D., et al. 2004. Mental health 2003: The Indian scene. In *Mental health: An Indian perspective (1946–2003),* ed. S. P. Agarwal, 3–24. New Delhi: Directorate General of Health Services, Ministry of Health and Family Welfare, Government of India/Elsevier.

Inter-agency Standing Committee. 2007. *IASC guidelines on mental health and psychosocial support in emergency settings*. Geneva: Inter-agency Standing Committee.

Jacob, K. 2006. The diagnosis and management of depression and anxiety in primary care: The need for a different framework. *Postgraduate Medical Journal* 82 (974): 836–839.

Jacob, K. 2010. Indian psychiatry and classification of psychiatric disorders. *Indian Journal of Psychiatry* 52 (7): 104.

Jacob, K. 2011. Repackaging mental health programs in low- and middle-income countries. *Indian Journal of Psychiatry* 53:195.

Jacob, K., et al. 2007. Mental health systems in countries: Where are we now? *The Lancet* 370:1061–1077.

Kakuma, R., et al. 2011. Human resources for mental health care: Current situation and strategies for action. *The Lancet* 378:1654–1663.

Kapur, R. 2004. The story of community mental health in India. In: *Mental health: An Indian perspective, 1946–2003,* eds. S. P. Agarwaal and D. S. Goel et al., 92–100. New Delhi: Published for Directorate General of Health Services Ministry of Health & Family Welfare [by] Elsevier.

Lancet global mental health group. 2007. Scale up services for mental disorders: A call for action. *The Lancet* 370:1241–1252.

Lawn, J. E., et al. 2008. Alma-Ata 30 years on: Revolutionary, relevant, and time to revitalise. *The Lancet* 372:917–927.

Lund, C., et al. 2011. Poverty and mental disorders: Breaking the cycle in low-income and middle-income countries. *The Lancet* 378:1502–1514.

Mari, J., et al. 2009. Packages of care for schizophrenia in low- and middle-income countries. *PLoS Medicine* 6 (10): e1000165.

Mbuba, C. K., and C. R. Newton. 2009. Packages of care for epilepsy in low- and middle-income countries *PLoS Medicine* 6 (10): e1000162.

Mental health policy group. 2011. http://mhpolicy.org/. Accessed 4 Jan 2012.

Murthy, R. S. 2011. Mental health initiatives in India (1947–2010). *National Medical Journal of India* 24 (2): 98–107.
Oxford Policy Management. 2002. Evaluation of the Prime Minister's Lady Health Worker Programme.
Painuly, N., and P. Sharan. 2008. Effectiveness of training of non-mental health care providers in mental health in low- and middle income countries: A systematic review. *Primary Care Community Psychiatry* 13 (2): 83–89.
Patel, V., and G. Thornicroft. 2009. Packages of care for mental, neurological, and substance use disorders in low- and middle-income countries: PLoS medicine series. *PLoS Medicine* 6 (10): e1000160.
Patel, V., et al. 2009. Packages of care for depression in low- and middle-income countries. *PLoS Medicine* 6 (10): e1000159.
Prince, M. J., et al. 2009. Packages of care for dementia in low- and middle-income countries. *PLoS Medicine* 6 (11): e1000176 (V. Patel ed.).
Rohde, J., et al. 2008. 30 years after alma-ata: Has primary health care worked in countries? *The Lancet* 372:950–961.
Saraceno, B., et al. 2007. Barriers to improvement of mental health services in low-income and middle-income countries. *The Lancet* 370:1164–1174.
Saxena, S., and P. Sharan. 2003. Research for change: Supporting mental health research in India. *National Medical Journal of India* 16 (3): 122–125.
Saxena, S., and P. Sharan. 2004. Enhancing mental health research publication from low and middle-income countries. *Indian Journal of Psychiatry* 46(1): 72–78.
Saxena, S., and P. Sharan. 2010. Mental health resources and services. In *Mental and neurological public health a global perspective,* ed. V. Patel, et al., 418–430. New Delhi: Elsevier.
Saxena, S., P. Sharan, and B. Saraceno. 2004. Research for change: The role of scientific journals publishing mental health research. *World Psychiatry* 3 (2): 66–72.
Saxena, S., et al. 2006. The 10/90 divide in mental health research: Trends over a 10-year period. *British Journal of Psychiatry: The Journal Of Mental Science* 188:81–82.
Saxena, S., et al. 2007. Resources for mental health: scarcity, inequity, and inefficiency. *The Lancet* 370:878–889.
Sharan, P., et al. 2009. Mental health research priorities in low- and middle-income countries of Africa, Asia, Latin America and the Caribbean. *British Journal of Psychiatry: The Journal Of Mental Science* 195 (4): 354–363.
Sood, M., and P. Sharan. 2011. A pragmatic approach to integrating mental health in undergraduate training: The AIIMS experience and work in progress. *National Medical Journal of India* 24 (2): 108–110.
Thornicroft, G., et al. 2010. WPA guidance on steps, obstacles and mistakes to avoid in the implementation of community mental health care. *World Psychiatry: Official Journal of the World Psychiatric Association* 9 (2): 67–77.
Tomlinson, M., et al. 2009. Setting priorities for global mental health research. *Bulletin of the World Health Organization,* 87 (6): 438–446.
Trivedi, J. K., and A. Q. Jilani. 2011. Pathway of psychiatric care. *Indian Journal of Psychiatry* 53 (2): 97–98.
United Nations Statistics Division. 2011. Standard Country and Area Codes Classifications (M49). http://millenniumindicators.un.org/unsd/methods/m49/m49regin.htm. Accessed 3 Dec 2011.
Walley, J., et al. 2008. Primary health care: making alma-ata a reality. *The Lancet* 372:1001–1007.
WHO. 2001. *The world health report: 2001: Mental health: New understanding, new hope.* Geneva: WHO.
WHO. 2007. The optimal mix of services for mental health. Information sheet 2. In Mental health policy, planning & service development. http://www.who.int/mental_health/policy/services/en/index.html. Accessed 3 Jan 2012.
WHO. 2008. *mhGAP: Mental health gap action programme: Scaling up care for mental, neurological and substance use disorders*. Geneva: WHO.

WHO. 2009a. *Improving health systems and services for mental health*. Geneva: World Health Organization.
WHO. 2009b. *Mental health systems in selected low- and middle-income countries: A WHO-AIMS crossnational analysis*. Geneva: WHO.
WHO. 2010a. Best practices: Mental health service development. Mental health policy, planning & service development. http://www.who.int/mental_health/policy/services/en/index.html. Accessed 3 Jan 2012.
WHO. 2010b. *mhGAP intervention guide for mental, neurological and substance use disorders in non-specialized health settings: Mental health gap action programme (mhGAP)*. Geneva: World Health Organization.
WHO. 2011a. *Mental health atlas 2011*. Geneva: WHO.
WHO. 2011b. WHO Model Lists of Essential Medicines. 17th edition. 2011. http://www.who.int/medicines/publications/essentialmedicines/en/. Accessed 3 Jan 2012.
WHO, WONCA. 2008. *Integrating mental health into primary care: A global perspective*. Geneva: World Health Organization and World Organization of Family Doctors.
WHO/UNESCO/ILO/IDDC. 2010. *Community based rehabilitation guidelines; supplementary booklet; mental health and CBR (P3–20)*. Geneva: World Health Organization.
Wig, N. 2004. Vision 2020: Roadmap for the future-II. Mainstreaming mental health. In *Mental health: An Indian perspective, 1946–2003,* ed. S. P. Agarwal, et al. 341–343. New Delhi: Published for Directorate General of Health Services Ministry of Health & Family Welfare [by] Elsevier.

Chapter 16
Mental Health in the Context of East and West: Beyond Resources and Geographical Realities

Levent Küey and ElifKırmızı Alsan

16.1 Introduction

Transcultural comparisons taking the differences and commonalities into consideration in the fields of mental health and ill mental health have always been a focus of scientific interest. The 'East' and 'West' comparisons in this regard, could be the one most widely deliberated. The notion of 'East and West' as a conceptual tool signifies more than mere geographical realities. Human beings, while constructing their own history, are also re-constructed by the events and the conceptualizations of these developments, hence these conceptualizations reflect what the humanity has been experiencing. In other words, our conceptualizations both reflect and re-construct our realities. 'East and West', as a human-made conceptual construct, has evolved to signify many social, cultural, political, economical and psychological realities and meanings, beyond its geographical references.

This article, after briefly discussing the historical roots of the conceptualization of 'East and West', will be focused on the relations of mental health and ill mental health with this and related conceptualizations.

16.2 East and West: Historical Roots

The dualism of 'East and West' in its broader conceptualization, in fact, reflects one of many dualistic perspectives utilized to understand and set basis to intervene with the social life and the world. Dualisms as such seem to be as old as the humanity itself; goes well back to the time of first settlers, i.e., the transition of some

L. Küey (✉)
Istanbul Bilgi University, Istanbul, Turkey
e-mail: kueyl@superonline.com

E. Alsan
Department of Psychiatry, Kocaeli University, Kocaeli, Turkey

J. K. Trivedi, A. Tripathi (eds.), *Mental Health in South Asia: Ethics, Resources, Programs and Legislation,* International Library of Ethics, Law, and the New Medicine 58,
DOI 10.1007/978-94-017-9017-8_16

hunter-gatherer tribes to become first farmers in various parts of the world. This transition from a nomadic culture to a sedentary proto-urban culture took place around big rivers as the Tigris–Euphrates river system, Amazon, Indus River, Green River, and Nile River. The culture and life styles established in Mesopotamia, Andes, Mesoamerica, India, China, and Egypt, none of which is a part of the 'Western World' today, are considered as the cradle of human civilizations (Smith et al. 2009; Columbia Encyclopedia 2012).

Settling in a relatively defined geographical area and land gave way to the need of defending that life zone from the attacks of the out-siders, who were mainly, the nomadic communities in the eyes of the sedentary ones. City walls were built up. Cities with their inner organizations for managing the agriculture and related issues of sharing the products and wealth were also created behind these walls. "Neolithic Revolution", mainly characterized by the development of settled communities, and the "Urban Revolution", as stated by Childe and agreed upon by prominent twentieth century archaeologists (Smith et al. 2009), gave way to utilization of surpluses hence creating social classes and hierarchy, monumental sacred, administrative and public buildings and first examples of writing and documentation, in these more densed settlements (Smith et al. 2009; Gordon 1936). These developments meant leaving footprints for the next generations, i.e., the history was born. Of course such revolutionary changes also were reflected in and reshaped the minds of these newly emerging city people; the consciousness of 'insiders and outsiders' /'us and them' has emerged. Here, 'us' describing me and my tribe/family/city members and 'them' the Others, i.e., the savages, the barbarians, the non-believers and etc. Furthermore, since, humans are social beings, human existence and survival need a net of connections and interactions with their fellow beings. This need presented a basis for social groupings and in turn, laid down the ground for the categorization of ingroups and outgroups (Küey 2010).

These early city people had the possession of more commodities (food, equipment, and land and etc.) along with their children, families, and tribe members, to be protected. From a psychological standpoint, these people needed to feel themselves as a part of their immediate community to breed themselves not only physically but also emotionally with a feeling of security. This was a remedy for their essential need of self confidence in differentiating between what is to be considered as hostile and life threatening and what is to be defined as friendly and nurturing. In fact, this was a primal need of human beings experienced to a certain extent even by the earlier tribes of hunter-gatherers. 'Us and them', on one hand, served as basis for developing group identities, and on the other, served as first conceptual tools of discriminating 'the other'. The human history had shown that group identities were grassroots both for self-security and for "murderous identities" (Maalouf 2003).

Hence, the categorization of 'ingroup versus outgroup' or 'us versus them' or 'me versus the Other' had also been the source of discrimination and stigmatization throughout the human history. Especially, race, ethnicity, gender and religion based ignorance; prejudice and discrimination have been causing heart-rending human suffering in almost all societies across the world. Many human designed mass violence acts have been executed in the name of such group differences along with so-called justifications referring hierarchical status differences between group identities and cultures. At times, policies of cultural purification based on group identities and mobilizing masses, ended up with genocides. Besides its multiple socio-economic, cultural, and political reasons, the process of discrimination covers various dimensions of socio-psychological interactions. Meanwhile, in almost all societies, these multiple sources of discrimination gave way to that society to form its own specific 'scapegoats'. Paradoxically, the very same psychosocial phenomena leading to discrimination also sets the ground for 'sub-identities' or 'group identities' for their members. Human beings seeking for higher levels of social integration to survive and enhance their security and well being, internalized such social group identities to complement their personal identities. These personally incorporated social identities, which could be constructed on the differences in gender, age, ethnicity, belief systems, and etc., are also sources of perceived and/or personal discrimination (Küey 2010). Today, in our so-called global post-modern world, the concept of 'East and West' and Eurocentrism and Orientalism still serves (Ponting 2000) as another starting point of discrimination and reflects the limiting characteristics of our categorical dichotomous thinking.

In fact, some of the above mentioned variables on which the social identities were constructed, had been perceived as justifications of discrimination and stigmatization disregarding the inner heterogeneity of such social identities. Human history is full of depressing realities and devastating consequences of discrimination based on the marginalization of groups by any of the above mentioned variables (Küey 2010). Needless to add, discrimination is not only a fact of the past, but also, a reality of our daily social life today; we are suffering from the negative effects of various discriminations, such as the discriminations based on the notion of 'clash of civilizations/cultures' between East and West (Huntington 1996).

During the course of human history, as transactions and transportation increased mainly through wars and trade among these primary units of settlements, the connotations of 'insiders and outsiders' were broadened. Already available terms of geography, e.g., East and West turned out to become as just another dualistic form of defining the insiders and outsiders for the people of both sides. Consequently, a geographical reference point was overloaded with external meanings. East and West can describe the two different parts of a house, a campus, a country, and a planet. However this dualism can also describe many other things such as a warmer climate in the eastern part of a house in comparison to the west wing, or more mathematical work as in the eastern part of a university campus with engineering faculties in comparison to the western part with the social sciences, or a more liberal economy as in the former West Germany in comparison to the former East Germany or more importance on the individual values (more emphasis on rationality, more control over one's own life and choices, and priority of individual freedom over group dependency) in the Western countries in comparison to the Eastern countries where the collective values (interdependency, fatalism, and priority of group welfare over individual rights) are held superior to those of the individual.

Ironically, the human-made nature of this dualism makes the distinctions obsolete as one focuses away: that is the house with the east and west wings can be in the east part of a town, or the very same town can be located in the west part of the country which can be classified among the eastern countries of the globe. Similarly the infinite possibilities of nature can defy our classifications and there can be a house with a very cold east wing due to poor isolation or an Eastern country where individualism has flourished turning our biases into jokes. Likewise, Australia, in the 'east of the west' geographically, could be mainly considered as a part of the 'Western World'. Furthermore, east defined in North Americas and UK is usually sub-classified as near-east, middle-east and far-east, where the people of 'this middle-east', defines 'this far-east' as merely 'east'! How many "easts" does "west" have and vice versa?

It is a broadly accepted idea that social power relations determine the stratification of 'us' and 'them' groupings. In other words, whether a group is to be designated as the 'Other' and labelled with prejudice and discriminated will heavily depend on the *zeitgeist* of the current social powers. Here, a question arises: is the East-West dichotomy a part of the *zeitgeist* of the current dominating social powers? The issue of 'East and West' and related mental health issues could be considered in this context.

Today, in many scientific circles, 'East and the West' signify a relationship of power and domination. The West gives voice to the East, without allowing the East to speak for herself (Said 1994). The West has the power to define the East as 'the Other'. The East is described by Said as the place of Europe's greatest, richest and oldest colonies, the source of its civilizations and languages, its cultural contestant, and one of its deepest and most recurring images of 'the Other' (MacCallum 2002).

16.3 East and West: Mental Health Issues

This duality of East and West reflects itself onto mental health practices in a very similar manner. The diagnostic guidelines and classifications, the pharmacological and psychotherapeutical interventions, the biological model of mental disorders all stem from a Western heritage where the Eastern influence is allowed to enrich the various contents with its peculiarities in the form of cultural syndromes (APA 2000), biological variations (Chen et al. (2009), traditional medicine practices and local psychotherapeutical approaches (Stein 2009) that are doomed to be local due to their otherness. The question here is whether it is possible-and possibly better- to approach the issue of East and West in a non-dualistic manner within the framework of mental health?

There are many inequalities in terms of the resources between the East and the West. For example, the number of psychiatrists per 100,000 people was reported to be 16.5 in the United States (Scully and Wilk 2003); whereas, according to the report of South Asian Zone of the World Psychiatric Association there is around 1 psychiatrist per 3 million people which covers countries such as India, Thailand, Malaysia, Singapore, Indonesia and Philippines. To make matters even more complicated it is thought that over 70 % of the psychiatrists are in large urban centers where less than 20 % of the population stays. The result is that here are huge underserved areas of the region where modern mental health services are all but non-existent (Trivedi 2012). Similar inequalities between East and West exist for not only the human sources for mental health but also for mental health facilities, institutions and universities.

The fact that most of the current concepts in psychiatry have a Western origin also poses an area of inequality. Although some psychotherapeutic methods such as "incubation" or "temple sleep" date back to ancient Egypt, and the famous Islamic-Eastern scientist known to use psychotherapy in treatment and the Kitab al-Qanun fi al-tibb (The Canon on Medicine) by İbnSina (Avicenna) was translated into a number of other languages, including Persian, Latin, Chinese, Hebrew, German, French, and English, and was the major textbook of medicine in Europe for centuries, there was a period of decline for the East until the entrance of modern medicine in the eighteenth century (Mohit 2001). Doctors in European missions and colonial armies have led to the domination of the Western medicine in the East. This domination has given rise to the fact that most of the current concepts in psychiatry have a Western origin creating a further area of inequality (Mohit 2001).

For example, it has been criticized that the Western conceptualization of insight is overly biomedical and fails to allow for social constructions and culturally appropriate explanatory models of mental illness (Saravanan 2007). However it has also been noted that globalization has made the medical model available to non-Western populations as well (Saravanan 2007). Nevertheless the fact that attribution of illness to supernatural phenomena which is closely related with the cultural characteristics of the patient is associated with poorer insight (Saravanan 2007) perhaps points to the need for cultural adaptation of even the very basic concepts of psychiatry. As it is said, if "culture bound syndromes" are realities in the field of ill mental health, psychiatry seems to be "culture-bound" itself (Kleinman 1988).

The cultural structure and the related values of a given society modify the range, the content, and the expression of psychiatric symptoms. A sole Western or an Eastern point of view could make the mental health worker blind to the cultural differences reflected onto the symptoms. For example, in a study about Social Anxiety Disorder (SAD) in the West and in the East, the Western diagnostic term of Social Anxiety Disorder, characterized by excessive and persistent fear and avoidance of social or performance situations with feelings of self-consciousness and embarrassment, has been compared with the Eastern diagnostic term of Taijinkyofusho (TKS) where the patient presents with fear of offending or embarrassing others. TKS has been included in DSM-IV as a culture bound variation of SAD that is specific to Japan. TKS appears to differ from SAD in that it is characterized more by fear of offending others than fear of embarrassing oneself. Although the clinical approach, dating back to Kraepelin's work on the psychotic patients in Java, argues that psychiatric disorders are similar in different parts of the world, the anthropological position criticizes contemporary psychiatric nosology for committing a category fallacy in which Western classifications are incorrectly reified as universal natural kinds. It is just the right time to recall: diagnostic categories and classification systems of mental disorders are human-made conceptual constructs and mirror the current *zeitgeist* of the scientists; i.e., they are not value-free universal facts. The integrated approach could overcome this problem of dualism and implementation of a non-dualistic perspective makes it possible that although psychiatric classifications potentially have scientific validity, social constructs of disorders also affect our expression and experience of mental disorders. In terms of SAD and TKS it can be said that in the West, constructs of social anxiety are embedded in the concerns of the society with individual performance; whereas, in the East the expression and experience of social anxiety are more likely to reflect social concerns about disrupting harmony. It has been argued that these differences mirror the Eastern emphasis on social cohesion (Stein 2009). Nevertheless, it is hard to make generalizations about societies and in today's world, due to globalization and fast international exchange of information and interaction; the similarities between cultures are increasing. Therefore cultural sensitivity also should not make us blind to the individual differences because a patient with a Japanese heritage can present to the clinic with SAD symptomatology and display fear of embarrassing himself (Stein 2009).

We should be reminded of, at this point, the current discussions in the classification and diagnostic systems of mental disorders. While DSM system of the APA (2010), generated mainly as a "national system of a Western country" has been widely accepted in many different sociocultural contexts of the world, the ICD system of the WHO (WHO 1992) was re-adapted and revised according to the local needs of some countries. The Chinese (Yan-Fang 2002), the Cuban (Otero-Ojeda 2002) and the Latin American classification systems (Berganza et al. 2002) are good examples of these efforts. It seems that the discussions characterized by 'categorical vs. dimensional approaches' and 'prototypical vs. operational criteria' have some implications also for the East-West dichotomy, and will continue to be an issue of debate in the future of psychiatry.

Another main area of strong debate related to the East-West dualism is the psychotherapeutic interventions. Referring to the differences in explanatory approaches and formulation of mental disorders in East and West, various psychotherapeutic techniques are thought to be more effective in different cultural settings. One argument at this point covers the power relations. The power relationship in the East-West dualism where the Western definitions and values are held superior to the Eastern/local approaches limits the use of local therapeutic interventions. The local psychotherapeutical approaches to culture bound syndromes tend to be limited to their own geographical regions. For example the Morita psychotherapy which has been used in Japan for many years for TKS carries similarities with cognitive behavioural therapy such as exposure exercise; however perhaps the limited number of controlled studies and the fact that it has been used for a culture bound syndrome from a Western point of view, has limited its wide-spread use (Stein 2009).

The differences in family structures in Eastern and Western societies along with the so-called collectivistic and individualistic life styles have an effect on the practice of psychiatry and mental health policies. For example, the extended structure of families in Arab societies is associated with a high level of tolerance to assimilating family members with psychiatric illness in the community though the severely mentally ill have tended to be isolated by society and even at times subject to physical strain (Gordon and Murad 2005). Treatment in the community can be both enhanced and disadvantaged by the characteristics of society. For example in Palestine, the extended family system allows for the chronically mentally ill to be cared for at home. However there is no provision for any other community services outside of the family. The patient may have to remain at home even when this may not be good for his emotional well-being. Furthermore, the stigma related to mental illness could be causing delays in help seeking behaviour and obtaining the needed treatment due to the reluctance of the carers and family members (Gordon and Murad 2005).

Extended family support is of course an important social support system and resource. In many parts of the world, some psychiatric wards hospitalize the patients with mental disorders with the escort of their family members. There the question arises whether this exercise could always provide 'support' for the person with ill mental health or not. One could argue that this is not a supportive source in many occasions, because, frequently, there are inter familial conflicts and the family members are used as extra human resource to fill the gap of lacking hospital staff in many institutions. Furthermore, the lack of social support and welfare systems and low level of the ratio of the working women in the Eastern societies increase the importance of the family care givers (domestic care), especially the women, as wives, daughters, and mothers. But in fact, frequently these women themselves are in need of a great help and support for their burn-out. Taking care of a person with severe mental illness at home without any professional help is highly traumatizing for the family members, especially the women. Helplessness generating in such situations could contribute to further stigmatization and discrimination and even brutality against the person with mental illness.

Family support is idealized frequently both in West and East in different ways; in the West the social transition in the last two centuries caused a decline in family support and furthermore, the dissolution of the 'social state' in the last three decades had many psycho-social consequences causing an idealization of the importance of the family, and increase in the importance of the sub-groupings, peer support, friendship groups and such. In the East the transition towards industrialization and urbanization caused dissolution in the family support systems but without replacing it with alternative social support systems. People developed and relied on new systems of support in urban areas such as 'ghettos' of subgroups settling in big cities. However it should also be remembered that, from an explanatory point of view, at times it is the very family system that causes or triggers the psychiatric disorders at hand. We should also remember that traumatic events such as domestic violence and sexual and physical abuse most frequently take place in the extended family, itself.

16.4 Conclusions

Transcultural comparisons in the fields of mental health and ill mental health show vast inequalities and discrepancies between East and West.

Inequalities in resources shape the differences and conceptualizations in all ways of life and these inequalities reconstruct the differences, perceptions, and attitudes.

The notion of 'East and West' itself carries the load of inequality by definition, for the relationship between the East and the West seems to be a relationship of power. The West gives voice to the East, without allowing the East to speak for herself and has the power to define the East as 'the Other'.

Beyond the inequalities in resources, this dualism reflects itself onto mental health theory and praxis in a many different ways. The range, content, patterns and expression of psychopathology and psychiatric symptomatology; the explanatory models of mental disorders; the diagnostic and classification systems; the psycho-pharmacological and psychotherapeutical interventions; the role of family and psychosocial support systems; basic conceptualizations of mental health and disease are some of the main areas of diversity.

In this day of globalization, to improve the quality of psychiatric theory and praxis within a framework of high ethical standards, a paradigm focusing on the unity in diversity and embracing a non-dualistic approach to cultural differences would bring all the colours of humanity and therefore psychiatry onto the canvas.

References

American Psychiatric Association. 2000. *Diagnostic and statistical manual of mental disorders*. 4th ed, Text Rev. Washington, DC: American Psychiatric Association.

Berganza, C. E., J. E. Mezzich, and M. R. Jorge. 2002. Latin American guide for psychiatric diagnosis (GLDP). *Psychopathology* 35:185–190.

Chen, P. Y., S. C. Wang, R. E. Poland, and K. M. Lin. 2009. Biological variations in depression and anxiety between East and West. *CNS Neurosciences and Therapeutics* 15:283–294.

Columbia Encyclopedia. 2012. In Encyclopædia Britannica. http://www.britannica.com/EB-checked/topic/126960/Columbia-Encyclopedia.

Gordon, C. V. 1936. *Man makes himself*. London: Watts and Co.

Gordon, H., and I. Murad. 2005. Psychiatry and the palestinian population. *Israel journal of psychiatry and related sciences* 42 (2): 73–80.

Huntington, S. P. 1996. *The clash of civilizations and the remaking of world order*. New York: Simon and Schuster.

Kleinman, A. 1988. *Rethinking psychiatry: From cultural category to personal experience*. New York: Free Press.

Küey, L. 2010. Stigma, women, and mental health. In *Oxford textbook of women and mental health*, ed. Dora Kohen. Oxford: OUP.

Maalouf, A. 2003. *In The name of identity: Violence and the need to belong* (English translation of Les Identitésmeurtrières, 1998; translated by Barbara Bray). New York: Penguin Books.

MacCallum, E. J. 2002. Othering and psychiatry nursing. *Journal of Psychiatric and Mental Health Nursing* 9:87–94.

Mohit, A. 2001. Mental health and psychiatry in the middle East: Historical development. *Eastern Mediterranean Health Journal* 7 (3): 336–347.

Otero-Ojeda, A. A. 2002. Third cuban glossary of psychiatry (GC-3): Key features and contributions. *Psychopathology* 35:181–184.

Ponting, C. 2000. *World history—A new perspective*. London: Chatto & Windus.

Said, E. W. 1994. *Orientalism*. New York: Vintage Books, Random House Inc.

Saravanan, B., K. S. Jacob, S. Johnson, M. Prince, D. Bhugra, and A. S. David. 2007. Assessing insight in schizophrenia: East meets West. *British Journal of Psychiatry* 190:243–247.

Scully, J. H., and J. E. Wilk. 2003. Selected characteristics and data of psychiatrists in the United States, 2001–2002. *Academic Psychiatry* 27:247–251.

Smith, M. E. 2009. V. Gordon childe and the urban revolution: A historical perspective on a revolution in urban studies. *Total physical response* 80 (1): 3–29.

Stein, D. J. 2009. Social anxiety disorder in the West and in the East. *Annals of Clinical Psychiatry* 21 (2): 109–117.

Trivedi, J. K. WPA Zone 16 Report. Retrieved on July 9, 2012, from http://www.wpanet.org/detail.php?section_id=5&category_id=8&content_id=65.

World Health Organization. 1992. *The ICD-10 classification of mental and behavioural disorders: Clinical descriptions and diagnostic guidelines.* 1st ed. Geneva: World Health Organization.

Yan-Fang, Chen. 2002. Chinese classification of mental disorders (CCMD-3): Towards integration in international classification. *Psychopathology* 35:171–175.

Part IV
Mental Health Legislation

Chapter 17
A Mental Health Law for Low and Middle Income Countries

George Szmukler, Adarsh Tripathi, Rowena Daw and John Dawson

17.1 Introduction

Mental health law in its current form in western countries evolved in the context of nineteenth century asylum psychiatry. It was a solution created to deal with a social problem—how to manage persons unable to participate in rational exchange in a social contract, who at the same time could not be held responsible for their actions. A system of legal relations was established involving family, doctors, administrators and the judicial system. The detained person fell into a 'special subordinate legal status', becoming subject to a regime of (principally) medical paternalism (Unsworth 1987). Being a mentally ill person was virtually synonymous with being in a psychiatric institution, and until the early twentieth century, there was no 'voluntary' status. Detention and treatment became increasingly regulated by legal statute in Europe and the United States in the latter years of the nineteenth century, resulting in complex statutes such as the Lunacy Act 1890 in England.

Many new forms of psychiatric treatment (especially medications), and a more optimistic outlook on their value, emerged in the mid twentieth century in

---

To appear in: 'Mental Health in South Asia; Ethical, Legal and Allied Issues' (Ed JK Trivedi & A Tripathi)
Springer series in International Library of Ethics, Law and the New Medicine

---

A. Tripathi (✉)
Department of Psychiatry, King George's Medical University,
Shahmeena Road, Chowk, Lucknow, UP, India
e-mail: dradarshtripathi@gmail.com

G. Szmukler
Institute of Psychiatry, King's College, London, UK

R. Daw
Legal consultant; formerly Head of Policy at MIND; later,
Head of Policy at Royal College of Psychiatrists

J. Dawson
University of Otago, Dunedin, New Zealand

J. K. Trivedi, A. Tripathi (eds.), *Mental Health in South Asia: Ethics, Resources, Programs and Legislation,* International Library of Ethics, Law, and the New Medicine 58,
DOI 10.1007/978-94-017-9017-8_17

some countries. In combination with the emergence of the 'welfare state', where benevolent social interventions often used informal methods of control, this produced legislation such as the Mental Health Act 1959 in England, which gave doctors much greater discretion over detention and involuntary treatment, and encouraged voluntary admission to hospital without formal legal process. Recognition that formal protections were still required led to greater opportunities for compulsory patients to appeal for discharge to courts or tribunals.

A growing emphasis on human or civil rights, initially in the US, during the 1960s and 70s, spread in the following two decades to Europe and Australasia. Together with well publicised criticisms of psychiatric practice emanating from the 'anti-psychiatry' movement, and fiscal pressures arising from the need to refurbish and maintain decaying hospitals, this led to statutes with tighter criteria for involuntary admission, usually centred on 'dangerousness' rather than 'need for treatment', and greater attention to 'due process'. This legalistic emphasis in mental health law reduced medical discretion and increased procedural regulation of involuntary psychiatric treatment. This trend has affected legislation in most western countries, even if there has subsequently been a turning away from strict 'dangerousness' criteria in some states in the late twentieth century (Appelbaum 1994). In another major development in mental health law, reflecting the shift of treatment to the community, involuntary outpatient treatment or 'community treatment orders' were introduced in a number of jurisdictions.

All South Asian countries except Bhutan have dedicated mental health legislation for the regulation of mental health services in the region. In Bhutan, only specific sections of the penal code are dedicated to mental health (Jacob et al. 2007). India, Pakistan and Bangladesh inherited a form of the British-era Lunacy Law in 1912 which, despite becoming clearly outdated, persisted for a long time unaffected by the above mentioned developments around the world. India replaced this law with the Mental Health Act of 1987 (Sarkar 2004), Pakistan replaced it with the Pakistan Mental Health Ordinance in 2001 (Government of Pakistan 2001), Bangladesh has submitted a draft of the Mental Health Act to the proper authority as of 2002, but it is yet to be implemented and the Indian Lunacy Act of 1912 remains in force (WHO 2006a). Sri Lanka and Nepal have also enacted regulatory mental health legislations (WHO 2006b; de Silva and Hanwella 2010). Although there are many advances in such laws compared to their predecessors, many aspects are criticised, for example, maintenance of the legalistic flavour of the previous laws, lack of mention of community care, and lack of adequate regard for autonomy and human rights issues (Narayan et al. 2011; Abbasi 2008; Trivedi 2002).

17.2 'Mental' Disorder and 'Physical' Disorder: Decision-making Capacity[1]

Mental health law has developed separately from the law applied to the rest of medicine. However, with psychiatry's full acceptance into the medical fold, the question can be asked why a difference should remain today.

[1] The term 'capacity' is commonly used for the capability to make treatment decisions characterized by features such as the ability to understand and retain information relevant to the decision, to

Informed consent (or its converse, refusal), as understood in general medical practice, has played a subsidiary, or more commonly, no role in relation to involuntary treatment in psychiatry. The 'exceptionalism' of mental health legislation resides in large part in this absence—an absence that is understandable if we recall the social context of the nineteenth century origins of this kind of law.

There are some jurisdictions, especially in the US, where decisions about detention and involuntary treatment are separated, and where the patient has the right to refuse treatment even though involuntarily hospitalised. The patient's capacity to refuse is separately determined, and only if the patient is found incompetent or an immediate emergency exists, can non-consensual treatment be given. There are two difficulties here: first, what is the justification for detention in a hospital, which is presumably to serve a health interest, if treatment cannot be given; and, second, in what sense can a patient who is *detained* in hospital consent 'voluntarily' to treatment? In light of these difficulties, this practice is hard to justify.

In some jurisdictions, where decisions about detention and treatment are not separated, the patient's inability to consent (usually vaguely specified) may still be an implicit criterion for involuntary treatment, but it is nearly always the risk of harm that is determinative (Dawson and Kampf 2006). Major international instruments—the UN Principles for the Protection of Persons with Mental Illness, the European Convention on Human Rights, and the 2004 Council of Europe Recommendations—concerning the human rights and dignity of persons with mental disorder, affirm the primacy of risk in relation to involuntary hospitalisation.

But when considered in the context of general medicine, in our view, current practices of involuntary treatment, with or without a separation of detention and treatment, are frankly discriminatory because they apply less favourable rules, unnecessarily, to those with mental disorders (Campbell and Heginbotham 1991; Szmukler and Holloway 1998). Hence our proposal that decision-making capability, and the respect for the 'autonomy' or self-determination of the person that this serves, should be at the heart of a determination that involuntary treatment is justified (Dawson and Szmukler 2006; Szmukler et al. 2010).

Before examining this issue further, we need to briefly touch on a complex area. In the nineteenth century, a person in the asylum was more or less fully under the supervision of the doctor who could then effectively make decisions for the patient concerning virtually all of his or her affairs—medical, social, and so on; or decision-making powers were automatically transferred to others: e.g., to appointed managers of the person's property. More recently, in many western jurisdictions, the control of such matters, unrelated to treatment, has been removed from regulation by the Mental Health Act, and is now dealt with under some form of 'guardianship' or 'capacity' legislation, which may be administered through a special court or tribunal. For example, the Mental Health Act 1983 in England first narrowed the powers conferred on guardians appointed under the 1959 Act, to treatment and

use, weigh and appreciate that information, and so on. The UN Convention on the Rights of Persons with Disabilities places great emphasis on another usage, 'legal capacity' (Article 12). Thus to avoid confusion in use of the word 'capacity' we use the term 'decision-making capability' to describe the capabilities described above and presented in Clause 3 of the Model Law (at 1.7, below).

care-related issues, and then in 2005 a Mental Capacity Act was passed that deals comprehensively with decision-making about other interests of such persons, including their property.

The point here is that impaired decision-making capability, which is nowadays assessed separately for each specific decision required of the person, is the key justification for these guardianship schemes. And these schemes, whether by legislation or case law, usually cover the non-consensual treatment of persons with an impairment or disturbance of mental functioning when their condition is 'physical' (non-psychiatric) and that treatment is considered by doctors to be in the person's best interests: an example would be the evacuation of a subdural haemorrhage in someone who is confused and resisting following a head injury sustained in an accident. It must be added though, that the boundaries and interrelationship between the respective roles of civil commitment and 'guardianship' legislation, where the two co-exist, can be ambiguous.

17.3 Capacity-based (Mental) Health Law

Why do we need capacity-based law in relation to a person's health interests? We accept that patients with physical disorders, provided they have decision-making capability, can make decisions that may be seriously detrimental to their health or safety. In contrast to the necessary role of impaired decision-making in patients with 'physical disorder', under conventional mental health legislation capacity plays little or no role in decisions to initiate psychiatric treatment against a patient's wishes. The criteria for the involuntary treatment of 'mental disorders' thus fail to respect the autonomy of the patient. The patient's reasons for rejecting the treatment are not determinative (as they must be in determining decision-making capability), nor is the question of whether treatment is in the 'best interests' of the patient—from the perspective of the patient, not the clinical team—central to the legal decision. The key considerations in mental health law are the presence of a mental disorder and risks to the patient's health or safety, or to the safety of others. For persons with physical disorders with decision-making capability, their personal values are given dominion. Those with mental disorder are not accorded this privilege.

There seems to be an underlying assumption in legislation of this kind that 'mental disorder' necessarily entails mental incapacity, so the question does not need to be asked, and that the wishes and preferences of a person with a disordered mind are not to be taken seriously in determining where their best interests lie. But research has shown that even among the most ill patients, those admitted to acute psychiatric wards, 40–60 % retain capacity (Okai et al. 2007; Owen et al. 2008). By failing to respect the person with a mental disorder's autonomy; by not presuming decision-making capability unless there is reason for doubt; by assuming that mental disorder entails impaired decision-making; and by enshrining these prejudices in legislation that applies uniquely to those with 'mental disorder', current mental health legislation discriminates against those with mental disorders.

Just as discriminatory as those aspects of mental health legislation that refer to the safety of the patient are those referring to the 'protection of other persons'. People with mental disorders are unusual in being liable to detention (in hospital) because they are assessed as presenting a risk of harm to others, but before they have actually committed an offence (or where there is strong evidence that they have, for example, when detained on remand). This constitutes a form of preventive detention that is selective.

The further clarify this point, let us imagine two overlapping populations of people—first, all people with a mental disorder and second, all people who present a risk to others. Some people with a mental disorder will also fall into the group who present a risk. Let us imagine that we can accurately assess the risk presented by an individual.[2] Now take everyone in the community who presents a level of risk above an unacceptable threshold of risk. A small proportion of these will be people with a mental disorder. Conventional mental health law allows those with mental disorder—only—to be detained on the basis of risk alone, not the much larger remainder of the group (who are not mentally ill) who present an equal level of risk. How can this be justified? There is no evidence that risk is easier to assess in those with mental disorder, or that violence is more predictable in this group. In the absence of antisocial personality disorder or drug abuse, violence is at most only modestly more frequent in people suffering from a psychosis (Coid et al. 2006; Elbogen and Johnson 2009). Even if this were not so, selective detention on the basis of risk would not be justified. Could treatability be a justification? If risky people with mental disorders were treatable perhaps that would be a good reason for involuntary treatment. Lengthy hospitalisations, lasting many years if not decades, in secure hospitals, suggest this is not the case. And it is possible, probably likely, that psychosocial interventions, such as anger management or controlled drinking programmes for risky persons without a mental disorder could prove more effective in reducing the total number of violent incidents in the community than treating the relatively small number of patients with mental disorder.

One cannot escape the conclusion that the 'protection of others' criterion in mental health legislation is discriminatory. If preventive detention is to be allowed for the mentally disordered solely on account of their risk to others, if we are to avoid discrimination, it should be permitted for all of us. Fairness demands that all those presenting an equal level of risk to others should be equally liable to detention. This of course would produce a generic dangerousness or preventive detention regime that many would find unacceptable. But the principle of non-discrimination requires that either we have generic legislation applicable to all of us, or we have no preventive detention for anyone, including those with mental disorder.

This does not mean that a person's dangerousness is unimportant. If it is reliably linked to an individual's mental disorder for which there is treatment, and that person lacks capacity, their involuntary treatment is justified if in their best interests. The risk

[2] In fact, even the best available risk assessment instrument, under the best research conditions, will be wrong more than 9 times out of 10 in predicting whether a person with a mental disorder will commit a serious act of violence, those acts occurring in about 1 % of patients with a psychosis in any year (Large et al. 2011).

they present might weigh substantially in the best interests assessment. What if the patient *has* decision-making capability? If involuntary 'treatment' is to be imposed in these circumstances, no health interest is necessarily served—the person with capacity is usually regarded as the best judge of what is in their best interests. Protection of the public becomes the sole interest. Why then is health legislation appropriate?

It is also a dangerous situation. There is a tendency for societies to try to expand the 'mental disorder' category to include people with various undesirable or threatening behaviours. This potential is clearly demonstrated in proposals under which the application of a state-defined 'diagnosis' (such as 'dangerous severe personality disorder') and an ascription of risk could justify someone's detention, even in the absence of a previous violent offence or the possibility of effective treatment (Maden and Tyrer 2003).

Persons with mental disorder do not receive the protections from preventive detention that the rest of us do. Mental health legislation denies such protection, thus reinforcing an underlying stereotype that the mentally ill are inherently dangerous.

17.4 The Fusion Law Proposal

It is for these reasons that we propose a single comprehensive statute that 'fuses' the strengths of capacity-based regimes with those of civil commitment regimes (Dawson and Szmukler 2006). In this statute, there would be no need for specific reference to 'mental disorder'. Nor would there be any need for a separate legal regime directed solely to involuntary psychiatric treatment. We propose a single, comprehensive involuntary treatment regime based on a decision-making capability test, like the one found in section 2 of the Mental Capacity Act 2005 for England and Wales, based on the ability of a person to understand, recall, process, and weigh relevant treatment information, and their ability to communicate treatment decisions[3].

We recognise, however, that the strength of such a guardianship regime, like the MCA, in giving due weight to the patient's autonomy when capacity is retained, is counterbalanced by a number of weaknesses in such schemes. These lie in the lack of sufficient attention to powers governing emergency treatment, forced treatment and detention in hospital. These are exactly the areas in which civil commitment schemes, such as the MHA, are strong. The use of force, and the detention and involuntary treatment of objecting patients, are clearly authorised and regulated by mental health legislation. Our proposed law thus fuses the strengths of capacity-based and commitment laws. This thinking has led to drafting of a model law giving practical expression to this proposal (Szmukler et al. 2010).

[3] The assessment of decision-making capability may involve, especially in difficult cases, a need to also consider beliefs, values and emotional factors. For further discussion see Tan, A. Stewart, R. Fitzpatrick & T. Hope. Competence to make treatment decisions in anorexia nervosa: thinking processes and values. Philosophy, Psychiatry, Psychology, 13,4 (2006): 267; Banner and Szmukler (2014) Radical Intepretation and decision-making capacity, Journal of Applied Philosophy. (in press).

The model law operates on a number of levels, and is divided into several Parts. In all cases, the patient must lack decision-making capability to consent to be covered by the regime, but other features of the situation then determine precisely which Parts of the model law apply. These distinguishing features include whether the patient: objects to their treatment or care; is effectively detained for treatment purposes, even if they do not object; or requires 'serious medical treatment', of a kind already subject to special regulation under English law. In addition, certain provisions cover aspects of forensic mental health care.

The first level of the model law states general principles.

The second part provides a 'general authority' to give care and treatment to an incapacitated patient, when that patient does not object, or when emergency intervention is required. These rules cover much the same ground as the older common law justification of necessity.

A third part of the model law then regulates certain 'serious medical treatments' and some other medical decisions for people who lack decision-making capability, such as withdrawal of life support and non-therapeutic sterilisation. Much of the detail here could be covered by regulations issued under the authority of the statute.

(In the full version of the model law (Szmukler et al. 2010), a fourth part deals with persons who lack decision-making capability and who do not object to placement in a hospital or care home, and treatment, but who are 'deprived of liberty', in its clearest sense meaning that the person is unable to leave. Included in this group are long-term residents in institutional care who acquiesce to being there, but may not be subject to any processes checking on their living conditions or the treatment they might be receiving. A set of safeguards is set out for this group. The protections are important, and one would hope that safeguards along these lines could be implemented, but at this stage it is not clear how this can best be done, even in western jurisdictions. We have thus omitted this part in the modified law, but it should nevertheless be kept in mind and reviewed at a later stage).

The fifth part of the modified model law then establishes a regime for the compulsory treatment of objecting patients who lack the decision-making capability to consent, regardless of its cause. This regime is not based on criteria of mental disorder and dangerousness, but on the patient's impaired decision-making and best interests. It would cover all forms of compulsory treatment, of both 'physical' and 'mental' disorders. This Part includes many familiar features of a civil commitment regime, including powers of immediate detention of the person; compulsory examination and certification of the person by qualified health professionals; a period of assessment; mandatory consultation with the patient's family and other carers; and automatic access to independent review of compulsory status. The process would culminate in a Compulsory Treatment Order, with an initial life of 6 months, based on a prepared treatment plan.

The sixth part of the modified law concerns the special position of forensic patients, and indicates how they could be managed under a regime based principally on the concepts of impaired decision-making and best interests, not mental disorder and dangerousness. A limited compromise of impaired decision-making principles may be required in some instances, to promote the safety of other people.

17.5 An Adaptation of the Model Law for Low and Middle Income Countries

There is a range of specific factors—cultural, historical, political, economic—that must play a crucial role in shaping mental health law in low and middle income countries such as those of South Asia (see for example, (Kallivayalil et al. 2009; Kala 2004; Trivedi 2002). We suggest that the principles of the fusion proposal lend themselves to adaptation to the particular circumstances surrounding healthcare in a country. South Asia is one of the most heavily populated regions in the globe. Most of the countries in this region are developing countries. Some challenging features that have been characteristic of this region are poverty, illiteracy, unemployment, natural disasters, conflicts and terrorism. Existing health care facilities in this region are inadequate for such a vast population, especially so for mental health care. We propose that the checks and balances in the Model Law can be tailored to accommodate the limitations placed by the resources available. Compared to conventional mental health laws, there is scope for greater involvement of non-medical practitioners who nonetheless have necessary expertise in mental health care, while independent authorisation of involuntary treatment could be conducted by a credible, respected single person member of a mental health Tribunal type of body with appropriate legal training, but not necessarily a judge. A highly significant role for 'substitute decision makers' (or 'facilitators'[4]) should comport well with societies in which family ties are strong.

17.6 Model Law

The model law that we present is not an attempt to produce a full statute. There are many details that are not covered. There is also no mention of 'positive' rights, such as rights to a minimum standard of health care and social care, employment, and so on. We have focused on involuntary treatment. Space does not permit an examination as to whether the model law is consistent with the UN Convention on the Rights of Persons with Disabilities. We would argue that the criteria for involuntary treatment are 'disability neutral' and thus non-discriminatory against people with a disability.

[4] The role of the 'substitute decision maker' in the sense we are using the term here refers to a person who acts according to the patient's 'best interests', from the patient's point of view; that is, makes, in association with others who know the patient well, the best interpretation of what the person would have decided were they to have retained capacity and found themselves in their current circumstances. As the term 'substitute decision maker' entails some ambiguity, a case can be made for using the term 'facilitator' as suggested by Bach and Kerzner (2010) which would make it clear that role of this person is to give expression to the ill person's will and preferences—when supported decision making has proved inadequate—and to do so until the person has recovered the capability to make decisions, with or without supports.

A caveat, though obvious, is worthy of mention. Even a perfect law cannot compensate for poor healthcare or inadequate resources. Under such circumstances the law will be bypassed or short-circuited, or totally ignored.

We conclude this chapter by presenting a draft statute that has been adapted to low or middle-income countries, but that can, of course, be modified further—for example with regard to the personnel involved and their professional backgrounds provided that their training has been appropriate, or duration of the various orders—while observing the basic principles.

17.7 Outline of Adapted Model Law[5]

Part I Principles
Part II General provisions
Part III Serious medical treatment
(Part IV Deprivation of Liberty—Omitted)
Part V Compulsory provision of care and treatment
Part VI Forensic provisions
Part VII The Mental Capacity Tribunal
Part VIII Patient safeguards

Part I Principles
1. Principles of the Act
The following principles apply for the purposes of the Act:

1. A person must be assumed to have decision-making capability unless lack of decision-making capability is established.
2. A person is not to be treated as unable to make a decision unless all practicable steps to help the person to do so have been taken without success, including all attempts at supported decision making[6].
3. A person is not to be treated as unable to make a decision merely because the person makes an unwise decision.
4. An act done, or decision made, under the Act for or on behalf of a person who lacks decision-making capability must be done, or made, in his or her best interests, except as otherwise specified.
5. Before the act is done, or the decision is made, regard must be had to whether the purpose for which it is needed can be as effectively achieved in a way that is less restrictive of the person's rights and freedom of action.

[5] The numbering of clauses and sub-clauses follows that in the full 'model law' in Szmukler et al. (2010). For example, Part IV has been entirely omitted. The original numbering has been retained so the reader may compare versions and see how the law might be modified according to the circumstances surrounding healthcare in a particular state.

[6] It would be in keeping with the UN CRPD to add clauses that clarify the nature of supported decision-making and how it could be implemented and supported.

6. All powers shall be exercised, and all services provided without any direct or indirect discrimination on the grounds of disability, age, gender, sexual orientation, race, colour, language, religion or national, ethnic or social origin and any differences on these grounds should be respected.
7. Any compulsory detention or treatment of a person under the Act should be matched by a reciprocal duty to provide treatment and support that is likely to provide a health benefit to that person.
8. Family members, friends or partners, who provide care to patients on an informal basis, should receive respect for their role and experience and have their views and needs taken into account.

Part II General Provisions

2. Scope of the Act

Except as otherwise provided the Act applies to persons who because of an impairment or disturbance in the functioning of the mind[7] lack the decision-making capability at the material time to make a decision relating to their care or treatment.

3. Definition of Decision-making Capability

1. For the purposes of the Act a person ("P") is unable to make a decision and lacks decision-making capability if unable:

   a. to understand the information relevant to the decision
   b. to retain that information
   c. to use, weigh or appreciate that information as part of the process of making the decision, or
   d. to communicate the decision (whether by talking, using sign language or any other means).

2. The fact that P is able to retain the relevant information for a short period only does not prevent P from being regarded as able to make the decision.

4. Definition of Best Interests[8]

1. In determining what is in the best interests of a person ("P"), the decision-maker must consider all the relevant circumstances including whether it is likely that P will at some time have decision-making capability in relation to the matter, and if so when that is likely to be.

[7] It is arguable whether the specification of an 'impairment or disturbance in the functioning of mind' is necessary. It might be claimed that this identifies a category of persons with a 'disability' and that by introducing such a 'status' the proposed model law may contravene what some authorities' view as a requirement under the UN CRPD that criteria for non-consensual treatment must be based on 'disability-neutral' criteria. (See Szmukler et al. (2014) for further discussion of this subject). It is hard to imagine an impairment of decision-making capability that is not associated with some kind of disturbance of mind.

[8] See footnote 4 above as to what is meant by 'best interests'.

2. He or she must, so far as reasonably practicable, permit and encourage P to participate, or to improve P's ability to participate, as fully as possible in any act done, and any decision made, affecting P.
3. The decision-maker must consider, so far as is reasonably ascertainable:

   a. P's past and present wishes and feelings (and, in particular, any relevant written statement made by P with decision-making capability)
   b. the beliefs and values that would be likely to influence P's decision if he or she had decision-making capability, and
   c. the other factors that P would be likely to consider if able to do so.

4. The decision-maker must take into account, if it is practicable and appropriate to consult them, the views of:

   a. anyone named by P as someone to be consulted on the matter in question or on matters of that kind
   b. anyone engaged in caring for P or interested in P's welfare
   c. any substitute decision maker appointed by the person or appointed for the person by the Tribunal as to what would be in P's best interests.

5. Notwithstanding (4) above, if a clause of this Act requires the agreement of any person that provision shall apply.
6. The principle of best interests applies to all decisions and to those participating as carers (or advocates) in decisions made on behalf of P, unless otherwise specified.
7. For the purpose of this Act a substitute decision maker[9] is a person who has been appointed by the person ("P") or by the Tribunal to act on behalf of P for the purposes of making decisions in relation to the care or treatment of P.
8. If all the other factors above are met, a decision may be in P's best interests although it is not in accordance with P's present expression of wishes and feelings, and although P objects to the treatment.
9. In determining best interests an advance refusal of treatment made by P in accordance with clause 53 shall be binding upon a decision maker in accordance with the provisions of clause 54.
10. Where:

   a. under the Act, P's treatment is authorised only when it is in his or her best interests, and
   b. during the course of such treatment P poses a serious threat of harm to another person,

P may be provided with such treatment as is immediately necessary to prevent such harm occurring and is proportionate to the likely seriousness of that harm.[10]

---

[9] See footnote 4 above concerning the meaning of the term 'substitute decision-maker'. It could be replaced by 'facilitator'.

[10] For example, when P is admitted to hospital for treatment in his or her best interests, but is violent to others while there.

5. Further Definitions: Care and Treatment and Primary Carer

1. "Care or treatment" that may be provided under the Act includes actions in relation to medical treatment, nursing, psychological or care needs, habilitation and rehabilitation and specific welfare arrangements, and includes the use of restraint or seclusion in accordance with guidelines established by Regulations.
2. "A decision in relation to care or treatment" of the person ("P") includes a decision to admit the person to a hospital or care home for the purposes of care or treatment.
3. The "primary carer" is the person who has the closest day to day care of P or, in the absence of such a person, a person who has an ongoing concern for the well-being of P.
4. The primary carer shall act in the role of substitute decision maker until another person is appointed unless:

   a. P objects to the primary carer being appointed
   b. the primary carer is unable to act or is otherwise unsuitable.

5. If subclause (4) (a) or (b) applies the appropriate authority shall appoint an advocate to act in the role until the substitute decision maker is appointed.

6. General Authority

1. Subject to the other provisions of the Act, a person is authorised to do an act with respect to the welfare, care or treatment of a person who lacks decision-making capability ("P") if that act is in the best interests of P.
2. Nothing in this clause excludes a person's civil or criminal liability resulting from negligence in doing the act.
3. If that act involves the restraint of P the act must be a proportionate response to the likelihood of harm to P and the seriousness of that harm if the act is not done.
4. A person restrains P if he or she:

   a. uses or threatens to use force to secure the doing of an act which P resists, or
   b. restricts P's liberty of movement, whether or not P resists, or authorises another person to do any of those things.

5. This clause does not authorise the use of force on P to administer medication unless it is immediately necessary to prevent serious harm to P.
6. This clause does not authorise the provision of serious medical treatment, unless the requirements of Part III are met.
7. Except as provided by (3) above, this clause does not authorise the deprivation of P's liberty.
8. This clause does not authorise a person to do an act that is contrary to:

   a. a decision made by P in a valid advance directive, as provided by clauses 53 and 54
   b. the decision of a substitute decision maker acting within the scope of his or her authority.

7. Application of the Parts of the Act

The following Parts of the Act apply to certain decisions or acts:

1. Part III applies if the decision or act involves serious medical treatment.
2. Part V applies if a person ("P") objects to a decision or act that involves the provision of care or treatment to P, unless that decision or act is authorised by clause 6.

Part III Serious Medical Treatment

8. Application of this Part

Except as otherwise provided, this Part applies to every person receiving care or treatment under this Act or receiving treatment authorised by a substitute decision maker or the Tribunal under clause 49 or 50.

9. Requirements Before Serious Medical Treatment can be Provided

1. If a health or social care provider is proposing to provide, or secure the provision of, serious medical treatment for a person ("P") who lacks decision-making capability to consent to the treatment, the following provisions apply:

   a. The clinician in charge of P's care or treatment (the responsible clinician) shall consult:

      1. P, unless inappropriate or impracticable
      2. the substitute decision maker for P
      3. P's primary carer.

   b. Where no substitute decision maker has been appointed the responsible clinician shall apply to the Tribunal for the appointment of a suitable person to act.
   c. If the appointment of such a person is impracticable the primary carer of P shall be appointed to act in that role in accordance with clause 5(4).

2. Before serious medical treatment is provided to P the approved clinician must prepare a written care plan.
3. "Serious medical treatment" means treatment that is defined in Regulations.

10. Approved Doctor to Provide Second Opinion on Serious Medical Treatment Where There is a Disagreement as to P's Best Interests

1. In the event of a disagreement between the responsible clinician and the substitute decision maker, or person acting in that role under clause 5(4), that the treatment is in the best interests of a person ("P"), an approved doctor must examine P and give a second opinion.
2. In any case where it is proposed to provide serious medical treatment to P a request for an approved doctor to examine P and give a second opinion may also be made by:

   a. the substitute decision maker
   b. an advocate
   c. P's primary carer.

3. If agreement on whether the proposed treatment is in P's best interests still cannot be reached following the second opinion, the case will be referred to the Tribunal for a determination.

11. Treatment Urgently Required

1. If serious medical treatment needs to be provided to a person who lacks decision-making capability as a matter of urgency in order to save life or serious and imminent deterioration in health, it may be provided on the basis of the opinion of one medical practitioner despite the absence of a second opinion or a care plan.
2. This clause does not authorise a person to provide treatment that is contrary to:

   a. a decision made by P in a valid advance directive, as provided by clauses 53 and 54
   b. the decision of a substitute decision maker acting within the scope of his or her authority.

12. Serious Medical Treatment Requiring Approval by a Second Medical Opinion or the Tribunal

1. If a healthcare provider is proposing to provide, or secure the provision of:

   a. electroconvulsive therapy
   b. medication for mental disorder beyond the period of 3 months from the date of the first treatment provided under this Act
   c. other treatments prescribed in Regulations

for a person ("P"), who lacks decision-making capability to consent to the treatment, the agreement of an approved doctor qualified to give a second opinion on the treatment shall be obtained before the treatment proceeds.

The agreement to the proposed treatment of the approved doctor who gives the second opinion shall be recorded on an approved form.

2. Regulations shall provide for the period of time for which the approved form is in force.
3. In respect of treatment provided under clause (1) (b) above, the maximum period for which the approved form shall be in force is 6 months.
4. If a healthcare provider is proposing:

   a. to withhold or withdraw artificial nutrition or hydration from a person in a permanent vegetative state or a minimally conscious state
   b. organ or bone marrow donation by a person who lacks decision-making capability to consent
   c. non-therapeutic sterilisation of a person who lacks decision-making capability to consent
   d. other treatment prescribed by Regulations

an application shall be made to the Tribunal for a determination on the matter.

Part V Compulsory Provision of Care and Treatment

21. Application of this Part

a. This Part applies to a person ("P") if the following conditions are met:

   1. P has an impairment or dysfunction of the mind.
   2. P lacks decision-making capability to make a decision about his or her care or treatment.
   3. P needs care or treatment in his or her best interests.
   4. P objects to the decision or act that is proposed in relation to his or her care or treatment and that decision or act is not authorised by clause 6.
   5. The proposed objective cannot be achieved in an alternative less restrictive fashion.
   6. Treatment is available that is likely to alleviate or prevent a deterioration in P's condition.
   7. The exercise of compulsory powers is a necessary and proportionate response to the risk of harm posed to P or any other person, and to the seriousness of that harm, if the care or treatment is not provided.

b. If any of these conditions are no longer met P shall be discharged from compulsory powers.

22. Preliminary Examination

1. If the appropriate authority receives a reasonable request for a health assessment of a person ("P") from any person with a legitimate interest in P's welfare, and if the conditions in clause 21 appear to be met in P's case, it must, as soon as practicable after receiving the request, arrange for P to be examined by a registered medical practitioner or an approved mental health worker.
2. After examining P, if the registered medical practitioner or approved mental health worker considers it likely that all the conditions in clause 21 are met in P's case and that compulsory assessment is necessary, the registered medical practitioner or approved mental health worker may apply to the authority for the compulsory assessment and registration of P.
3. The appropriate authority shall:

   a. appoint a responsible clinician to be in charge of the assessment, care or treatment of P
   b. ensure P is advised, as far as practicable, of the availability of advocates
   c. register P as a compulsory patient
   d. provide P with a copy of the certificate of registration and appropriate information.

4. P may be detained in hospital for assessment and treatment under this clause for up to 72 h from the time of admission to hospital, or the responsible clinician may direct that P be assessed in the community if that would be safe and viable, provided no treatment is provided to P that is contrary to clause 28 below.
5. The substitute decision maker for P or the primary carer shall be consulted concerning P's assessment if practicable.

23. Powers of Entry and Inspection

An approved health or social care professional as provided in Regulations may at all reasonable times enter and inspect any premises (not being a hospital) in which a person who lacks mental decision-making capability is living, if he or she has reasonable cause to believe that the person is not under proper care.

24. Power to take a Person to a Place of Safety

1. If it appears to a justice of the peace or approved person (as designated by Regulations) on information on oath by an approved health or social care professional that there is reasonable cause to suspect that a person ("P") appears to lack decision-making capability to make decisions about his or her care or treatment and:

   a. has been, or is being, ill treated or neglected in any place, or kept otherwise than under proper control, or
   b. being unable to care for himself or herself is living alone and is in need of care and attention

The justice or approved person may issue a warrant authorising a police officer to enter the premises, by force if necessary, and if thought fit to remove P to a place of safety with a view to making proper arrangements for P's care.

2. If a police officer finds in a place to which the public have access a person ("P") who appears to be unable to make decisions about care or treatment, and who appears to be in immediate need of care or control, the constable may, if he or she thinks necessary to do so in the best interests of that person or for the protection of others, remove that person to a place of safety.
3. P may be detained under this clause for a period not exceeding 24 h for the purpose of being examined by a registered medical practitioner under clause 22.

25. Conveyance to Hospital

1. A person:

   a. for whom an application for compulsory assessment has been made by a registered medical practitioner
   b. who has been registered as a compulsory patient under clause 22(3)
   c. who is lawfully recalled to hospital by the responsible clinician

may be taken by an authorised person and conveyed to a hospital, or to another designated place of assessment, at any time within the following 72 h.

2. The range of persons who may be authorised to exercise this power shall be designated by Regulations.

26. Application in Respect of Patient Already in Hospital

If, in the case of a person ("P") who is receiving treatment in a hospital, it appears to a nurse or other approved healthcare worker of the prescribed class:

a. that P is suffering from an impairment or dysfunction of the mind of such a degree that it is necessary for P's health or safety or for the protection of others for P to be immediately restrained from leaving the hospital; and
b. that it is not practicable to secure the immediate attendance of a practitioner for the purpose of furnishing a report under clause 22

the nurse or other approved healthcare worker may record that fact in writing, and in that event P may be detained in the hospital until a medical practitioner has arrived and examined P, provided that the maximum period for which P may be detained under this clause is 24 h.

27. Initial Assessment

1. At the end of 72 h after the person ("P") is registered as a compulsory patient, he or she must be discharged from compulsory care, unless a health or social care professional as provided in Regulations has examined P and made a report to the appropriate authority that the conditions in clause 21 are likely to be met in P's case and that it is appropriate for P to be subject to initial assessment under this Part.
2. The report under the above clause shall not be provided by the registered medical practitioner or approved health worker who provided the initial report under clause 22.
3. If both examiners agree that the conditions appear to be met, P may be detained and given care or treatment for a further 28 days, or the responsible clinician may direct that P be assessed in the community as provided below, but no treatment may given to P contrary to clause 28 below.
4. Each examiner must give an opinion as to whether it is appropriate for P:

   a. to be detained in a hospital while the assessment is carried out, or
   b. to be assessed in the community.

5. In considering whether it is appropriate for the assessment to take place in the community they shall each consider:

   a. P's views on being assessed in the community
   b. whether P can be safely and effectively assessed in the community
   c. whether care or treatment for P can be provided safely and effectively in the community.

6. If, during the period of assessment, the responsible clinician considers:

   a. a person under community assessment requires care or treatment in hospital
   b. a person in hospital could be adequately assessed in the community

the responsible clinician may direct such a change in the place of assessment.

7. If at any time during the period of initial assessment the responsible clinician considers there are no longer reasonable grounds to believe that P meets the conditions in clause 21, the responsible clinician shall immediately discharge P from compulsory assessment.

28. Treatment Within the Period of Preliminary Examination and Initial Assessment

1. During the period of preliminary examination and initial assessment the person ("P") may not be provided with medical treatment to which he or she objects unless:

    a. it is covered by the general authority established by clause 6
    b. it is necessary to save life or prevent serious and immediate deterioration in P's health or to protect another person from harm
    c. where serious medical treatment is to be provided that is covered by clause 12, the requirements of that clause are satisfied.

2. Any treatment provided under (1) (a) or (b) shall not be given in the community but P shall be conveyed to hospital for the purposes of treatment.

29. The End of the Initial Assessment Period

1. Before the end of the 28 day period of initial assessment the responsible clinician may apply to the Tribunal for a Compulsory Treatment Order.
2. If no application is made to the Tribunal the authority for the compulsory assessment of P shall lapse at the end of 28 days.
3. A Compulsory Treatment Order may be made, following a full hearing, by an appointed member of the Tribunal.
4. An application may not be made for a Compulsory Treatment Order under (1) without an initial Assessment Order having first been made, unless:

    a. a substitute decision maker has been appointed for P and he or she does not object
    b. P has on a previous occasion been admitted to hospital lacking decision-making capability and needing treatment in his or her best interests and was, immediately prior to registration under this Part, being treated as a voluntary patient in the same health service.

34. Compulsory Treatment Orders

1. The responsible clinician may apply to the Tribunal during the compulsory assessment period for a Compulsory Treatment Order.
2. The application shall be based on:

    a. the written recommendation of a registered medical practitioner and another health or social care professional that the conditions in clause 21 are met
    b. a draft care plan.

3. if the Tribunal finds that the conditions in clause 21 are met it may make an order considered appropriate in the circumstances.
4. The Tribunal may specify the kinds of conditions that can be imposed by the responsible clinician on the person ("P") within the community. These conditions may include:

    a. where P may reside
    b. where and when P shall attend for treatment

c. restrictions or limits that can be imposed on P's conduct or freedom of movement, provided that any such restrictions must be proportionate to the harm that is likely to occur if they are not imposed.

5. The Tribunal shall authorise the care plan, subject to any amendments it requires, although amendments to the treatment provisions may only be made with the agreement of the Responsible Clinician and the medical member of the Tribunal.
6. Once it has been authorised by the Tribunal, the care plan provides sufficient authority for authorised persons to provide the care or treatment described in the plan, including the authority to detain P in hospital, and to return P to hospital if he or she is absent without permission, when hospital treatment is included in the plan.
8. The duration of the order shall be specified by the Tribunal but shall not exceed 6 months.
9. P or P's substitute decision maker may make one application to the tribunal for review of the terms of, or discharge from, the Compulsory Treatment Order, at any time while the order is in force.
10. Notwithstanding (9), P or P's substitute decision maker may apply to the Tribunal for review of, or discharge from, the Compulsory Treatment Order if P is returned from community treatment to detention in hospital for treatment for more than 72 h.

35. The Care Plan

1. The care plan to be approved by the Tribunal must include:

   a. a description of the medical treatment to be provided to the person ("P") while the plan is in force, provided no treatment may be authorised contrary to Part III
   b. such other information relating to the care of P as may be prescribed in Regulations.

2. In preparing a plan for P, the responsible clinician must consult the following persons about the treatment proposed:

   a. P, unless inappropriate or impracticable
   b. P's substitute decision maker
   c. the primary carer of P.

3. The responsible clinician must send a copy of the plan to:

   a. P
   b. P's substitute decision maker
   c. the primary carer of P

as soon as practicable after the plan is in force.

4. The responsible clinician may amend the care plan with the agreement of the substitute decision maker at any time while it is in force.

5. If there is disagreement between the responsible clinician and the substitute decision maker as to:

    a. a change to the treatment
    b. a condition of a community treatment order
    c. a change in the location of treatment from community to hospital or from hospital to community

and it cannot be resolved between them, an opinion shall be sought from a approved doctor qualified to give a second opinion and the change shall not be instituted without his or her agreement.

36. Renewal and Termination of Compulsory Treatment Orders

1. If at any time during the life of a Compulsory Treatment Order the responsible clinician considers the conditions in clause 21 are no longer met, the responsible clinician shall discharge P from the Order.
2. The responsible clinician may, before the Compulsory Treatment Order has expired, apply to the Tribunal for a new order under Clause 34.
3. If no new order has been made, the order shall lapse at its conclusion and P shall be immediately released from compulsory treatment.

37. Community Treatment

1. Before deciding whether the person ("P") may reside in the community under a Compulsory Treatment Order the responsible clinician shall be satisfied that:

    a. compulsory care in the community is compatible with safe and effective care
    b. appropriate services are available in the community
    c. P has been consulted as far as practicable, and P's views carefully considered, as to whether community treatment should proceed
    d. any carers have been consulted and their views considered
    e. the SDM has been consulted and his or her view considered.

2. If the conditions in (1) (a) or 1(b) cease to apply P shall be recalled by the responsible clinician to hospital or discharged from compulsory care.

38. Power to Recall to Hospital

1. The responsible clinician may recall to hospital a person ('P") under compulsory treatment in the community, if the responsible clinician considers:

    a. P requires medical treatment in hospital; and
    b. there would be a risk of harm to the health or safety of P or to other persons if P were not recalled to hospital for that purpose.

2. The responsible clinician may also recall P to hospital if P fails to comply with a condition specified under clause 34(6) above.
3. The lawful recall of P to hospital shall be sufficient authority for an authorised person to take and convey P to hospital and for P to be detained there in accordance with the provisions of the Act.

39. Treatment Without Consent Under Assessment Order or Compulsory Treatment Order

The consent of a person ("P"), who is the subject of an Assessment Order or a Compulsory Treatment Order, shall not be required for the provision to P of any treatment, given by or under the direction of P's responsible clinician:

a. that is covered by the general authority established by clause 6
b. that is included in the care plan approved by the Tribunal, or in a lawfully amended care plan
c. that needs to be provided as a matter of urgency in order to save P's life or serious and imminent deterioration in P's health.

Part VI Forensic Provisions

40. Remand on bail or to hospital for a report on mental condition

1. An accused person (P) charged with a criminal offence may be remanded by the court on bail or, if it would be impractical for P to be assessed on bail, to hospital for up to 28 days for a report to be prepared on his or her mental condition, where:

    a. the court is satisfied on the evidence of a medical practitioner that there is reason to suspect that P has an impairment or dysfunction of mind, and
    b. P consents to the exercise by the court of this power, or
    c. if P lacks decision-making capability to consent, the court is satisfied on the evidence of a medical practitioner that an assessment is in P's best interests.

2. P and P's substitute decision maker may request a second medical opinion as to whether the conditions in (1) are met.
3. P may appeal to the Tribunal against the order, and if the Tribunal is satisfied that any of the conditions in (1) are not met, it shall discharge P from assessment.

41. Remand to Hospital for Treatment

1. An accused person (P) charged with a criminal offence may be remanded by the court on bail or, if it would be impractical for P to be treated on bail, to detention in hospital for treatment for his or her mental condition, provided that:

    a. the court is satisfied on the evidence of a medical practitioner and an approved health or social care professional that P has an impairment or dysfunction of the mind
    b. P needs care or treatment for his or her health or safety or the safety of another person
    c. where P has decision-making capability he or she consents to the exercise by the court of this power, or where P lacks decision-making capability the court is satisfied on the evidence of a medical practitioner that treatment is in P's best interests
    d. treatment is available which is likely to alleviate or prevent a deterioration in P's condition.

2. P and P's substitute decision maker may request a second medical opinion as to whether the conditions in (1) are met.
3. The duration of the order shall not exceed 6 months.

42. Due Process and Treatment During Remand Under Clause 40 or 41

1. The accused person ("P") must be represented by a lawyer when a court makes a remand order under clause 40 or 41.
2. P may apply to the Tribunal for discharge from a remand order, and if the Tribunal is satisfied that any of the conditions in clause 40(1) or 41(1) are not met, as required, P shall be discharged from the order and returned to the court.
3. P and P's substitute decision maker may request a second medical opinion, to be placed before the court or Tribunal, as to whether the conditions in clause 40(1) or 41(1) are met.
4. During a remand for assessment or treatment under clause 40 or 41, P may not be provided with medical treatment to which he or she objects, unless:

    a. it is covered by the general authority established by clause 6, or
    b. there are reasonable grounds to believe P lacks the decision-making capability to consent to treatment, and

    1. (i) the treatment is necessary to save life or prevent serious and immediate deterioration in P's health, or to protect another person from harm
    2. (ii) where serious medical treatment covered by clause 12 is to be provided, the requirements of that clause are satisfied.

5. That a person has been remanded for assessment or treatment does not prevent an application being made for that person's compulsory assessment or treatment under Part V.

43. A Hospital Order with a Concurrent Sentence

1. If a person is convicted of a criminal offence punishable by imprisonment the court may, after determining the sentence for the offence, make an order that the person ("P") be detained in hospital, if the court is satisfied on the evidence of a medical practitioner and a social or health professional that the following conditions are met:

    a. P has an impairment or dysfunction of the mind
    b. P needs care or treatment in his or her best interests or to protect the safety of another person
    c. treatment is available that is likely to alleviate or prevent deterioration in P's condition
    d. if P has decision-making capability, he or she is willing to accept treatment, or if P lacks decision-making capability, treatment is in his or her best interests.

2. P may be treated under this order despite his or her objection so long as P lacks decision-making capability to consent to treatment, provided that where serious medical treatment covered by clause 12 is to be provided, the requirements of that clause must be satisfied.

3. While P is under treatment in a hospital under this provision P's sentence will continue to run.
4. If any of the conditions in (1) are no longer met P shall be admitted to prison to complete the sentence or, if the sentence is in the community, to the terms of the sentence.
5. P may apply to the Tribunal for discharge from a treatment order made under this clause, and if any of the conditions in (1) are no longer met the Tribunal shall discharge P from the order, and P shall then be admitted to prison to complete his or her sentence or, if the sentence is in the community, the terms of the sentence.
6. Where at the end of P's sentence the conditions in (1) continue to apply:

    a. if P lacks decision-making capability to consent to treatment P shall be deemed to be subject to a compulsory treatment order made under clause 34
    b. if P has decision-making capability and is willing to consent to treatment he or she shall be treated as a voluntary patient.

44. Treatment Order Without a Concurrent Sentence

1. Where a person ("P") is convicted of an offence punishable by imprisonment the court may make a treatment order without sentencing P if:

    a. P has an impairment or dysfunction of the mind
    b. P lacks decision-making capability to make a decision about his or her care or treatment
    c. P needs care or treatment in his or her best interests
    d. treatment is available that is likely to alleviate or prevent deterioration in P's condition
    e. the care or treatment cannot safely be provided in a less restrictive manner
    f. (f) the court is of the opinion, having regard to all the circumstances of the case, including the nature of the offence and the character and antecedents of the offender and other methods of dealing with P, that an order under this clause is the most suitable method of disposing of the case.

2. A treatment order made under this clause deems P to be subject to a compulsory treatment order made under clause 34 and the associated provisions of the Act shall apply.

45. Treatment Order for a Person Found Not Guilty by Reason of Insanity or Unfit to Plead

1. If a person ("P") accused of an offence punishable by imprisonment is found not guilty by reason of insanity or unfit to plead for the offence, the court may:

    a. if it considers it necessary in the interests of P or to protect the safety of another person, order P's detention in hospital for treatment under this clause
    b. make a Compulsory Treatment Order for P under clause 34, if the conditions for such an order are met
    c. make no order, if P is subject to a sentence imposed on another charge
    d. if it would be safe to do so, direct P's immediate release.

2. Where the court orders P's detention in a hospital under clause (1) (a) the consent of P shall not be required for the provision of treatment to P by or under the direction of his or her responsible clinician, for the duration of the order, where the following conditions are met:

    a. where serious medical treatment covered by clause 12 is to be provided, the requirements of that clause are satisfied
    b. where P lacks the decision-making capability to consent, the treatment is in P's best interests or necessary to protect the safety of others
    c. where P has the decision-making capability to consent, the responsible clinician is satisfied that:

        1. P needs treatment in his or her own interests or for the protection of another person from harm
        2. P is suffering from an impairment or dysfunction of the mind that contributed significantly to the acts or omissions that constituted the offence
        3. treatment is available that is likely to alleviate or prevent a deterioration in P's condition and is likely to reduce the risk of the recurrence of those acts or omissions.

3. P may apply to the Tribunal for discharge from an order under clause (1) (a) at intervals specified by Regulations.
4. Where, on such an application, the Tribunal is satisfied that the order is no longer necessary in the interests of P or for the protection of others, or that continuation of the order would be disproportionate to the seriousness of the offence with which P was charged, the Tribunal shall:

    a. exercise the power under clause 34 to make a Compulsory Treatment Order for P, if the conditions for such an order are met;
    b. direct P's discharge from the order.

46. Transfer from Prison to Hospital

1. A person ("P") serving a custodial sentence may be transferred from prison to hospital if one medical practitioner and one health or social care professional certify that the following conditions are met:

    a. P has an impairment or dysfunction of the mind
    b. P needs care or treatment in hospital
    c. treatment is available which is likely to alleviate or prevent a deterioration in P's condition
    d. if P retains decision-making capability he or she is willing to accept treatment
    e. if P lacks decision-making capability, treatment is in his or her best interests.

2. If any of the conditions in (1) are no longer met P shall be admitted to prison to complete the sentence.
3. P may apply to the Tribunal for discharge from treatment in hospital under this clause, and if any of the conditions in (1) are no longer met the Tribunal shall discharge P and direct that P be admitted to prison to complete his or her sentence.

47. Care Plans for Patients Detained Under Part VI

1. A preliminary care plan shall be prepared for a person remanded to hospital for assessment under clause 40 as if that person was subject to an Assessment Order under clause 33.
2. A care plan shall be prepared for a person ordered to undergo treatment in hospital under clauses 41, 43, 44, 45 and 46, as if that person was subject to a Compulsory Treatment Order under Part V.

Part VII The Mental Capacity Tribunal

48. The Mental Capacity Tribunal

1 There shall be a Mental Capacity Tribunal that shall consist of two divisions:

   a. the Primary Division
   b. the Appeal Division.

2 The Primary Division shall hear all cases in the original jurisdiction except matters reserved to the Appeal Division.
3 Hearings before the Primary Division may, as provided in the Act, be conducted by a single legal member of the Tribunal.
4 The Appeal Division shall have the powers of a court and shall be presided over by a judge.
5 The Appeal Division shall have jurisdiction over:

   a. appeals from the Primary Division
   b. any other specified matters.

Part VIII Patient Safeguards

49. Appointment of person to act as substitute decision maker

1. P may appoint a substitute decision maker to have authority to make decisions about P's care or treatment in the event that P has lost the capacity to make the decisions.
2. P must have reached 18 and have decision-making capability to make the gift of power.
3. The gift of power must be in writing and be witnessed.
4. Subject to (5) the authority includes the power to consent to or to refuse treatment by a person providing care (subject to an advance refusal by P).
5. The power conferred on a substitute decision maker shall not authorise:

   a. giving or refusing consent to life sustaining treatment unless the power expressly so provides
   b. the provision of treatment contrary to Parts III or V of the Act
   c. the provision of treatment to which the patient objects unless that treatment is authorised by clause 6.

50. Powers of the Tribunal to Make Decisions and Appoint Substitute Decision Makers

1. If a person ("P") lacks decision-making capability in relation to any matter concerning his or her care or treatment the Tribunal may:

a. by making an order, make the necessary decision or decisions on P's behalf, or
b. appoint a person (the "substitute decision maker") to make decisions on P's behalf in relation to a specified matter or matters.

2. The Tribunal shall not authorise the provision of treatment contrary to Parts III or V of the Act.

References

Abbasi, Y. 2008. Mental health ordinance 2001-is it really being used? *Journal of Pakistan Medical Association* 58 (10): 578–580.

Appelbaum, P. S. 1994. *Almost a revolution: Mental health law and the limits of change*. New York: Oxford University Press.

Bach, M., and L. Kerzner. 2010. A new paradigm for protecting autonomy and the right to legal capacity. Law Commission of Ontario. http://www.lco-cdo.org

Campbell, T., C. Heginbotham. 1991. *Mental illness: Prejudice, discrimination and the law*. Vermont: Dartmouth.

Coid, J., M. Yang, A. Roberts, S. Ullrich, P. Moran, P. Bebbington, and N. Singleton. 2006. Violence and psychiatric morbidity in a national household population-a report from the British household survey. *American Journal of Epidemiology* 164:1199–1208.

Dawson, J., and A. Kampf. 2006. Incapacity principles in mental health laws in Europe. *Psychology, Public Policy and Law* 12:310–331.

Dawson, J., G. Szmukler. 2006. Fusion of mental health and incapacity legislation. *British Journal of Psychiatry* 188:504–509.

de Silva, V., Hanwella, R. 2010. Mental health in Sri Lanka. *Lancet* 376:88–89.

Elbogen, E. B., S. C. Johnson. 2009. The intricate link between violence and mental disorder: Results from the national epidemiologic survey on alcohol and related conditions. *Archives of General Psychiatry* 66:152–161.

Jacob, K. S., P. Sharan, I. Mirza, M. Garrido-Cumbrera, S. Seedat, J. J. Mari, S. Saxena. 2007. Mental health systems in countries: Where are we now? *Lancet* 370:1061–1077.

Kala, A. K. 2004. Why we should opt for a new mental health act, and not tinker with the old one? *Indian Journal of Psychiatry* 46:96–98.

Kallivayalil, R. A., J. K. Trivedi, and A. Tripathi. 2009. Social factors and forensic psychiatry in India. *Current Opinion in Psychiatry* 22:482–487.

Large, M. M., C. J. Ryan, S. P. Singh, M. B. Paton, and O. B. Nielssen. 2011. The predictive value of risk categorization in schizophrenia. *Harvard Review of Psychiatry* 19:25–33.

Maden, T., P. Tyrer. 2003. Dangerous and severe personality disorders: a new personality concept from the United Kingdom. *Journal of Personality Disorders* 17:489–496.

Narayan, C. L., M. Narayan, and D. Shikha. 2011. The ongoing process of amendments in MHA-87 and PWD Act-95 and their implications on mental health care. *Indian Journal of Psychiatry* 53:343–350.

Okai, D., G. Owen, H. McGuire, S. Singh, R. Churchill, and M. Hotopf. 2007. Mental capacity in psychiatric patients: Systematic review. *Britissh Journal of Psychiatry* 191:291–297.

Owen, G. S., G. Richardson, A. S. David, G. Szmukler, P. Hayward, and M. Hotopf. (2008). Mental capacity to make decisions on treatment in people admitted to psychiatric hospitals: cross sectional study. *British Medical Journal* 337:a448.

Pakistan, G. o. (2001). Mental health ordinance, 2001. http://www.emro.who.int/mnh/whd/Pakistan-Ordinance.pdf. Accessed 10 Feb 2012.
Sarkar, J. (2004). A new mental health act for India: An ethics based approach. *Indian Journal of Psychiatry* 46:104–114.
Szmukler, G., and F. Holloway. (1998). Mental health legislation is now a harmful anachronism. *Psychiatric Bulletin* 22:662–665.
Szmukler, G., R. Daw, and J. Dawson. 2010. A model law fusing incapacity and mental health legislation. *Journal of Mental Health Law* 20:1–140 (Special Issue Ed.).
Szmukler, G., R. Daw, and F. Callard. (2014). Law on mental health consistent with the UN convention on the rights of persons with disabilities. *International Journal of Psychiatry and the Law* (in press).
Trivedi, J. K. (2002). The mental health legislation: an ongoing debate. *Indian Journal of Psychiatry* 44:95–96.
Unsworth, C. (1987). *The politics of mental health legislation*. Oxford: Clarendon Press.
World Health Organisation. (2006a). *AIMS Report on Mental Health System in Bangladesh.* WHO and Ministry of Health & Family Welfare, Dhaka.
World Health Organization. (2006b). *WHO-AIMS Report on Mental Health System in Nepal.* WHO and Ministry of Health, Kathmandu, Nepal.

Chapter 18
Mental Health Legislation in South Asian Countries: Shortcomings and Possible Solutions

Mohan Isaac

18.1 Introduction

Nearly a quarter (1.64 billion) of the global population of over 6.8 billion lives in South Asia. India with a population of over 1.2 billion is the second most populous country in the world. Pakistan (184 million) and Bangladesh (164 million) have the sixth and seventh largest population in the world. Besides India, Pakistan and Bangladesh, the other countries which constitute South Asia are Afghanistan, Bhutan, Maldives, Nepal and Sri Lanka. All these eight countries are members of the South Asian Association for Regional Cooperation (SAARC) which was formed in 1985. While seven of these countries, were founding members, Afghanistan joined the association in 2007. Being former British colonies, all these countries except Afghanistan, Bhutan and Nepal are also member states of the Commonwealth of Nations. These countries of South Asia share not only a historical and cultural legacy but a common legal heritage as well. Legislation, including mental health legislation is always influenced by the specific historical, political, social, cultural and economic context in which it is enacted. While mental health legislations vary widely across the world, due to the common legal heritage, the current scenario of mental health legislation is largely similar in all the countries of the South Asian region. Broadly, the shortcomings as well as the possible solutions too are similar. To understand the current shortcomings in mental health legislation and their possible solutions, it is essential to provide a background description of South Asia and an overview of the current health and mental health care systems in these countries.

M. Isaac (✉)
Fremantle Hospital, Level 6, W Block, Fremantle, WA 6160, Australia
e-mail: Mohan.Isaac@uwa.edu.au

J. K. Trivedi, A. Tripathi (eds.), *Mental Health in South Asia: Ethics, Resources, Programs and Legislation,* International Library of Ethics, Law, and the New Medicine 58,
DOI 10.1007/978-94-017-9017-8_18

18.2 Socio-economic and Cultural Background of South Asia

South Asia is the poorest region on the earth after Sub-Saharan Africa. It is also one of the most densely populated regions of the world. Although in recent years, some countries in South Asia, have registered very impressive economic growth, world's largest population of poor, hungry and illiterate live here. The World Bank (2010) classifies economies according to their gross national income (GNI) per capita, into low income (US$ 995 or less), middle income (sub divided into lower middle income, US$ 996—US$ 3945 and upper middle income, US$ 3946—US$ 12,195) and high income (US$ 12,196 and above) countries. According to this classification, Afghanistan, Bangladesh and Nepal are low income countries and all the other South Asian countries except Maldives fall under the category of lower middle income group. Maldives which has the smallest population (396, 334) amongst all the South Asian countries had a per capita GNI of US$ 3970 in 2009 which made it move up to an upper middle income country.

Although countries in the South Asian region have very many similarities such as, a British colonial past, an under developed industrial base; agriculture based economies, very low standards of living and similar problems of inadequate resources and capacities, it is also a region of great contrasts. The countries of this region do not constitute a very homogenous group. There are striking differences between the countries as well as between different regions within the countries. They show widely varying profiles of development. The Gross National Income (GNI) may not always provide a complete picture of a country's overall development. The United Nations Development Programme (UNDP) has developed a composite index called the Human Development Index (HDI) to better capture the complex relationship between a country's income and human progress. HDI which was created for the UNDP by Mahboubul Huq, an influential Pakistani economist and former finance minister of Pakistan, indicates average progress of a country in human development. It is a measure of three dimensions of human development namely living a long and healthy life (measured by life expectancy), being educated (measured by adult literacy and school enrolment at the primary, secondary and tertiary level) and having a decent standard of living (measured by purchasing power parity, PPP, income). Since 1990, the UNDP has released the Human Development Report (HDR) annually. The HDR ranks all the countries in the world according to their HDI. Tables 18.1 and 18.2 provide some of the human development indicators and the HDI ranks for the countries of South Asia in 2010. Only Sri Lanka among the South Asian countries has achieved acceptable levels of under 5 mortality rate (per 1000 live births), adult literacy rate and life expectancy at birth.

About 70% of the population in South Asia live in rural areas and continue to depend largely on agriculture. The total male population is slightly larger than total the female population and majority of the population belong to the younger age groups. The literacy rates of the male population are slightly better than the female population. It is estimated that between 35 and 40% of the population in the region

Table 18.1 Human Development Indicators (HDI) in South Asian countries 2010

Country	Life expectancy at birth (yrs)	Expenditure on health, public (% of GDP)	Under 5 mortality (per 1000 live births)	Adult literacy rate (both sexes)	Human development index (HDI) rank
Afghanistan	44.6	1.8	257	28	155
Bangladesh	66.9	1.1	54	56.5	129
Bhutan	66.8	3.3	81	52.8	NA
India	64.4	1.1	69	68.3	119
Maldives	72.3	6.4	28	97.3	107
Nepal	67.5	2.0	51	60.3	138
Pakistan	67.2	0.8	89	54.2	125
Sri Lanka	74.4	2.0	15	90.8	91

Table 18.2 GDP per capita (2008 PPP—Purchasing Power Parity—US$) in South Asia

Country	GDP per capita (2008 PPP—Purchasing Power Parity—US$)
Afghanistan	1419
Bangladesh	1458
Bhutan	5532
India	3354
Maldives	5721
Nepal	1189
Pakistan	2625
Sri Lanka	4995

live below the poverty line of less than US$ 1.25 a day. A similar proportion of population do not have access to either safe drinking water or adequate sanitation. South Asia has one of the highest child malnutrition rates in the world.

South Asia is ethnically diverse. Numerous languages and dialects are spoken across the region. India has 22 officially recognized languages. Hinduism, Islam, and Buddhism are the major religions in South Asia. India, Pakistan and Bangladesh were under British rule till 1947. Nepal and Bhutan were protectorates of Great Britain. While most countries in the region currently are parliamentary democracies, Bhutan continues to be a constitutional monarchy. Pakistan is an Islamic republic. The region has witnessed long—running conflict and is considered to be one of the most violent places in the world. While an ethnic conflict of more than 25 years in Sri Lanka ended recently, Pakistan and Afghanistan continue to be violent places. Political instability is a chronic problem in some of the countries. Equally common are natural and man made disasters of various types and intensity, affecting large populations in all the countries of the region.

Although the countries in the region had formed the South Asian Association for Regional Cooperation (SAARC) more than 25 years ago, there has not been satisfactory integration between the countries due to a variety of political and socio economic reasons. The integration has been least trade wise; the trade between South Asian states is only 2 % of the combined GDP of the states. Along with rapid and impressive economic growth in some parts of the region, income inequalities and health inequalities too are disturbingly growing. A recent commentary on “the poor half billion in South Asia” notes that the rapid economic growth has created “two South Asia”—one dynamic, urbanized and globally integrated, and the other rural, impoverished and lagging (Ghani 2010). South Asia today is best known for its large scale poverty, increasing income inequality, frequent natural and man made disasters, chronic ethnic conflicts; violence and war, gross underdevelopment and uneven development, all factors which are important social determinants of a population’s quality of mental health (Isaac 2011).

18.3 Health and Mental Health Care Systems in South Asia

Health is comparatively a low priority for all the countries in the region as evidenced by the low spending on health, usually below 2 % of the GDP. Bhutan and Maldives which spent 3.3 and 6.4 % of their GDP on health are exceptions. Coverage and effectiveness of public health services are suboptimal in all countries in the region. Public health systems are dysfunctional and very vulnerable, due to lack of financial and human resources in most countries. Health systems are inequitable and are constrained by various factors such as low budgets, chronic shortage of motivated and adequately trained staff, irregular supply of essential drugs, lack of transportation facilities, non-functioning equipment and poor organization and management. Most people are burdened by out of pocket expenses as they have to depend on private health services. Large proportion of the population is not covered by any type of health insurance. In most countries of the region, an unregulated and highly commercialized private sector in health care is growing fast. India is considered to have the sixth most privatized health system in the world. A series of papers in The Lancet recently described the challenges affecting health care system in India which is reflective of problems all over the South Asian region. Summarizing the findings of the series, Reddy et al. (2011) noted that “India still has unacceptably high infant, child and maternal mortality rates, and high numbers of premature deaths that are attributable to chronic diseases. The burden of disability is further worsened by unrecognized and inadequately treated mental illness and the increasing toll of intentional and unintentional injuries. Several adverse social determinants together corrode the health of vulnerable populations, whereas risk factors like smoking, oral tobacco consumption and binge drinking of alcohol account for much death and disability”. The only exception to the poor state of public health systems and health and education indicators in the region is Sri Lanka.

Up until the time of independence from British rule in 1947, mental health services in the region were exclusively centred in the large mental hospitals, previously called asylums, located in some of the cities of the region—the undivided Indian sub-continent (Madras, Bombay, Calcutta, Poona, Ranchi, Lahore, Colombo etc.)—a reflection of the pre-chlorpromazine era asylum based psychiatric services in Britain and other western countries. Even after attaining independence, mental health services continued to be centred in asylums for several more years. In some countries, more asylums were established. Like asylums everywhere else in the world then, these were large, overcrowded, under funded and poorly staffed. Mental health services were marked by gross neglect due to a variety of reasons which included pervasive stigma, widespread misconceptions, grossly inadequate budgets and acute shortage of trained personnel. There were very few trained psychiatrists and other mental health professionals in any of the countries to develop alternate mental health services. Centres or facilities for training in any of the mental health disciplines were also non-existent. Few psychiatrists who worked in South Asian countries then were all trained in Britain (Neki 1973).

During the decades that followed independence, post graduate training facilities were created in few centres and small psychiatric units were set up in many of the general hospitals of the region (General Hospital Psychiatry Units). However, as a series of reports including the Lancet series on Global Mental Health (Jacob et al. 2007), the World Health Organizations Mental Health Atlas project of 2005 and 2011 (WHO 2005, 2011) and data obtained by the International consortium on mental health policy and services (Gulbinat et al. 2004) from Pakistan (Karim et al. 2004), India (Khandelwal et al. 2004) and Nepal (Regmi et al. 2004) show that the available mental health resources and services are grossly inadequate when compared to the enormous needs. Most of the psychiatric hospital beds are located in stand alone psychiatric settings such as mental hospitals. Majority of the services are largely located in urban areas. The country profiles also indicate that there are wide variations in the availability of different components of mental health services in different parts of the country. A recent review of the availability of resources for mental health in developing countries showed that resources were not only very scarce but were inequitably and inefficiently used (Saxena et al. 2007). Mental health continues to occupy a very low priority in terms allocation of funds, in the background of overall poor funding of around 1–2 % of the GDP for health. It is known that the proportion of persons with mental disorders who receive services in a country corresponds to the country's percentage of GDP (gross domestic product) spent on health care (Wang et al. 2007).

Most persons with serious mental disorders are looked after by their families in their homes and communities. Various traditional, religious and alternate methods of treatment are popular for all forms of mental disorders and are still extensively used all across the region. Ignorance, stigma and misconceptions about the causation and management of mental disorders continue to be widely prevalent. Consequently, even when mental health services are provided, their utilization is generally low.

Coupled with the severe shortage of trained mental health professionals such as psychiatrists is the poor quality of psychiatric training in medical schools of South Asia. Psychiatry is neither taught adequately nor examined at the medical graduation level except in very few medical schools in the region (Trivedi and Dhyani 2007). As a result, most primary care doctors and physicians have no skills or interest in diagnosing and managing mental and emotional disorders. The severe shortage of trained mental health professionals is further complicated by constant "brain drain" (Patel 2003).

During the past decade, a private sector in mental health care has grown steadily in the metropolitan areas of the region. However, this care is accessible to only a small section of the population who can afford the price which is often out-of-pocket expenditure. Specialized services such as child and adolescent mental health services, addiction services, community based and rehabilitation facilities and specialized counselling services have come up in many centres. Another sector which is increasingly playing a role in certain aspects of mental health care delivery in all the countries of the region, in particular in India is the not-for-profit, civil society organizations—both national and international non-governmental organizations (NGOs) (Tara and Patel 2010).

18.4 Current Status of Mental Health Legislation

It is in the context of the present sociocultural and economic situation and functioning of the health and mental health care systems described above that current status of mental health legislation be evaluated. Up until about two and half decades ago, the law dealing with mentally ill persons in the Indian sub continent was the Indian lunacy Act of 1912, enacted by the British colonial government for the entire British India and its protectorate territories. This law considered mentally ill persons as potentially dangerous and was primarily concerned with custodial aspects of treatment in mental hospitals such as rules and procedures for admission to and discharge from these institutions. It paid no attention to the human rights aspects of persons with mental illnesses. From the early 1950s, the infant psychiatric organizations of the sub continent made efforts to reform/revise the Indian Lunacy Act of 1912 or repeal it and enact new mental health laws. But nothing much occurred till quite recently and various forms of abuse and human rights violation of persons with mental illnesses continued in all countries of the region.

In India, an enquiry into the conditions in mental hospitals all over the country by the National Human Rights Commission in 1999 noted that "...38 % of the hospitals still retain the jail like structure that they had at the time of inception, patients are referred to as inmates and persons in whose care the patients remain through most of the day are referred to as warders". The Commission also noted that "... the deficiencies in the areas described so far are enough indicators that the rights of the mentally ill are grossly violated in mental hospitals" (National Human rights Commission 1999). In 2001, 28 chained mentally ill persons were burnt to death due to an accidental fire at a religious healing centre in Erwadi in Tamil Nadu state of South India (Srinivasa Murthy 2001). Media exposes and public interest litigations about the poor and scandalous situations in many mental hospitals coupled with assertive action by the courts and directives from the Supreme Court of India contributed to initiation of slow but positive mental hospital reforms in many centres in India (Murthy and Sekar 2008).

Table 18.3 Classification of countries based on the status of mental health legislation

Countries	Description
Bhutan, Maldives	No mental health legislation
Bangladesh, Sri Lanka	New legislation in draft form only, continue to follow provisions of archaic legislation
Afghanistan, India, Nepal, Pakistan	New legislation, enacted during the past two/three decades
India, Nepal	Legislation, currently being revised
All countries	Legislations do not reflect international standards and lags behind universally accepted values and principles

The current status of mental health legislation in all the countries of the South Asia region is now available from a variety of sources. These include the World Health Organizations (WHO) Atlas Project (WHO 2005, 2011). WHO obtained comprehensive country profiles of mental health services using the WHO Assessment Instrument for Mental Health Systems (Saxena et al. 2007), an assessment tool for mental health systems designed for low and middle income countries consisting of six domains which included "legislative framework" (WHO-AIMS Afghanistan, Bhutan, Bangladesh, Maldives, Nepal 2006, 2007). Information about mental health legislation in Pakistan, India and Nepal is available also from the International consortium on mental health policy and services (Gulbinat et al. 2004; Karim et al. 2004; Khandelwal et al. 2004; Regmi et al. 2004). In addition, Gilani et al. (2005); Jha and Adhikari (2009) and Weerasundare (2011) have commented about mental health legislation in their respective countries, Pakistan, Nepal and Bangladesh. Tables 18.3 and 18.4 provides a classification of countries based on the current status of their mental health legislations and other details regarding year of enactment and the status of implementation.

While Bhutan and Maldives, both countries with small populations do not have dedicated mental health legislation, some legal provisions concerning mental health are covered in other laws related to disability, welfare and general health legislation. The penal code of Bhutan has provisions that allow the treatment of patients with mental illness involuntarily. A draft mental health act is awaiting approval and enactment since 2002 in Bangladesh. Bangladesh which was part of Pakistan until it attained independence in 1971 did not have a traditional mental hospital. Sri Lanka continues to rely on an archaic legislation—Lunacy Ordinance of 1873—with minor modifications made over fifty years ago (Weerasundera 2011). In Pakistan, the Pakistan Mental Health Ordinance came into effect in February 2001 bringing about significant changes to the provision of care and treatment for persons with mental disorders as well as management of their property and other related matters. In Pakistan, ordinances hold as much value as any act of Parliament (Gilani et al. 2005). Afghanistan enacted mental health legislation in 1997 and the Ministry of Public Health is planning to revise the current mental health policy as well as legislation to formulate a new legislation (WHO-AIMS Report 2006). The Indian lunacy Act 1912 was repealed and a new Mental Health Act was passed by the Indian parliament in May 1987. Nepal established its Mental Health treatment and Protection Act in 2006. Both India and Nepal have recently initiated a process of

Table 18.4 Current status of mental health legislation in South Asia

Country	Presence of MH Legislation	When drafted/ enacted	Current status/Comments
Afghanistan	Yes	1997	New Act planned (WHO-AIMS 2006)
Bangladesh	Yes	2006	Only in draft form. Yet to be approved and enacted (Rabbani 2012)
Bhutan	No	N/a	Legal provisions concerning mental health, in other laws (welfare, disability, health) (WHO MH Atlas 2011)
India	Yes	1987	Currently, a new mental health care act is being developed, draft available, yet to be finalized and placed before the Parliament (Narayan et al. 2011)
Maldives	No	N/a	Legal provisions concerning mental health, in other laws (welfare, disability, health) (WHO MH Atlas 2011)
Nepal	Yes	2006	Mental health (treatment and protection) act 2063(2006), current plans for reform of the law(Mental health worldwide 2011)
Pakistan	Yes	2001	Mental health ordinance, 2001 (holds as much value as any act passed by Parliament (Gilani et al. 2005)
Sri Lanka	Yes	2000	Remains as a draft for over a decade, pending passing by Parliament and enactment (Weera-sundera 2011)

reviewing and reforming their mental health legislation to make them consistent with the provisions of the United Nations Convention on the Rights of Persons with Disabilities (UNCRPD) which came into effect in May 2008.

The latest mental health atlas of WHO notes that worldwide, only 59 % of the people live in a country where there is dedicated mental health legislation. While dedicated mental health legislation is present in 77 % of high income countries, in comparison only 39 % of low income countries have dedicated mental health legislation. Mental health legislation is less frequent in the African and South East Asian regions of the world (WHO 2011).

18.5 Shortcomings of Mental Health Legislation

According to WHO definitions, mental health legislation may cover a broad array of issues including access to mental health care and other services, quality of mental health care, admission to mental health care facilities, consent to treatment, freedom from cruel and inhuman and degrading treatment, freedom from discrimination, the enjoyment of a full range of civil, cultural, economic, political and social rights, and provision for legal mechanisms to promote and protect human rights (e.g. review bodies to oversee admission and treatment to mental health facilities, monitoring bodies to inspect human rights conditions in facilities and complaints mechanisms) (WHO 2011). Going by this broad scope and coverage of mental health legislation, all the dedicated mental health acts in the region have an institutional and custodial philosophy and lack human rights and community based mental health care approach. Besides ambiguities and lacunae in the acts themselves, there are major flaws in their implementation too.

Despite the existence of these acts, horrifying conditions and extreme form of abuses continue in psychiatric institutions of the region (Patel et al. 2012; Sharma 2012; Drew et al. 2011). In a powerful cover story which appeared in popular and widely circulated Indian news weekly (The Week) on the state of mental health care in India, Sharma (2012) described the conditions in many mental hospitals in India thus:

> "…The stench from the lavatory next to the dormitory is nauseating.", "…about 400 patients share 250 beds. Patients at a severe stage of mental illness are locked up in 4+5 feet cells, with an Indian style closet- they eat sitting next to it. And to kill body lice, say a hospital employee, patients are stripped and sprayed with insecticides meant to kill cockroaches. The pathetic and horrible condition is compounded by inhumanity", "Patients live in stinking barracks. The cells have no fans, even as the temperature soars over 40 degrees Celsius. Patients are forced to sleep on the dirty floor, as there are no beds in most wards", "Evidence of chaining patients, clinical abuse and active neglect are seen".

While one of the fundamental aims of mental health legislation is to protect the rights of persons with mental disorders, in most parts of the South Asian region, mentally ill people continue to be vulnerable to various types of abuse and violation of their rights. In some instances, instead of protecting the rights of persons with mental disorders, legislation may take away their rights. The mere existence of mental health legislation in itself does not guarantee against rights abuses. More than legislation is needed to protect the human rights of persons with mental disorders (Irmansyah et al. 2009). Inability to access adequate and affordable mental health services, inhumane and degrading treatment and being subjected to stigma and discrimination in the community are still widely rampant in most low and middle income countries including countries of the South Asian region (Drew et al. 2011). Lack of adequate processes for involuntary admission and treatment, dehumanizing and abusive conditions of treatment, arbitrary use of seclusion and restraint, lack of periodic review process, lack of proper diagnostic processes, and lack of formal process for determining capacity or competence are some of the major shortcomings with existing legislation in the region. The current legislations in the region do not reflect the international standards and lag far behind universally accepted values and principles. Considering the reality of the situation in large parts of South Asia including Nepal, Jha and Adhikara (2009) observe that *"mental health act for a low income country is like an elephant at a poor man's home"*. They wonder: *"a country where even essential medical and psychiatric facilities are not available, how can it afford a piece of legislation which requires enormous manpower and systems in place to implement and maintain the spirits of the mental health act?"*

Due to the neglect or low priority assigned to all issues related to mental health in all countries of South Asia, mental health legislation and its reform too take a back seat. The arduous path taken by the draft Indian Mental Health Act to replace the Indian Lunacy Act of 1912 is an example which exemplifies the difficulties of mental health reform and new legislation in the region. The efforts to amend or pass a new mental health act and repeal the Indian lunacy Act of 1912 were initiated within the first few years after India became independent and when the Indian psychiatric Society was formed during the late 1940s. An ad hoc committee of the Indian Psychiatric Society prepared a draft "Indian Mental Health Act" as early as the early 1950s (Banerjee 2001). However, due to numerous reasons including lack of political will, lack of interest or demand from any of the stakeholders and inability of the professionals and experts to reach a consensus, it took more than thirty years and many revisions for the Mental Health Act to be ultimately passed by both houses of the Indian Parliament and obtain the assent of the President of India in May 1987. Again, there were further long delays for the implementation of the much awaited new mental health act. The Government of India notified the Act only in 1990 and it took three more years, until 1993, for the Mental Health Rules to be framed. Even after the rules were framed, the Act remained largely non-functional and actual implementation was delayed in most states in the country until the Supreme Court of India issued directives to the federal and state governments for the implementation of the Act, The Supreme Court directives came following a Public Interest Litigation related to the fire accident at a religious treatment centre in Erwadi in Tamil Nadu state in South India which killed many chained mentally ill persons (Srinivasa Murthy 2001; Kala 2004). The reasons for non-implementation of the new legislation were many, "unrealistic minimum standards for mental hospitals, restrictive licensing requirements, complete exclusion of government institutions and the traditional health sector from the anvil of the Act, and divergent perspectives among psychiatrists, government and the legal point of view" (Jacob et al. 2007).

Within a decade after passing of the mental health act 1987, criticisms came up about various provisions of the Act. "Incongruities", "defects", "lacunae", "imperfections", "absurdities", "discriminations", "irrationalities" and difficulties in the implementation of the Act were pointed out. The shortcomings and the need for change were highlighted by numerous experts and authors (Antony 2000, 2010; Banerjee 2001; Trivedi 2002; Kala 2004; Murthy 2010; Narayan et al. 2011). A consensus grew at various levels for either amending the Act 1987 or drafting a new mental health act. With the adoption of the United Nations convention of on rights of people with disability (UNCRPD) in December 2006, India being a signatory to the convention was obliged to review and revise legislations related to persons with all disabilities including mental disabilities. The Government of India initiated the process of drafting a new mental health act in 2010. After countrywide consultation with different stake holders, a new draft mentalhealth care bill (The Mental Health care Bill 2011) is now available in India, pending finalization and presentation to the Parliament. The new bill is drafted taking into consideration the basic principles of mental health care law (WHO 2005); Murthy (2010) and Narayan et al. (2011) have extensively reviewed and critiqued the shortcomings of the 1987 act and the innovative changes in the new draft bill.

18.6 Challenges, Issues and Possible Solutions

There are numerous challenges in the application of international values, principles and standards to national contexts of low and middle income countries. In most South Asian countries there are no constitutional guarantees for proper health services. Socioeconomic rights such as right to education, right to health, social welfare, housing, employment etc. are not adequately addressed. There are also difficulties relating to peoples civil and political rights (rights pertaining to democratic processes) in some countries where they are hampered the existing form of government, unstable governments, political corruption etc. In most parts of South Asia, in general, people value family networks and community based approaches to individual, rights based value of personal autonomy. It is in the background of these socioeconomic and cultural realities of the region that legislation in the field of mental health and implementation of legislation be considered. In some countries, there is also the law-policy debate—which first, a mental health policy for the country or a new mental health act?

Mental health legislation cannot be considered in isolation. It becomes relevant only in the larger context of mental health service provision. In all the countries of the South Asian region, there is large treatment gap in the field of mental health care. The urgent need is to quickly enhance the scale and quality of mental health services at all levels (Eaton et al. 2011). All governments will have to recognize services for mental as an important priority and sanction adequate funds to develop services. Mechanisms will have to be created at federal, state and district levels to plan, implement and oversee feasible mental health care delivery programmes suitable to the culture and the country. Based on past experiences from each country and evidence generated internationally from low and middle income countries, feasible and effective country specific as well as region specific programmes will have to be developed taking into consideration the limited number of trained and motivated human resources available in each country. There is also a need to meaningfully address the severe shortage of trained mental health personnel in the region. Reform of mental health legislation will have to go hand in hand with increased resource allocation and improvement of services. Family members of the mentally ill, recovering mentally ill themselves and civil society organizations will have to be involved in the process of review and revision of legislation.

From the currently predominant institution (mental hospital) based and centralised care, public mental health services will have to increasingly become community based and integrated with existing system of primary health care. There is need for strengthening primary health care as well as primary care mental health services. The quality and spread of state run primary health services need great improvement. The main goal will be making basic mental health services accessible to all so that the current high levels of treatment gap can be steadily reduced. Capacity building for mental health will have to be undertaken on war footing. Training positions in psychiatry and allied disciplines will have to be increased. Content, quality and duration of training in mental health in undergraduate medical, nursing and para-medical education will need to be considerably improved.

The mental health Gap Action Programme (mhGAP) launched recently by the World Health Organization (WHO 2010) aims at scaling up services for mental, neurological and substance use disorders in low and middle income countries. The mhGAP has developed an intervention guide for mental, neurological and substance use disorders for use in non-specialist settings. The intervention guide presents integrated management of priority conditions using protocols for clinical decision-making. The priority conditions included are: depression, psychosis, bipolar disorders, epilepsy, developmental and behavioural disorders in children and adolescents, dementia, alcohol use disorders, drug use disorders, self-harm/suicide and other significant emotional or medically unexplained complaints. The guide is developed for use by health-care providers working in non-specialized health-care settings and needs adaptation for each country/region setting. Adaptation, pilot testing and use of this guide in countries of the region is likely to help substantially to reduce the treatment gap as well as human rights violation of the mentally ill.

References

Antony, J. T. 2000. A decade with mental health act, 1987. *Indian Journal of Psychiatry* 42:347–355.

Antony, J. T. 2010. On drafting a new mental health act. *Indian Journal of Psychiatry* 52:9–12.

Banerjee, G. 2001. The law and mental health: an Indian perspective Mental Health Reviews. http://www.psyplexus.com/excl/lmhi.html. Accessed 22 June 2012.

Drew, N., M. Funk, and S. Tang. 2011. Human rights violation of people with mental and psychosocial disabilities: An unresolved global crisis. *Lancet* 378:1664–1675.

Eaton, J., L. McCay, M. Semra, et al. 2011. Scale up services for mental health in low income and middle income countries. *Lancet* 378:1592–1603.

Ghani, E. 2010. *The poor half billion in South Asia: What is holding back lagging regions?* New York: Oxford University Press.

Gilani, A. I., U. I. Gilani, P. M. Kasi, and M. M. Khan. 2005. Psychiatric health laws in Pakistan: From lunacy to mental health. *PLoS Medicine* 2 (11):e317.

Gulbinat, W., R. Manderscheid, F. Baingana, et al. 2004. The International consortium on mental health policy and services: Objectives, design, and proper implementation. *International Review of Psychiatry* 16:5–17.

Irmansyah, I., Y. A. Prasetyo, and H. Minas. 2009. Human rights of persons with mental illness in Indonesia: More than legislation is needed. *International Journal of Mental Health Systems* 3:14. doi:10.1186/1752-4458-3-14.

Isaac, M. 2011. Mental health services in South Asia: Past, present and future. South Asian *Journal of Psychiatry* 2:4–12.

Jacob, K. S., P. Sharan, I. Mirza, et al. 2007. Mental health systems in countries: Where are we now? *Lancet* 370:1061–1077.

Jha, A., and S. R. Adhikari. 2009. Mental health services in Nepal: Observations, objections and outlooks for the future. *Journal of Nepal Medical Association* 48:185–190.

Kala, A. K. 2004. Why we should opt for a new mental health act and not tinker with the old one?. *Indian Journal of Psychiatry* 46:96–98.

Karim, S., K. Saeed, M. H. Rana, et al. 2004. Pakistan: Mental health country profile. *International Review of Psychiatry* 16:83–92.

Khandelwal, S. K., H. P. Jhingan, S. Ramesh, et al. 2004. India: Mental health country profile. *International Review of Psychiatry* 16:126–141.

Mental Health Worldwide. 2012. Review of Nepal mental health act. http://mentalhealthworldwide.com/2011/11/review-of-nepal-mental-health-act/. Accessed 25 July 2012.

Murthy, P. 2010. The mental health act 1987—quo vadimus? *Indian Journal of Medical Ethics* 7:152–156.

Murthy P., and K. Sekar. 2008. A decade after the NHRC quality assurance initiative: Current status of government psychiatric hospitals in India. In *Mental health care and human rights national human rights commission and national institute of mental health and neuroscience,* eds. D. Nagaraj and P. Murthy. New Delhi: NHRC.

Narayan C. L., M. Narayan, and D. Shikha. 2011. The ongoing process of amendments in MHA—87 and PWD Act—95 and their implications on mental health care. *Indian Journal of Psychiatry* 53:343–350.

National Human Rights Commission. 1999. *Quality assurance in mental health. national human rights in mental health.* New Delhi: National Human Rights Commission.

Neki, J. S. 1973. Psychiatry in South-East Asia. *British Journal of Psychiatry* 123:257–269.

Patel, V. 2003. Recruiting doctors from poor countries: The great brain robbery? *BMJ* 327:926.

Patel, V., A. Kleinman, and B. Saraceno. 2012. Protecting the human rights of people with mental disorders: A call to action for global mental health. In *Mental health and human rights: Vision, praxis and courage,* eds. M. Dudley, D. Silove, and F. Gale. New York: Oxford University Press.

Rabbani, G. 2012. *Draft mental health act 2006.* Bangladesh: Personal communication.

Reddy, K., V. Patel, P. Jha, et al. 2011. Towards achievement of universal health care in India by 2020: A call to action. *Lancet* 377:760–768.

Regmi, S. K., A. Pokharel, S. P. Ojha, et al. 2004. Nepal: Mental health country profile. *International Review of Psychiatry* 16:142–149.

Saxena, S., G. Thornicroft, and M. Knapp. 2007. Resources for mental health: Scarcity, inequity and inefficiency. *Lancet* 370:878–889.

Saxena, S., A. Lora, M. van Ommeren, et al. 2007. WHO's assessment instrument for mental health systems; collecting essential information for policy and service delivery. *Psychiatric Services* 58:816–821.

Sharma, G. 2012. Damned lives and statistics; the horrid state of mental health care in India can drive one insane. *The Week,* pp. 40–46, May 27.

Srinivasa, Murthy R. 2001. Lessons from the erwadi tragedy for mental health care in India. *Indian Journal of Psychiatry* 43:362–366.

Thara, R., and V. Patel. 2010. Role of non-governmental organizations in mental health in India. *Indian Journal of Psychiatry* 52:389–395.

The mental health care Bill. 2011. http://mohfw.nic.in/WriteReadData/l892s/6420662643 DRAFT%20THE%20MENTAL%20HEALTH%20CARE%20BILL.pdf. Accessed 8 Aug 2012.

The World Bank. 2010. GNI per capita, Atlas method (current US$). http://data.worldbank.org/indicator/NY.GNP.PCAP.CD.

Trivedi, J. K.. The mental health legislation: An ongoing debate. *Indian Journal of Psychiatry* 44:95–96.

Trivedi, J. K., and M. Dhyani. 2007. Undergraduate psychiatric education in South Asian countries. *Indian Journal of Psychiatry* 49:163–165.

United Nations Development Programme (UNDP). 2010. Human development reports. http://hdr.undp.org/en/data/profiles/.

Wang, P. S., A. Aguilar-Gaxiola, J. Alomo, et al. 2007. Use of mental health services for anxiety, mood and substance disorders in 17 countries in the WHO world mental health surveys. *Lancet* 370:841–850.

Weerasundera, R. 2011. Mental health legislation in Sri Lanka: The time for change is now. Sri Lanka *Journal of Psychiatry* 2:43–44.

World Health Organization. 2005a. *Mental health atlas 2005 world health organization*. Geneva: World Health Organization.
World Health Organization. 2005b. *WHO resource book on mental health human rights and legislation*. Geneva: World Health Organization.
WHO-AIMS report on mental health system in Bhutan. 2006a. *WHO and ministry of health & family welfare*. Bangladesh: WHO-AIMS.
WHO-AIMS report on mental health system in Bhutan. 2006b. *WHO and ministry of health*. Maldives: WHO-AIMS Report on mental health system in Bhutan.
WHO-AIMS report on mental health system in Bhutan. 2006c. *WHO and ministry of health*. Nepal: WHO-AIMS Report on mental health system in Bhutan.
WHO-AIMS report on mental health systems in Afghanistan. 2007. *WHO and ministry of public health*. Afghanistan: WHO.
WHO-AIMS report on mental health system in Bhutan. 2007. *WHO and ministry of health*. Bhutan: WHO.
World Health Organization. 2010. *mhGAP Intervention guide world health organization*. Geneva: World Health Organization.
World Health Organization. 2011. *Mental health atlas 2011 world health organization*. Geneva: World Health Organization.

Chapter 19
Mental Health Legislation: Evolution and Implementation in South Asian Countries

Vijoy K Varma, Harischandra Ghambeera and Nitin Gupta

19.1 Introduction

Major Psychiatric disorders impair the insight of the individual into their illness. The lack of insight affects the decision making abilities resulting in a reduction of the capacity to protect their own interest. The stigma attached to mental illnesses also affects the persons with mental illnesses and their family members adversely, leading to discrimination and marginalization. Stigma affects not only the individuals and their families; it also affects the establishment of mental health services to provide equitable services (Sartorius and Schulze 2005). In order to overcome these vulnerabilities of persons with mental disorders and establishment of optimal services, different countries throughout the world have proposed and enacted legislations which are called "mental health legislations". Thus mental health legislation is essential for the protection of human rights of mentally ill people and establishment of decent minimum mental health service standards. In addition, it also helps in providing legal framework for addressing critical issues of psychiatric patients, integration of psychiatrically ill patients with the community, establishing uniform services and providing quality care accessible equitable to everybody are some other functions that can be achieved by mental health laws. New and comprehensive legislation will certainly be helpful in prevention of mental illnesses and promotion of mental health. Establishing equitable services throughout any country will obviously be

V. K. Varma (✉)
Indiana University School of Medicine, Indianapolis, IN, USA

Postgraduate Insititute of Medical Education & Research, Chandigarh, India
e-mail: vijoyv@frontier.com

H. Ghambeera
National Institute of Mental health, Angoda, Sri Lanka
e-mail: hgambheera@gmail.com

N. Gupta
Government Medical College & Hospital, Sector 32, Chandigarh 160030, India

J. K. Trivedi, A. Tripathi (eds.), *Mental Health in South Asia: Ethics, Resources, Programs and Legislation,* International Library of Ethics, Law, and the New Medicine 58,
DOI 10.1007/978-94-017-9017-8_19

helpful in early detection and treatment and thereby minimizing the duration of untreated psychosis (DUP). An important objective in establishing an equitable service throughout the country would be prevention of relapses and thereby minimizing the cognitive deterioration due to longstanding mental illnesses.

Although South Asian societies have a rich tradition of concepts of mental health and illness, historically modern mental health services were started by the respective colonial powers in these countries, primarily for their own personnel, particularly those of the armed forces and the 'urban poor'. This was accompanied by equally inappropriate mental health legislation (Varma et al. 1984). It was only subsequently and indirectly that these became available to the indigenous population. However, there had been a preoccupation with the 'right to freedom' in the mental health planning and legislation. This had taken precedence over 'right to treatment'. But, it is heartening to note that the latter is increasingly being seen as the more important consideration. Unfortunately, mental health services in most of the countries in South Asia operated under an informal system, and are still laboring under antiquated laws framed under the colonial powers, with recent attempts to modernize it. However, such attempts have often been fragmented, inadequate and controversial (Gilani et al. 2005; Weerasundera 2011; Kala 2012).

19.2 Historical Background

Some of the countries in the region, namely India (now three independent countries, India, Pakistan, and Bangladesh) and Sri Lanka have been invaded by various nations in history. Those colonial powers influenced heavily the changes in health systems, life styles, governance and administration. South Asian countries inherited many characteristics from those cultures during their periods of colonial administration. These countries were predominantly under the British rule during last few centuries. In addition to many administrative rules and regulations, countries in South Asia also inherited mental health legislation from British rulers.

History of mental health care in South Asia dates back to several centuries. From the sorcery of the devil dance, the practice of Ayurvedic Medicine and ancient traditions of local religions such as Hinduism, Buddhism and Islam and the influence of the countries that colonized South Asia have brought diverse combinations of influence. However the British rulers developed a western model of mental health services in their colonies in South Asia as well as in other regions for the benefit of their own people. The mental health services in Britain were based around Lunatic asylums in that era and these were extended to their colonies (Carpenter 1988).

19.3 Evolution of British Mental Health Legislations

Vagrancy Act enacted in 1774 provided provisions to detain undesirable people away from the society, protect property from the people who did not own property and prevent traveling abroad. The Act had been enacted with a special reference to insane people (Hamilton 1983).

Mentally ill people in the United Kingdom were kept either in their family home or in a Madhouse which is entirely a private enterprise. Madhouses detained insane people purely on commercial basis with little or no medical intervention. People were detained in atrocious conditions without a scientific selection procedure. In order to set out a legal framework for regulating "madhouses" (private imprisonment) the Parliament of Great Britain enacted the Madhouses Act in 1774.

The intended purpose with regard to licensing and regulating the private mad houses had not been served by the 1774 mad house Act and the need of an act with better and effective provisions arose. The Act had been repealed by the 1828 Act with other provisions thought to be more effective and some provisions for improving the treatment of insane persons. This Act has been amended in 1832 with more effective and purposeful provisions.

Mentally ill offenders were allowed to walk free from court acquitted on the grounds of insanity before the enactment of Criminal Lunatic Act in 1800. There was no provision to detain them for the safety of others unless incarcerated under the provisions of Vagrancy Act 1744. Enabling the detention of mentally abnormal offender on the grounds of dangerousness the British parliament enacted the Criminal lunatic Act in 1800 (Hamilton 1986).

The county Asylum Act was established for the first time in 1808 and revised twice in 1828 and in 1845. The county asylum act set forth the provisions for building asylums and monitoring. The Lunacy Act was passed through the parliament at the same time with the amended County Asylum In 1845. The lunacy act 1845 provided provisions to establish the Lunacy Commission. These two acts were interdependent and they consolidated the lunacy law in England.

The 1890 Lunacy Act was a major consolidating Act that remained the core of English and Welsh Lunacy Legislation until it was repealed by the 1959 Mental Health Act. The major change associated with the Lunacy Act 1890 regarding admission was that it states patients apart from those who were dealt with by the court of Chancery (chancery lunatics) should not be detained without judicial order from a Justice of Peace specializing in such "reception orders (Hamilton 1983).

The 1930 Mental Treatment Act modernized, without replacing, the Lunacy Laws. A major change it made was to provide provision for voluntary admissions and psychiatric outpatient clinics and 'asylums' became 'mental hospitals'.

The idiots also were incarcerated in asylums together with lunatics until the Idiot's act 1886 distinguished lunatics from Idiots. The act was intended to provide for facilities for the care, education and training of idiots and imbeciles (Idiots Act 1887).

The Lunacy Act 1890 was repealed following World War II by the Mental Health Act 1959. This Act abolished the Board of Control, and aimed to provide informal treatment for the majority of people suffering from mental disorders, whilst providing a legal framework such that such people could, if necessary, be detained in hospital against their will. It also aimed to make local councils responsible for the care of mentally disordered people who did not require hospital admission (Hamilton 1983).

However, like its predecessors, the 1959 Act did not provide clarity as to whether a legal order to detain a mentally disordered person in hospital also empowered the hospital to impose medical treatment against the person's wishes. It had become clear by the 1970s that a specific legal framework for medical treatments such as psychiatric medications, electroconvulsive therapy and psychosurgery was needed in order to balance the rights of detained persons with society as a whole.

19.4 Our Laws Belong More to the Past

The English East India Company initially arrived in Asia for trading and in early 17th century it started its operation in India (India, Pakistan and Bangladesh). In the 18th century with the declining power of Mughal Empire, the company started gaining the military and political power in India starting from Bengal. The East India Company invaded the western coastal areas of the Island, Sri Lanka (formerly Ceylon) in 1796 and became a crown colony in 1802. In 1815 the entire island came under British crown with the occupation of Kandyan Kingdom.

The first mental health legislations introduced in India was the Lunacy Act 1858 (Somasundram 1987) and in Sri Lanka it was Mental Disease ordinance 1873 (GOSL 1960). They provided guidelines for setting up mental asylums and procedural checks for admission and treatment on the patients with a view "to segregate those who by reasons of insanity were troublesome and dangerous to the other fellow citizens. At the initial stages the primary objectives of the asylums were to provide services to their own people but not the native people of the Indian subcontinent.

The Lunacy Act 1845 and the County asylum Act 1845 was in force in the United Kingdom governing mental health services and giving legal rights for incaserating mentally ill patients during the period they introduced Mental Health legislations in India and Sri Lanka. Even though occupational therapy and other moral treatment have been introduced, asylums in India and Sri Lanka remained primarily designed for custodial care and detention rather than treatment.

The Lunacy Act 1858 of India was amended in 1912 brought the mental hospital under the charge of Civil Surgeons instead of the Inspector Generals of Prison as in the earlier times. For the first time, psychiatrists were appointed and the control of such asylums handed over to the central government. Unfortunately, colonial conditions of the time provided certain Character to the Mental Health policies in India of those days. Further, the names of all asylums were also changed to mental hospitals.

Out of seven countries in South Asia, there are no mental health legislations ever enacted in three countries namely Bhutan, Nepal and Maldives. Nepal Psychiatric society has drafted new legislation exactly a decade ago. This has not yet gone through the Parliament for the enactment. Bangladesh and Sri Lanka have laws enacted as far back as 1858 (amended in 1912) and 1872 respectively by British rulers in colonial times. In the light of development of Community psychiatric services and appropriate mental health legislations, Afghanistan, India and Pakistan have enacted new laws in 1997, 1987 and 2001 respectively.

Considering ages of enactment of mental health legislations, countries of South Asia belong to three distinct eras. Countries that do not have any written laws like Bhutan, Maldives and Nepal belong to the pre legislative era. Bangladesh and

Sri Lanka have laws enacted in the Asylum Era. Moving forward with the rest of the world, Afghanistan (WHO 2006), India (GOVT of India 1987) and Pakistan GOVT of Pakistan (2001) have drafted and enacted new mental health Acts facilitating deinstitutionalization and establishment of community Psychiatric services.

People with mental disorders are, particularly vulnerable to abuse, exploitation and violation of rights. They have been denied access to equal services when compared to other illnesses due to stigma attached to mental illnesses. Mental health Legislation remedied the situation to a great extent and facilitated the establishment of comparable and accessible services. In the circumstances mentally ill patients in Bhutan, Maldives and Nepal may be subjected to all the possible adversities mentioned above in the absence of the mental health legislations and a reasonable and decent minimum service. While there are no legislations, there are no decent mental health services established for approximately 31 million people in these three countries (Nepal 29, 959, 364, Bhutan 725, 940, Maldives 315, 885- 2010: World Bank). However, the Bhutan Penal Code, which has been enacted in 2004, contains clauses to safeguard the interests of people with mental illness in the criminal justice system, including the provision of adequate treatment (WHO 2006).

There is lack of community psychiatric services in Nepal. There are 17 community-based psychiatric inpatient units (i.e. general hospitals and teaching hospitals) available in the country, with a total of 1.00 beds per 100,000 populations. There is also one mental hospital with a total of 0.2 beds per 100,000 populations. The majority of mental health service users are treated in outpatient facilities. Both human resources and infrastructure in psychiatric services in Nepal are scarce and unevenly distributed in urban and rural areas. Further analysis of service indicates that there is only 0.13 Psychiatrists available for 100,000 populations. In the absence of the mental health legislations all the mentally ill people are forced to be admitted voluntarily, perhaps with the consent of relatives (WHO 2006). The new laws have been drafted by Nepal Psychiatric Society and handed over to the relevant authorities for enactment.

Hence, it is imperative and important to highlight and recognize the role of the World Health organization (WHO) in the development and implementation of mental health legislation in various countries; especially from the SAR; across the world.

19.5 Role of the World Health Organization (WHO) in Mental Health Legislation

The World Health Organization (WHO) has long been aware that mental health legislation is an important component of mental health programs, and WHO Expert Committees have repeatedly drawn attention to the need to consider such legislation in the various fields of mental health. Thus, the problems of legislation affecting psychiatric treatment were discussed by the WHO Expert Committee on Mental Health as early as 1955 (WHO, 1955a). The WHO conducted a comparative study of legislation in 37 countries on hospitalization of mental patients (WHO, 1955b). A newer and more extensive survey of mental health legislation of a substantial number (47) of Member States was published in 1978, which included a special section on mental health law in developing countries: "Many developing countries were formerly colonies and, during the colonial period, the imperial powers simply transferred the system of large, centralized mental hospitals that had developed in Northern Europe and North America during the preceding century, an institutional response almost wholly inappropriate to the rural/agrarian communities of those countries. …" (Curran and Harding 1978). They further pointed out that "all countries can learn from the way in which others have tackled the difficult issues in the field."

As per the WHO survey conducted about a quarter of a century back (Curran and Harding 1978), the "basic statutory structure" of mental health legislation from the standpoint of its suitability for developing countries should include the following:

- It (mental health legislation) should enable advantage to be taken of all potential sources of help, e.g., police, social work agencies, families, traditional healers.
- It should avoid complex procedures that waste the time of scarce trained personnel and are often difficult for the patients to understand.
- It should ensure that patient's rights are protected irrespective of their educational level.
- It should not create completely separate mental health services but rather aim at their integration in general health services, and should stimulate the involvement of other sectors, e.g., education and social welfare.
- It should be in line with national policy and require treatment of priority conditions.

In another WHO study (Varma et al. 1984), select mental health professionals from 10 developing countries of South-East Asia were sent a specially prepared questionnaire. These countries comprised 7 out of 11 countries of South-East Asia Region of the World Health Organization (SEARO), namely, Bangladesh, Burma (now Myanmar), India, Indonesia, Nepal, Sri Lanka and Thailand; and also included 3 neighbouring countries, namely Iran, Pakistan and Malaysia, on account of socio-cultural and historical similarities.

The questionnaire covered various aspects of mental health legislation pertaining to enactment, responsible agencies, licensing and registration, admission & discharge procedures, civil rights and criminal liability, legislative provisions for patients and treatments, processes of improvement or modification of the legislation.

It was seen that all countries (apart from Iran and Thailand) had some form of major enactment in the field of mental health legislation. The licensing and registration of institutions for the care of the mentally ill was compulsory in all the countries, except Bangladesh and Nepal. Voluntary admissions were possible and various types of judicial commitment procedures exist in all the countries. In Malaysia and Thailand, hospital staff could also order admission. As for statutory bodies for judicially committed patients and for their discharge, no such body existed in Indonesia, Iran, Nepal or Thailand. Regarding civil rights of judicially-committed patients, a loss of such rights are usually based on the merits of each case. No legal provisions existed in any of the countries governing particular forms of treatment. Regarding alcohol dependence, no legal provisions existed, but legislation existed in Burma (Myanmar) and Malaysia for drug dependence.

In summary, the findings demonstrated that where mental health legislation did exist, it was out of date and often based on the inappropriate legislation of a former colonial power. It was realized that new legislation is needed. But it was also appreciated that it was in the course of preparation in some countries, and there was a willingness to discuss it further.

19.6 Strategies for Implementation of Mental Health Legislation

We present below a few key strategies for consideration. It needs to be remembered that this list is neither exhaustive nor comprehensive. It is based upon the available literature with the authors coupled with their professional experience-cum-expertise in the field of mental health. These can be-

[1] *Streamlined delivery of mental health care in the community* (Weerasundera 2011): This can potentially force the law to keep in pace; especially by highlighting the gap between the available care and lagging, though available, legislative options.
[2] *Political will* (Weerasundera 2011): It is no surprise that political will has a key role to play in the implementation (or more significantly, non-implementation) of legislation. This is reflected from our discussion about the state of play of legislation in various SAR countries. Hence, it is imperative that key political people/leaders are identified and linked, as far as possible, with the process of implementation and/or modification of the concerned legislation.
[3] *Involvement of various relevant stakeholders* (Weerasundera 2011): Over the years, looking at the examples of legislative development from the west, it can be seen that involvement of patients/service users, caregivers, NGOs and other voluntary organizations, self-help groups, media, lay public etc have become an integral (if not mandatory) part of the movement of furthering the mental health movement and care. A most recent example from the west would be the Revised Mental Health Act (1983, revised in 2007) of the United Kingdom (http://www.dh.gov.uk/health/2011/07/mental-health-act/; last accessed 31st October 2012). Various relevant stakeholders participate and help to enhance awareness and facilitate a robust implementation of relevant legislative processes.
[4] *Assess the performance over the course of years following implementation* (Gilani et al. 2005): This helps to see if the particular piece (or whole) of legislation is working well and/or it merits changes. Periodic reviews by team of experts helps in being able to benchmark the performance arising out of the available legislation and standards.

[5] *Complying with International Standards* (Gilani et al. 2005): e.g. The Universal Declaration of Human Rights, and the Declaration of the Rights of the Mentally Retarded. These laws serve the role of a prescriptive function, protective function, instrumental function, and interpretative function (WHO 2001b). Although these laws cannot be incorporated directly and/or fully into the domestic (individual country) laws, yet they provide us with certain standards which policy makers should strive to achieve (Gilani et al. 2005).
[6] *Generating Awareness*: This can be achieved by various methods, especially by carrying out Promotional Campaigns and Activities on Mental Health (WHO-AIMS 2006b, 2007a) on a regular basis (WHO-AIMS 2006a).
[7] *Collection of Information*: This is required in order to improve the mental health system and provide a baseline for monitoring the change e.g. through WHO documents and the process of WHO-AIMS.
[8] *Identification of Mental Health Needs*: These require identification under the aspects of promotional/preventive measures, care and after-care, and creation of awareness as the overall objectives of legislation can possibly be achieved by co-operation and coordination between these three broad aspects (WHO 2001b). The mental health needs can be identified through two streams i.e. medical/public health needs (e.g. promotion of mental health, prevention of mental disorders, access to mental health services, quality of care etc.) and civil/legal needs (human rights, involuntary admission and treatment, regulatory mechanisms, care and custody etc.) (WHO 2001b). A detailed analysis of these needs will feed into various aspects and framework of the legislation to be modified/formulated/implemented.
[9] *Collaborative Initiatives*: Directly (through own efforts) or indirectly (with aid of other agencies e.g. WHO), initiatives at collaboration can help in review of progress, collation of information, identifying areas of good practice etc which can help to pave the way forward regarding implementation. An excellent example of this would be the Regional Workshop on Mental Health Legislation held in Sri Lanka in 2001 (WHO 2001b).

19.7 Conclusions

There are vast variations in mental health legislations of South Asian countries and also the wide variability in its implementation across the various SAR countries. Hence, there can be no single solution that fits best for all; neither can one apply a potential solution for one country onto another country. What probably is required is a detailed analysis of the situation at hand followed by a pragmatic analysis and finally a potential combination of various strategies to achieve the ultimate goal i.e. full and functional implementation of mental health legislation in that particular SAR country.

Therefore, while looking at options of how to implement mental health legislation in the SAR region, it should be ensured that whatever professional resources are available are not compromised, and that all contribute to their maximum at their levels of expertise. While safeguarding against clear abuse of patients and clearly wrong treatment options, the legislation should facilitate mental health services. The legislation will need to examine the contributions and needs of the various professional segments, such as government hospitals, private sector and the traditional indigenous healers. A logical and optimal compromise of conflicting claims and needs of the various sectors will have to be effected. Legislation should regulate, but not impede, in the work of general hospital units or of private nursing homes. In this way, legislation can provide for the underpinning under which mental health services can progress in the region. Finally, the rich traditional legacy of the countries in the region should not be lost sight of, but exploited to the maximum advantage of all concerned.

Implementation should hence be viewed as just one segment of the kaleidoscopic puzzle related to the intricacies of mental health care, services and especially mental health legislation. Without an in-depth understanding of the wider public health issues, civil and criminal legislative processes, implementation of mental health legislation shall remain a dream 'to be achieved' rather than 'actually achieved'.

References

Carpenter, J., and N. Mendis. 1988. *The history of mental health care in Sri Lanka*. Colombo: Marga Publications.

Curran, W. J., and T. W. Harding. 1978. *The law and mental health: Harmonizing objectives*, 5. Geneva: WHO.

Gilani, A. I., U. I. Gilani, P. M. Kasi, et al. 2005. Psychiatric health laws in Pakistan: From lunacy to mental health. *PLoS Medicine* 2 (11): 1105–1109.

GOSL (Government of Sri Lanka). 1960. *Government of Ceylon legislative enactments; Mental disease ordinance*. Colombo: Government printer.

Government of Pakistan. 1912. Lunacy act 1912.

Hamilton, J. R. 1983. Mental health act 1983. *British Medical Journal (Clinical Research Ed.)* 286 (6379): 1720–1725.

Hamilton, J. R. 1986. Insanity legislation. *Journal of Medical Ethics* 12:13–17.

Idiots Act. 1886 (1887). *British Journal of Psychiatry* 33:103–108

Kala, A. K. 2012. Contemporary debates about mental health legislation: A summer and a winter of discontent. In *Community mental health in India*, ed. B. S. Chavan, N. Gupta, P. Arun, A,. Sidana, and S. Jadhav, 161–168. New Delhi: Jaypee Publishers.

Mendis, P. 1997. Act for the protection of the rights of persons with disabilities in Sri Lanka. *Asia and Pacific Journal on Disability* 1:6–9.

Sartorius, N., and H. Schulze. 2005. *Reducing the stigma of mental illness*. Cambridge: Cambridge University Press.

Somasundram, O. 1987. The Indian lunacy act 1912: A historic background. *Indian Journal of Psychiatry* 29 (1): 3–14.

The Ministry of Health, Sri Lanka. The Draft Mental Health Act. http://203.94.76.60/Act/Act_Mental_Health_Act_Final_Draft_English.pdf. Accessed 31 Oct 2012.

Varma, V. K., S. K. Verma, and T. W. Harding. 1984. Mental health legislation in ten Asian developing countries: The perceived need for change. *International Digest of Health Legislation* 35 (2): 283–304.

Weerasundera, R. 2011. Mental health legislation in Sri Lanka: the time for change is now. *Sri Lankan Journal of Psychiatry* 2 (2): 43–44.

WHO Expert Committee on Mental Health. 1955a. Fourth Report. Legislation affecting psychiatric treatment. Geneva (WHO Technical Report Series, No. 98).
WHO/Ministry of Health Bhutan. 2006. The WHO-AIMS (Assessment Instrument for Mental health Systems) report on mental Health systems in Bhutan WHO and Ministry, Bhutan.
WHO/Ministry of Public Health, Afghanistan. 2006. The AIMS report on Mental Health System of Afghanistan, Kabul, Afghanistan.
WHO/Ministry of Health and Population Nepal. 2006. The WHO-AIMS report on Mental Health systems in Nepal. WHO and the Ministry, Nepal.
WHO-AIMS Report on Mental Health System in Afghanistan. 2006a. WHO and Ministry of Public Health, Kabul, Afghanistan.
WHO-AIMS Report on Mental Health System in Nepal. 2006b. WHO and Ministry of Health, Kathmandu, Nepal.
WHO-AIMS Report on Mental Health System in Bangladesh. 2007a. WHO and Ministry of Health & Family Welfare, Dhaka, Bangladesh.
WHO-AIMS Report on Mental Health System in Bhutan. 2007b. WHO and Ministry of Health, Thimpu, Bhutan.World Health Organization. 1955b. Hospitalization of mental patients. A survey of existing of existing legislation. *International Digest of Health Legislation* 6:3–100.
World Health Organization. 2001a. The World Health Report. Mental Health: New Understanding, New Hope. Geneva, WHO.
World Health Organization. 2001b. Meeting the Needs of Persons with Mental Disorders through Legislation. WHO Regional Office for South-East Asia, New Delhi.

Chapter 20
Mental Health Legislation: Comparison of South Asian and Western Countries

Jitendra Kumar Trivedi, Adarsh Triapthi and Sujit Kumar Kar

20.1 Introduction

A relationship between psychiatry and law is ever growing and last few decades have seen many important advances in this field around the world. The pace of growth however, remained dissimilar in different countries. Law or legislation is an essential and integral part of the society to regulate everything made to happen around in order. A country's strength of legislation decides its evolution and progress. Almost all countries of the world have legislation for health which decides the national priorities on health, national budget to be spend on health and many other health related issues. In majority of the countries, legislation related to health places emphasis on physical health and mental health is often ignored. Progressively, many countries have formulated or are planning to formulate legislations, policies or programs in order to empower the mental health system. Despite globalization movements, there is disparity from country to country in this aspect. There is a clear division between Western countries which are developed and Eastern Countries which are developing or underdeveloped. Other way, American, European and Australian countries are developed countries, most Asian countries are typically developing and African countries are mostly underdeveloped.

20.2 Mental Health Legislations

Persons suffering from mental illnesses face stigma, discrimination and marginalization in most parts of the world, which is a brutal violation of human rights (The Mental health context, WHO 2003). Mental health legislations are the laws

A. Triapthi (✉) · J. K. Trivedi · S. K. Kar
Department of Psychiatry, King George's Medical University,
Shahmeena Road, Chowk, Lucknow, UP, India
e-mail: dradarshtripathi@gmail.com

J. K. Trivedi, A. Tripathi (eds.), *Mental Health in South Asia: Ethics, Resources, Programs and Legislation,* International Library of Ethics, Law, and the New Medicine 58,
DOI 10.1007/978-94-017-9017-8_20

formulated to protect the rights of mentally ill patients (Mental Health Legislation and Human Rights, WHO 2003). Mental health legislation intends to strengthen the goals, objectives, values and principles related to mental health (Trivedi et al. 2007).

It avails basic mental health services, reintegrates mentally ill persons in the community as well as helps in developing mental health plan and policy for the country (Mental Health Legislation and Human Rights WHO 2003; Mental Health Atlas 2011; WHO 2011). As a whole mental health legislation facilitates mental health awareness, promotes mental health, makes the mental health services accessible, affordable & available and also focuses on rehabilitation of mentally ill (Mental Health Legislation And Human Rights, WHO 2003).

Currently most mental health legislations across the world focus mainly on some important factors like human rights, high quality mental health care, strengthening the administration, expanding the budget, participation of clients and mental health service management (Bertolote 1995).

In the survey conducted by World Health Organization in 2001, it was found that 31 % of world's population (25 % of countries of the world) was covered with mental health legislation (Mental Health Legislation and Human Rights, WHO 2003; Mental Health Resources in the World: WHO 2001). Currently as per the data of 2011, Mental health legislation is existing in more than 90 % of developed (high income) countries where as it is less than 40 % (around 36 %) in underdeveloped and developing (low income) countries (Mental Health Atlas 2011; WHO 2011). In just above 40 % of countries having mental health legislation, the legislation is either introduced or revised after the year 2005 (Mental Health Atlas 2011; WHO 2011). Approximately 6 % of countries don't have mental health legislation or any law in any form to protect the rights of mentally ill (Mental Health Atlas 2011; WHO 2011). When the data of 2005 is compared with that of 2011, it is found that mental health legislation covers 94 % of world's population in comparison to 83 % of world's population in 2005 (Mental Health Atlas 2011; WHO 2011; Mental Health Atlas WHO 2005).

There is no defined boundary of South Asia, although a region naming South East Asia region (SEAR) is defined in the mental health map of World Health Organization. Similarly the term "Western countries" is also very non-specific. But more or less it includes the American (both south and north) and European countries. It is very difficult to have a head to head comparison. So the comparison is being done considering the major countries of each region.

20.3 Current Status of Mental Health Legislation in South Asia

South Asia is one of the heavily populated regions in the globe. Most of the countries in this region are developing countries. There is no definitely defined boundary of South Asia. The World Health Organization has given a definite boundary to South-East Asian region. There are eleven member countries enlisted under the World Health Organization, regional office for South East Asia naming "India", "Bhutan", "DPR Korea", "Bangladesh", "Sri Lanka", "Nepal", "Myanmar", "Maldives", "Timor—Leste", "Indonesia" and "Thailand" (WHO regional office for South East Asia, SEARO Website 2011). South Asia includes important countries like India, Sri Lanka, Nepal, Bhutan, Bangladesh, Myanmar, Maldives, Indonesia, Afghanistan and Pakistan. China, which exactly not a part of South Asia but has a great influence in this region, is hence considered in the comparison.

The challenging features which are characteristic of this region are heavy population, poverty, illiteracy, unemployment, natural disasters, wars and terrorism. Existing health care facilities in this region seem to be inadequate for such a vast population. But since last few decades there is rapid expansion of the health care facilities. Mental health care delivery is disproportionately poor in comparison to general health care facility. Seven out of ten South East Asian countries currently have a dedicated mental health policy. But India being a rapidly advancing and highly influential country which roughly comprises more than two third of population of South East Asian region, does not have a dedicated mental health policy. The current mental health scenario of the south Asian countries is described below.

20.3.1 India

Among the south Asian countries, India is the most populated country. As per the 2011 census, approximately 17.5% of world populations reside in India which amounts to 1,210 million (Census of India 2011). India launched the mental health legislation in 1987 in the name of mental health act. In last twenty five years, (i.e. since 1987) India has started mental health programs and attempts have been made to expand the programs across the country, but the success of the programs is still questionable. This doubtful success of mental health program can be explained with reasons like—inadequate administrative support (improper and poor budgeting), unrealistic proposals, critical licensing procedure, weak legislation and lack of political commitment (Jacob et al. 2007; Antony et al. 2000; Danda et al. 2004). The government of India has increased funding towards mental health in its national five year plans which is summarized in the table below (WHO Atlas mental health resources in the world 2005; Agarwal et al. 2004; Goel et al. 2004; Planning Commission Government of India 2006; Government of India Ministry of Health and Family Welfare 2009).

National five year plan	Funding amount in rupees (crores)
Ninth five year plan (1997–2002)	28
Tenth five year plan (2002–2007)	190
Eleventh five year plan (2007–2012)	408

Following the Erwady tragedy in 2001, there is increased emphasis on human rights of mentally ill as well as strengthening of mental health legislation in India.

Other than the mental health act (1987), in India there are few more legislations related to mental health like—The Narcotic Drugs and Psychotropic Substances Act (1985), The Persons with Disabilities Act (1995), The Consumer Protection Act (1986), The Protection of Human Rights Act (1993), The Children Act (1960), The Juvenile Justice Act (2001), The National Trust for Welfare of Persons with Autism, Cerebral Palsy, Mental Retardation and Multiple Disabilities Act (1999), Section 309 of Indian Penal Code (attempted suicide as a punishable offence) and Section 377 of Indian Penal Code (Homosexuality as a punishable offence) (Khandelwal et al. 2004; Kala and Kala 2007, http://rehabcouncil.nic.in/pdf/nationaltrust1999.pdf).

New mental health related legislation is proposed in the name of "The Mental Health Care Act". This has been cleared by the Union Cabinet on June 14, 2013, and once approved by the Parliament, will repeal the Mental Health Act 1987. The Mental Health Care Bill 2012 makes significant strides over the Mental Health Act 1987 and proposes to bring about protection and empowerment of persons with mental illness. The new mental health care bill (2013) has focused more on the rights of mentally ill individuals (Kala 2013). The bill also proposed changes in the procedure of involuntary admission. Unmodified Electro-Convulsive Therapy is planned to be banned as per the proposed bill. The mental health care bill attempted to clarify the definition related to mental health and illness (Kala 2013).

20.3.2 Bangladesh

The total population of Bangladesh is approximately 140 million. The life expectancy of males and females in Bangladesh are 64 and 65 years respectively (Bangladesh Bureau of Statistics 2002). Mental health legislation and mental health policy/program exist in Bangladesh but there is poor mental health care delivery system (less than one psychiatrist and psychiatric nurse per 100,000 population) (Jacob et al. 2007).

20.3.3 Bhutan

The total population of Bhutan is estimated to be 6.37 lakh in 2006 with life expectancy of 66 years for males and 66.2 years for females respectively (WHO regional office for South East Asia, SEARO Website 2011). Though there is mental health policy/program in Bhutan but there is no mental health legislation. For every 100,000 population, there exists less than one psychiatrist and psychiatric nurse (Jacob et al. 2007).

20.3.4 Indonesia

The total population of Indonesia was 205.8 million in 2000 with life expectancy of males being 69 years in males in 2005 (WHO regional office for South East Asia, SEARO Website 2011).Both mental health legislation and policy exists in Indonesia. Indonesia has a better scenario of mental health care delivery system in comparison to majority of countries of this region. In Indonesia, there are approximately 0.2 psychiatrists and 0.9 psychiatric nurses for every 100,000 population (Jacob et al. 2007).

20.3.5 Maldives

The total population of Maldives is approximately three lakhs in 2006 (WHO regional office for South East Asia, SEARO Website 2011). Maldives don't have any mental health legislation or policy. There are 0.36 psychiatrists for every 100,000 populations (Jacob et al. 2007).

20.3.6 Myanmar

The total population of Myanmar was 55.4 million in 2005–2006 (WHO regional office for South East Asia, SEARO Website 2011). In Myanmar there exist mental health legislation, mental health policy and for every 100,000 populations, 0.2 psychiatrists and 0.6 psychiatric nurses are present (Jacob et al. 2007).

20.3.7 Nepal

The total population of Nepal was 25.8 million in 2006 (WHO regional office for South East Asia, SEARO Website 2011). The life expectancy of males was 60 years where as that of females was 61 respectively in 2001 (WHO regional office for South East Asia, SEARO Website 2011). Mental health policy and mental health legislation exist in Nepal. Nepal has approximately 0.1 psychiatrist and almost equal number of psychiatric nurses for 100,000 populations (Jacob et al. 2007).

20.3.8 Sri Lanka

The total population of Sri Lanka was 19.25 million in 2003 (WHO regional office for South East Asia, SEARO Website 2011).The life expectancy was 70.7 years in males and 75.4 years in females respectively in the year 2001. Mental health policy as well as mental health legislation present in Sri Lanka. In Sri Lanka, psychiatric nurses are almost nine times of that of psychiatrists and are major part of mental health professionals (0.2 psychiatrists and 1.8 psychiatric nurses for every 100,000 populations). There exist 1.8 mental health beds for every 10,000 population (Jacob et al. 2007).

20.3.9 Afghanistan

The total population of Afghanistan was 281.5 lakhs with life expectancy of males being 47 years and females being 50 years respectively by 2009 (Global Health Observatory 2009, Afghanistan, WHO). Afghanistan has mental health legislation but the existing mental health care delivery system is very weak. There is scarcity of mental health professionals (less than one psychiatrist/psychiatric nurse per 100,000 populations) as well as beds for mentally ill (Jacob et al. 2007).

20.3.10 Pakistan

The total population of Pakistan was approximately 180 million by 2009 with life expectancy of males and females being 62 years and 64 years respectively (Global Health Observatory 2009; Pakistan, WHO). In Pakistan, both mental health legislation and policy exist with 0.2 psychiatrists and 0.08 psychiatric nurses for every 100,000 population to deliver mental health services (Jacob et al. 2007).

20.3.11 China

The total population of China was 1.37 billion in 2010 with life expectancy of >73 years (http://www.gov.cn/). China has no specific mental health legislations but there exists mental health policy. Recently in the year 2011, the government of China implemented laws in order to protect the rights of mentally ill people.

Mental health legislations in most of the developing and underdeveloped countries compel the patients or their care givers to a particular treatment. It usually happens as these countries lack adequate manpower and mental health care facilities. So only few treatment options are left from which the patient or the care giver have to prefer one. Affordability is another major factor which decides the treatment option. As most of the countries in this region are low to middle income countries, the available best treatment option cannot be availed to most part of the population. So there is difficulty in preserving the basic mental health rights through legislation.

Country	% of GDP spent for health (%)	MH policy/program	Mental health legislation
India	4.8	Yes	Yes
Sri lanka	3.5	Yes	Yes
Nepal	5.3	Yes	Yes
Thailand	3.3	Yes	No
Myanmar	2.8	Yes	Yes
Bangladesh	3.4	Yes	Yes
Bhutan	3.1	Yes	No
Indonesia	3.1	Yes	Yes
Maldives	6.2	No	No
Afghanistan	6.5	Yes	Yes
Pakistan	2.4	Yes	Yes
China	5.6	Yes	No

Source: http://www.thelancet.com/series/global-mental-health

20.4 Current Status of Mental Health Legislation in Western Countries

Western countries comprise of the European countries as well as countries of North and South America. Most of the Western countries are developed countries. Health services are considered as national priority and high quality of general health service is provided by the governments. Similarly mental health services are also well organized and much emphasis is given in delivering the mental health service to people at need. In developed countries (most of the western countries) delivery of mental health services are superior to that of South Asian countries due to major reasons like—

- Adequate mental health manpower
- Adequate funding for mental health
- High quality health care
- Good facility for training of mental health professionals
- Professionals stick to the internationally accepted norms and procedures
- Protection of human rights
- Highly qualified policy makers
- Good review system
- Strong legislation

The mental health policies and legislations of western countries are multidimensional and more advanced. Frequent amendments are done in the mental health policies and legislations of western countries as these countries have a good review system. The policy and legislations related to mental health undergo continuous and regular scrutiny.

20.4.1 Argentina

The total population of Argentina was 40.276 million in 2009. The life expectancy of males was 72 years and that of females was 79 years respectively (Global Health Observatory 2009; Argentina, WHO). Argentina has mental health legislation, mental health policy as well as program. The mental health delivery system is stronger as evidenced by more than ten psychiatrists per 100,000 population and 6 mental health beds for 10,000 populations (Jacob et al. 2007).

20.4.2 Hungary

The total population of Hungary was 9.99 million in 2009. The life expectancy of males was 70 years and that of females was 78 years respectively (Global Health Observatory 2009; Hungary, WHO). Hungary has mental health legislation and mental health policy with 9 psychiatrists and 19 psychiatric nurses for every 100,000 populations. Also 9.6 mental health beds are present for every 10,000 population (Jacob et al. 2007).

20.4.3 Brazil

The total population of Brazil was 193.73 million in 2009. The life expectancy of males was 70 years and that of females was 77 years respectively (Global Health Observatory 2009; Brazil, WHO). Mental health legislation as well as mental health policy/program exist in Brazil. There are approximately 5 psychiatrists per 100,000 populations and approximately 2.5 mental health beds for 10,000 populations (Jacob et al. 2007).

20.4.4 Denmark

The total population of Denmark was 5.47 million in 2009. The life expectancy of males was 77 years and that of females was 81 years respectively (Global Health Observatory 2009; Denmark, WHO). Denmark has one of the well-developed mental health care infrastructures. Strong mental health legislation, mental health policy and availability of 7.1 mental health beds for every 10,000 population with 16 psychiatrists and 59 psychiatric nurses to deliver care for 100,000 populations, reflects the quality of care in Denmark (Jacob et al. 2007).

20.4.5 Canada

The total population of Denmark was 33.57 million in 2009. The life expectancy of males was 79 years and that of females was 83 years respectively (Global Health Observatory 2009; Canada, WHO). Canada has well-structured mental health legislation and mental health policy/program. There is adequate and well organized mental health care delivery system. There are 19 mental health beds for 10,000 population with 12 psychiatrists and 44 psychiatric nurses for 100,000 populations (Jacob et al. 2007). Ontario's Brain's law is followed across the country. But each state province of Canada can frame its own legislation. The legislations related to mental health in Canada and its provinces focus on human right protection, treatment and admission procedures, community treatment and criminal responsibilities.

20.4.6 Italy

The total population of Italy was 59.87 million in 2009. The life expectancy of males was 79 years and that of females was 84 years respectively (Global Health Observatory 2009; Italy, WHO). There are 9.8 psychiatrists and 33 psychiatric nurses available to serve every 100,000 populations. Mental health legislation and policy exist in Italy (Jacob et al. 2007).

20.4.7 Ireland

The total population of Ireland was 4.51 million in 2009. The life expectancy of males was 77 years and that of females was 82 years respectively (Global Health Observatory 2009; Ireland, WHO). Ireland has mental health legislation, mental health policy and 9.43 mental health beds/10,000 populations with adequate number of mental health professionals (6.82 psychiatrists and 136 psychiatric nurses for every 100,000 populations) to provide service (Jacob et al. 2007).

20.4.8 Greece

The total population of Greece was 11.16 million in 2009. The life expectancy of males was 78 years and that of females was 83 years respectively (Global Health Observatory 2009; Greece, WHO). Greece has mental health legislation, mental health policy with developed mental health care delivery system (8.7 mental health beds/10,000 populations, 15 psychiatrists and 3 psychiatric nurses for every 100,000 population) (Jacob et al. 2007).

20.4.9 Germany

The total population of Germany was 82.16 million in 2009. The life expectancy of males was 78 years and that of females was 83 years respectively (Global Health Observatory 2009; Germany, WHO). The German mental health delivery system is a highly developed system of care (7.5 mental health beds/10,000 population, 11.8 psychiatrists and 52 psychiatric nurses for 100,000 populations) with separate policy for mental health and mental health legislation (Jacob et al. 2007).

20.4.10 United Kingdom

The total population of United Kingdom was 61.56 million in 2009. The life expectancy of males was 78 years and that of females was 82 years respectively (Global Health Observatory 2009; Great Britain, WHO). Regularly updated and organized mental health policy as well as legislation, adequate number of mental health beds (5.8 beds for every 10,000 populations) and sufficient mental health professionals (11 psychiatrists and 104 psychiatric nurses for 100,000 populations) are characteristic mental health care scenario in United Kingdom (Jacob et al. 2007). Scotland is one of the member countries of United Kingdom, which has a well-developed mental health legislative system. The Mental Health (Care and Treatment) (Scotland) act, 2003 is strong legislation which has given much emphasis on areas like autonomy, individual rights & capacity (Darjee and Crichton 2004). In United Kingdom, there are multiple provinces covered by different mental health legislation. England and Wales, Scotland and Northern Ireland are governed by different mental health legislations but the ideologies behind the legislations are almost similar (Mental Health Atlas Jacob et al. 2005, WHO).

In UK, mental health is a priority for the government. Important mental health legislations in United Kingdom are—Mental health Act (1983), National Health Service(NHS) and Community Care Act (1990), Mental Health (Patients in the Community) Act (1995), Disability Discrimination Act (1995), Human Rights Act (1998), Mental Health Bill (2002), Mental Health (Care and Treatment) (Scotland) Act (2003), Mental Health Bill (2003), Mental Capacity Act (2005), Mental Health Bill (2006), Mental Health Act (2007), Health and Social Care Act (2012) (Darjee and Crichton 2004; National Health Service and Community Care Act 1990; http://www.parliament.uk/documents/post/pn204.pdf; The Mental Capacity Act 2005).

Most of the recent mental health legislations focus on rights of the mentally ill persons as well as public protection and risk management. The latest legislation related to mental health was passed in March, 2012 in the name of "Health and Social Care Act, 2012" and it was intended to maintain parity between mental and physical health (http://www.centreformentalhealth.org.uk/policy/legislation.aspx).

20.4.11 United States of America

The total population of United States of America was 314.66 million in 2009. The life expectancy of males was 76 years and that of females was 81 years respectively (Global Health Observatory 2009; USA, WHO). The United States of America has policy and legislations for mental health as well as adequate (7.7mental health beds/10,000 populations, 13.7 psychiatrists and 6.5 psychiatric nurses for 100,000 populations) and high quality mental health services (Jacob et al. 2007). In United States of America, much emphasis is given to protect the rights of mentally ill and challenged population. The legislations framed in American region in this regard are—American Declaration of the Rights and Duties of Man (1948), American Convention on Human Rights (1978), Additional Protocol to the American Convention on Human Rights in the area of Economic, Social and Cultural Rights (1988), The Mental Health Parity Act (1996), The civil commitment laws, Inter-American Convention on the Elimination of all Forms of Discrimination against Persons with Disabilities (1999), The Children's Health Act (2000) (Jacob et al. 2007).

Much emphasis is given to maintain privacy and confidentiality with regard to patient's mental health related information in the laid legislations.

20.4.12 France

The total population of France was 62.34 million in 2009. The life expectancy of males was 78 years and that of females was 85 years respectively (Global Health Observatory 2009; France, WHO). A well-developed system of mental health care (22 psychiatrists and 98 psychiatric nurses for 100,000 populations) with existence of mental health legislation and policy shows the quality of mental health service (Jacob et al. 2007).

20.4.13 Russia

The total population of Russia was 140.87 million in 2009. The life expectancy of males was 62 years and that of females was 74 years respectively (Global Health Observatory 2009; Russia, WHO). In Russia, for every 100,000 population there are 13.3 psychiatrists and 50 psychiatric nurses. There is well framed mental health policy and legislation (Jacob et al. 2007).

20.4.14 Jamaica

The total population of Jamaica was 2.72 million in 2009. The life expectancy of males was 69 years and that of females was 74 years respectively (Global Health

Observatory 2009; Jamaica, WHO). Jamaica has 5 mental health beds for every 10,000 populations, 1.6 psychiatrists and 8 psychiatric nurses for every 100,000 populations and mental health policy as well as legislation (Jacob et al. 2007).

Other Western countries like Uruguay, Venezuela, Equador, Finland, Greece, Iceland, and Cuba also have mental health legislation. Uruguay's mental health legislation has more focus on human right issues and promotion of mental health, at the same time there is little or no emphasis on regulation of mental health services. Venezuela's legislation on mental health emphasizes on promotion of mental health, protection of human rights, regulation of mental health care delivery, advocacy of mental health and admission procedures. Mental health legislations of Equador, Iceland and Cuba also focus in these areas. In Greece, mental health legislations focuses more on community based psychiatry. Though Iceland has no separate mental health act, but the existing legislations related to mental health have focus on the human right as well as social security issues and admission procedures.

In the WHO European Ministerial conference on mental health, held at Helsinki (Finland) in January 2005, the European countries declared to modify and upgrade the mental health system of their countries in accordance with the constitutional structure for facing challenges (WHO European Ministerial Conference on Mental Health: Facing the Challenges, Building Solutions, 2005). The proposed initiatives also emphasized on a better mental health policy and mental health legislation. The European ministry committed to bring following changes in the mental health legislation.

1. Enforcement of mental health legislation/policy that protects the human rights
2. Coordinating responsibilities for framing and implementation of mental health legislation/policy
3. Introduction of anti-discrimination legislation
4. Development of effective mental health policy and legislation by providing experts to the governments

There was also emphasis on improving the quality of mental health care by better service, education, tackling stigma as well as addressing other sensitive issues related to mental health (e.g.- suicide, violence etc) (WHO European Ministerial Conference on Mental Health: Facing the Challenges, Building Solutions 2005).

Country	% of GDP spent for health (%)	MH policy/program	Mental health legislation
USA	15.2	YES	YES
Canada	9.9	YES	YES
UK	8	YES	YES
Germany	11.1	YES	YES
France	10.1	YES	YES
Italy	8.4	YES	YES
Russia	5.6	YES	YES
Denmark	9	YES	YES
Ireland	7.3	YES	YES
Hungery	8.4	YES	YES
Jamaica	5.3	YES	YES

Source: http://www.thelancet.com/series/global-mental-health

20.5 Comparison of Mental Health Legislation: South Asian Countries Versus Western Countries

20.5.1 Statistical comparison

When the statistical data of western countries is compared with that of south Asian (particularly South East Asian countries) there is gross variation in mental health care delivery as well as in the legislative system.

As per the data from *"WHO Mental Health Atlas 2011"*, 88 % of population of American countries and more than 90 % of population of European countries covered by dedicated mental health policy whereas only 32 % population of South East Asian countries are covered with a dedicated mental health policy and it is because of India, which is the most populated country of this region which don't have a dedicated mental health policy (Mental Health Atlas 2011; WHO 2011). Other south Asian countries like Pakistan and Afghanistan have mental health policy. China also has mental health policy. When the mental health legislation of American & European countries are compared with South East Asian countries, there is not much difference between the two groups, so far the percentage of population covered with because ~80 % of American, approximately 81 % of European and 76 % of South East Asian population are covered by mental health legislation. But when the number of countries of these regions is compared, 60 countries out of 84 countries in the region of America and Europe have mental health legislation at the same time only 4 out of 10 countries of South East Asia have mental health legislation (Mental Health Atlas 2011; WHO 2011). Among other major countries of South Asia Pakistan and Afghanistan have mental health legislations. China doesn't have specific legislation for mental health (Jacob et al. 2007) but in 2011, the government of China had taken initiatives to develop legislation for the same.

20.5.2 Comparison of the Legislative Structures

The World Health Organization has developed a check list for mental health legislation which is very important for maintaining the basic standard of mental health legislation (WHO Checklist on Mental Health Legislation 2009). It also aims at helping countries to formulate draft as well as to review the adequacy and comprehensiveness of their mental health legislation (WHO Checklist on Mental Health Legislation 2009). The WHO checklist on mental health legislation has 27 important key points which is as below (The Mental health context, WHO 2003; WHO Checklist on Mental Health Legislation 2009).

1. Preamble and objectives
2. Definitions
3. Access to mental health care
4. Rights of users of mental health services
5. Rights of families and other care givers
6. Competence, capacity and guardianship

7. Voluntary admission and treatment
8. Non-protesting patients
9. Involuntary admission (when separate from treatment) and involuntary treatment (where admission and treatment are combined)
10. Involuntary treatment (when separate from involuntary admission)
11. Proxy consent for treatment
12. Involuntary treatment in community settings
13. Emergency situations
14. Determinations of mental disorder
15. Special treatments
16. Seclusion and restraint
17. Clinical and experimental research
18. Oversight and review mechanisms
19. Police responsibilities
20. Mentally ill offenders
21. Discrimination
22. Housing
23. Employment
24. Social security
25. Civil issues
26. Protection of vulnerable groups
27. Offences and penalties

A number of definitions are used in the mental health legislation. Broader the definition, more likely the chance of it being abused and narrower the definition, more risk of it excluding persons as normal (Report of the Expert Committee: Review of the Mental Health Act 1983, Department of Health 1999; Fisteina et al. 2009). Most of the western countries follow broader definitions and the south Asian countries follow the same as the mental health legislations adopt the definitions of western countries. Countries like Canada and United Kingdom follow "disability" (phenomenon that impair mental functioning) approach, India and Pakistan follow broad "disorder" (diagnosis of particular syndromes) approach whereas Sri Lanka follow "unsoundness of mind" approach (Fisteina et al. 2009).

Now days, much emphasis is given to the autonomy of the patients with mental illnesses. Most of the mental health legislations pay little consideration on the human rights (particularly autonomy) (Jones 2005). Society's preference and inadequate financial resources for the empowerment of human rights are the major reasons of this ignorance (Chodoff 1984; Dhanda 2005). This is one of the reasons of difference of mental health legislation among different countries of the world (Fisteina et al. 2009; Appelbaum 1997; Bartlett and Watchirs 2005; Gray and O'Reilly 2001; Zinkler and Priebe 2002). Mental health legislations of Western countries give much emphasis to human rights as compared to that of developing South Asian countries (Fisteina et al. 2009). The European council does not consider ethnicity, religious, political, cultural or philosophical beliefs and practices as a ground for involuntary treatment but one of the country of European council, north Ireland

and also Pakistan excludes sexual preferences, identity or practice and substance use as a ground for involuntary treatment (Fisteina et al. 2009). Most of the states of Canada as well as India reject intellectual disability as a reason for involuntary treatment (Fisteina et al. 2009). Few of the states of Canada, Scotland, England and Wales have mental health legislation which aims at prevention of mental illness as well as improving the quality of life whereas most of the south Asian countries as well as European countries have mental health legislation with no therapeutic aim for detention and it is done for public interest.

The approach towards persons with mental illnesses is different in mental health legislation of countries across the world. In most Western countries (Canada, United Kingdom etc), the mental health legislation follows broad "safety" approach to narrow "safety" approach (i.e.: detention is recommended in order to prevent immediate or impending physical harm to patient or others and also to prevent serious deterioration or psychological harm to patient or others) (Fisteina et al. 2009). In most of the developing countries of south Asia (India, Sri Lanka, Pakistan etc), the mental health legislation recommends detention of the mentally ill persons for the betterment of their health and functional ability (Fisteina et al. 2009). In United Kingdom, there exists laws (Mental Health (Care and Treatment) (Scotland) act 2003 and Mental Capacity Act 2005) in order to determine the capacity of mentally ill person for treatment or admission to hospital (The Mental Capacity Act 2005; Fisteina et al. 2009). In south Asian countries like Sri Lanka if the person's mental capacity is not adequate to make rational choices for treatment, then he/she can be hospitalized or treated. In Indian mental health legislation as well as legislation of European countries, treatment of the mentally ill persons is permitted without capacity assessment. In Pakistan and Jamaica, capacity assessment is not required for admission but for treatment is done when the capacity is impaired. In Sri Lanka there is no review or appeal system for legislation related to mental health but the legislation of India and Pakistan allow appealing even if there is no independent legal review system. In Jamaica, Canada, countries under council of Europe, there is regular independent automatic legal review system (Fisteina et al. 2009). Mental health legislation facilitates mental health care through primary health centers, govt. hospitals in order to make the mental health care facility accessible to all (World Health Report, WHO 2005). This is made possible through adequate funding and implementing programs and policy.

Mental health legislation also protects and respects the rights of users in terms of making illness and human right related information accessible, maintaining confidentiality and by providing humane treatment facility (World Health Report, WHO 2005). Mental health legislations of Western countries have much emphasis on protecting the rights of users as well as making the mental health care accessible to all in comparison to South Asian countries. Mental health legislation also protects the offenders and prisoners with mental disorders and avails mental health care facility to them. Most of the Western as well as South Asia countries have emphasis on this on their legislation. Special treatments (Electroconvulsive therapy, Psychosurgery, Sterilization operations etc), seclusion and restraints for mentally ill patients are usually decided by the medical boards. Clinical and experimental research in psychiatry is also coming under the legislative framework. Most of the Western countries and some of the South Asian countries emphasize on this in their mental health legislations. Another important responsibility of mental health legislation is providing protection to vulnerable groups of population i.e.—children & adolescents, women and elderly population. Eliminating discrimination of mentally ill persons and gender inequality are also dealt by the mental health legislation. Less importance is given to these areas in the mental health legislations of most of the developing South Asian countries as compared to the developed Western countries.

Parameters of mental health legislation	Western countries	South Asian countries
Preamble and objective	+++	+++
Clear definitions	++	++
Access to mental health care	+++	++
Rights of users of mental health services	+++	+
Rights of families and other care givers	++	+
Competence , capacity and guardianship	++	++
Voluntary admission and treatment	+++	++
Involuntary admission	+++	++
Proxy consent for treatment	++	++
Involuntary treatment in community settings	++	+
Emergency situations	++	++
Determination of mental disorder	+++	++
Special treatments (sterilization, interventions)	+++	+
Clinical and experimental research	+++	++
Oversight and review mechanisms	+++	++
Police responsibilities	+++	++
Mentally ill offenders	+++	+++
Considering Discrimination	++	+
Housing	++	++
Employment	++	+
Social security	++	+
Civil issues	+++	++
Protection of vulnerable groups (minors and women)	++	+
Offences and penalties	++	+

Number of "+" symbols signifies the strength of consideration in mental health legislation:
"+" absent or not very much considered in most considered
"++" considered by some of the countries
"+++" considered by most of the countries

Similarly WHO had also developed "Mental Health Care Law: Ten Basic Principles" in 1996 to guide countries developing mental health legislation. The ten basic principles are (Jacob et al. 2007; Mental Health Care Law: Ten Basic Principles, WHO, 1996) –

1. Promotion of mental health and prevention of mental disorders
2. Access to basic mental health care
3. Mental health assessments in accordance with internationally accepted principles

4. Provision of least restrictive type of mental health care
5. Self-determination
6. Right to be assisted in the exercise of self-determination
7. Availability of review procedure
8. Automatic periodic review mechanism
9. Qualified decision-maker
10. Respect of the rule of law

The differences in mental health legislation of Western countries and South Asian countries are summarized as below in the table.

Basic Legislative Principles	Western countries	South Asian countries
Promotion of mental health and prevention of mental disorders	+++	++
Access to basic mental health care	+++	+
Mental health assessments in accordance with internationally accepted principles	++	+
Provision of least restrictive type of mental health care	++	–
Self-determination	++	+
Right to be assisted in the exercise of self-determination	++	+
Availability of review procedure	++	+
Automatic periodic review mechanism	++	–
Qualified decision-maker	++	+
Respect of the rule of law	++	++

+++ present in most countries
++ present in many countries
+ present in few countries
– absent

20.6 Future Needs

Comparison of mental health legislations and mapping them for different countries is very important because by this always there will be a scope to improve. The counties will know own deficits and will get an idea to handle its problems. Many standard guiding principles are developed by WHO which is helpful in formulating mental health legislation. More than 75 % countries of South Asia and more than 80 % of Western countries have mental health legislation (Mental Health Atlas 2011; WHO 2011). The need of the future is to—

- Increase public awareness on mental health
- Aid countries (which don't have mental health legislation) in developing mental health legislation
- Strengthening the existing mental health legislations

20.7 Conclusion

Currently the center of attention is on strengthening of mental health legislations and preserving the rights of mentally ill, in both Western and South Asian countries. The World Health Organization is doing regular surveys and setting guidelines for framing mental health legislation. The World Health Organization (WHO) is trying to strengthen mental health care delivery system as well as the legislations governing mental health in developing and undeveloped countries. The Western countries have advanced mental health legislation whereas the mental health legislation is either an adoption or a modified version of the western one in the developing nations. The socio-cultural, economic backgrounds as well as national priorities are different in western and south Asian countries. Due to these understandable variations, the mental health legislations differ in these two groups of countries. But scarcity of resources, insufficient funding, inefficient system and the socio-culturo-geographic variations are the foremost causes of difference in mental health legislations. It is believed that mental health awareness is the first step to initiate global mental health movement. So the Western countries and WHO should cooperate in creating mental health awareness and developing proper mental health legislation in South Asian countries.

References

Agarwal, S. P., R. L. Ichhpujani, S. Shrivastava, and D. S. Goel. 2004. Restructuring the national mental health program. In *Mental health: An Indian perspective, 1946–2003,* eds. S. P. Agarwal, D. S. Goel, R. L. Ichhpujani, R. N. Salhan, and S. Shrivastava, 119–122. New Delhi: Directorate General of Health Services, Ministry of Health and Family Welfare.

Antony, J. T. 2000. A decade with the mental health act 1987. *Indian Journal of Psychiatry* 42:345–355.

Appelbaum, P. S. 1997. Almost a revolution: An international perspective on the law of involuntary treatment. *Journal of the American Academy of Psychiatry and the Law* 25:135–147. [PubMed: 9213286].

Bangladesh Bureau of Statistics. SVRS. 2002.

Bartlett, P., and H. Watchirs. 2005. Introduction to special edition on law and rights. *International Journal of Law and Psychiatry* 28:97–98.

Bertolote, J. M. 1995. Legislation related to mental health: A review of various international experiences. *Revista de Saúde Pública* 29 (2): 152–156.

Centre for mental health London United Kingdom. http://www.centreformentalhealth.org.uk/policy/legislation.aspx. Accessed 20 Oct 2012.

Census of India. 2011. http://censusindia.gov.in/2011-common/censusdataonline.html. Accessed 5 May 2012.

Chodoff, P. 1984. Involuntary hospitalization of the mentally ill as a moral issue. *American Journal of Psychiatry* 141:384–389. [PubMed: 6703103].

Danda, A. D., S. Goel, and R. K. Chadda. 2004. Law and mental health: Common concerns and varied perspectives. In *Mental health: An Indian perspective, 1946–2003,* eds. S. P. Agarwal, D. S. Goel, R. L. Ichhpujani, R. N. Salhan, and S. Shrivastava, 170–188. New Delhi: Directorate General of Health Services, Ministry of Health and Family Welfare.

Darjee, R., and J. Crichton. 2004. New mental health legislation. *British Medical Journal* 329:634–635.
Department of Health. London. 1999. Report of the Expert Committee: Review of the Mental Health Act 1983.
Dhanda, A. 2005. The right to treatment of persons with psychosocial disabilities and the role of the courts. *International Journal of Law and Psychiatry* 28:155–170. [PubMed: 15862872].
Draft Mental Health Care Act. 2010. www.prsindia.org/…/DraftMentalHealthCareAct. Accessed 30 Dec 2011.
Draft Mental Health Care Bill. 2012. http://mohfw.nic.in. Accessed 18 Oct 2012.
Fisteina, E. C., A. J. Hollanda, I. C. H. Clarea, and M. J. Gunn. 2009. A comparison of mental health legislation from diverse commonwealth jurisdictions. *International Journal of Law and Psychiatry* 32 (3): 147–155.
Global Health Observatory. 2009a. World Health Organization. http://www.who.int/countries/afg/en/. Accessed 28 Feb 2012.
Global Health Observatory. 2009b. World Health Organization. http://www.who.int/countries/pak/en/. Accessed 28 Feb 2012.
Global Health Observatory. 2009c. World Health Organization. http://www.who.int/countries/arg/en/. Accessed 28 Feb 2012.
Global Health Observatory. 2009d. World Health Organization. http://www.who.int/countries/hun/en/. Accessed 28 Feb 2012.
Global Health Observatory. 2009e. World Health Organization; http://www.who.int/countries/bra/en/. Accessed 28 Feb 2012.
Global Health Observatory. 2009f. World Health Organization. http://www.who.int/countries/dnk/en/. Accessed 28 Feb 2012.
Global Health Observatory. 2009g.World Health Organization. http://www.who.int/countries/can/en/. Accessed 28 Feb 2012.
Global Health Observatory. 2009h. World Health Organization; http://www.who.int/countries/ita/en/. Accessed 28 Feb 2012.
Global Health Observatory. 2009i. World Health Organization. http://www.who.int/countries/irl/en/. Accessed 28 Feb 2012.
Global Health Observatory. 2009j. World Health Organization. http://www.who.int/countries/grc/en/. Accessed 28 Feb 2012.
Global Health Observatory. 2009k.World Health Organization. http://www.who.int/countries/deu/en/. Accessed 28 Feb 2012.
Global Health Observatory. 2009l.World Health Organization. http://www.who.int/countries/gbr/en/. Accessed 28 Feb 2012.
Global Health Observatory. 2009m. World Health Organization. http://www.who.int/countries/usa/en/. Accessed 28 Feb 2012.
Global Health Observatory. 2009n. World Health Organization. http://www.who.int/countries/fra/en/. Accessed 28 Feb 2012.
Global Health Observatory. 2009o.World Health Organization. http://www.who.int/countries/rus/en/. Accessed 28 Feb 2012.
Global Health Observatory. 2009p.World Health Organization. http://www.who.int/countries/jam/en/. Accessed 28 Feb 2012.
Goel, D. S., S. P. Agarwal, R. L. Ichhpujani, and S. Shrivastava. 2004. Mental health 2003: The Indian scene. In *Mental health: An Indian perspective, 1946–2003,* eds. S. P. Agarwal, D. S. Goel., R. L. Ichhpujani, R. N. Salhan, and S. Shrivastava, 3–24. New Delhi: Directorate General of Health Services, Ministry of Health and Family Welfare.
Government of China. http://www.gov.cn/.
Government of India. Ministry of Health and Family Welfare (Department of Health) Approval letter No. V-15011/6/2007-PH (Pt2). Dated 24th April 2009.
Gray, J. E., and R. L. O'Reilly. 2001. Clinically significant differences among Canadian mental health acts. *Canadian Journal of Psychiatry* 46:315–321.
Jacob, K. S., P. Sharan, I. Mirza, M. Garrido-Cumbrera, S. Seedat, J. J. Mari, V. Sreenivas, and S. Saxena. 2007. Mental health systems in countries: Where are we now?; Global Mental

Health 4; Lancet Global Mental Health Group; Series on Global Mental Health. *Lancet*, Online September 4. http://www.thelancet.com/series/global-mental-health.

Jones, M. 2005. Can international law improve mental health? Some thoughts on the proposed convention on the rights of people with disabilities. *International Journal of Law and Psychiatry* 28:183–205. [PubMed: 15862874].

Kala, A. 2013. Time to face new realities: Mental health care bill-2013. *Indian Journal of Psychiatry* 55:216–219.

Kala, K., and A. K. Kala. 2007. Mental health legislation in contemporary India: A critical review. *International Psychiatry* 4:69–71.

Khandelwal, S., H. P. Jhingan, S. Romesh, R. K. Gupta, and V. K. Srivastava. 2004. India: Mental health country profile. *International Review of Psychiatry* 16:126–141.

Mental Health Care Act 2010 offers protection for the mentally ill. 27 December 2010. http://indiagovernance.gov.in/news.php?id=504. Accessed 30 Dec 2011.

Mental Health Care Law: Ten Basic Principles. (1996). Geneva, World Health Organization. http://whqlibdoc.who.int/hq/1996/WHO_MNH_MND_96.9.pdf.

National Health Service and Community Care Act. 1990. http://www.legislation.gov.uk/ukpga/1990/19/contents. Accessed 28 Feb 2012.

National Institute for Health and Family Welfare. Mental Health Act. 1987. http://www.nihfw.org/ndc-nihfw/html/Legislations/MentalHealthAct.htm. Accessed 1 Dec 2011.

Planning Commission Government of India. 2006. *Towards faster and more inclusive growth: An approach to the 11th five year plan*. New Delhi: Planning Commission, Government of India.

Reform of mental health legislation; postnote October 2003 Number 204. http://www.parliament.uk/documents/post/pn204.pdf. Accessed 28 Feb 2012.

The Mental Capacity Act. 2005. http://www.dh.gov.uk/en/Publicationsandstatistics/DH_063423.

The Mental health context. World Health Organization. 2003. (Mental health policy and service guidance package). Geneva http://www.who.int/mental_health/policy/services/3_context_WEB_07.pdf.

The National Trust for Welfare of Persons with Autism, Cerebral Palsy, Mental Retardation and Multiple Disabilities Act, 1999. http://rehabcouncil.nic.in/pdf/nationaltrust1999.pdf. Accessed 30 Dec 2011.

Trivedi, J., P. Narang, and M. Dhyani. 2007. Mental health legislation in south Asia with special reference to India: Shortcomings and solutions. *Mental Health Review Journal* 12 (3): 22–29.

WHO Checklist on Mental Health Legislation; World Health Organization. 2009 http://www.who.int/mental_health/policy/WHOLegislationChecklist.pdf. Accessed 28 Feb 2012.

WHO European Ministerial Conference on Mental Health: Facing the Challenges, Building Solutions; Helsinki, Finland, 12–15 January 2005. http://www.euro.who.int/__data/assets/pdf_file/0008/88595/E85445.pdf. Accessed 28 Feb 2012.

World Health Organization, regional office for South East Asia, SEARO Website, 2011. http://www.searo.who.int/EN/Section864/Section1007/Section1012.htm. Accessed 4 Feb 2012.

World Health Organization. 2005. *Atlas, mental health resources in the world 2005*. Geneva: WHO.

World Health Organization. Mental Health Atlas. 2005. Geneva: WHO.

World Health Organization. Mental Health Atlas 2011. Geneva: WHO.

World Health Organization. Mental Health Resources in the World. 2001.Geneva: WHO.

World Health Organization; Mental Health Legislation And Human Rights (Mental Health Policy and Service Guidance Package), World Health Organization, 2003.

World Health Report, WHO. 2005. http://www.who.int/whr/2005/en/index.html. Accessed 30 Jan 2012.

Zinkler, M., and S. Priebe. 2002. Detention of the mentally ill in Europe—A review. *Acta Psychiatrica Scandinavica* 106:3–8. [PubMed: 12100342].

Erratum

Mental Health in South Asia: Ethics, Resources, Programs and Legislation

Jitendra Kumar Trivedi • Adarsh Tripathi

King George's Medical University
Lucknow
India

J. K. Trivedi, A. Tripathi (eds.), *Mental Health in South Asia: Ethics, Resources, Programs and Legislation,* International Library of Ethics, Law, and the New Medicine 58, DOI 10.1007/978-94-017-9017-8_1

DOI 10.1007/978-94-017-9017-8_21

The publisher regrets to have made the following errors in the present version of the book:

1. The editor's name on the cover of the book should be "Adarsh Tripathi" instead of "A.K. Tripathi".
2. The "Response to the Reviewer's Comments" which is presently a part of the front matter should have been deleted.
3. The editor's name given at the end of the "Introduction" section of the front matter should be "Jitendra Kumar Trivedi" instead of "Jiterndra Kumar Trivedi".

The online version of the original book can be found at http://dx.doi.org/ 10.1007/978-94-017-9017-8

J. K. Trivedi, A. Tripathi (eds.), *Mental Health in South Asia: Ethics, Resources, Programs and Legislation,* International Library of Ethics, Law, and the New Medicine 58, DOI 10.1007/978-94-017-9017-8_21

MIX
Papier aus verantwortungsvollen Quellen
Paper from responsible sources
FSC® C105338
FSC
www.fsc.org

If you have any concerns about our products,
you can contact us on
ProductSafety@springernature.com

In case Publisher is established outside the EU,
the EU authorized representative is:
Springer Nature Customer Service Center GmbH
Europaplatz 3, 69115 Heidelberg, Germany

Printed by Libri Plureos GmbH
in Hamburg, Germany